Emerging Technologies in Hazardous Waste Management 8

Emerging Technologies in Hazardous Waste Management 8

Edited by

D. William Tedder

Georgia Institute of Technology
Atlanta, Georgia

and

Frederick G. Pohland

University of Pittsburgh
Pittsburgh, Pennsylvania

Springer Science+Business Media, LLC

Library of Congress Cataloging-in-Publication Data

Emerging technologies in hazardous waste management 8 / edited by D. William Tedder
and Frederick G. Pohland.
        p. cm.
    Includes bibliographical references and index.
    ISBN 978-1-4757-8677-4
    1. Hazardous wastes--Management--Congresses. 2. Hazardous
wastes--Purification--Technological innovations--Congresses. 3. Soil
remediation--Congresses. 4. Groundwater--Purification--Congresses. 5. Radioactive
wastes--Purification--Congresses. I. Tedder, D. W. (Daniel William), 1946- II. Pohland,
Frederick G., 1931- III. Emerging Technologies in Hazardous Waste Management
Symposium (1997: Pittsburgh, Pa.) IV. Emerging Technologies in Hazardous Waste
Management Symposium (1998 : Boston, Mass.) V. Computing in Science and
Engineering Symposium (1997 : Birmingham, Ala.)

TD1020 .E443 2000
628.5--dc21
                                                                    00-023975

ISBN 978-1-4757-8677-4   ISBN 978-0-306-46921-3 (eBook)
DOI 10.1007/978-0-306-46921-3

Proceedings of the Emerging Technologies in Hazardous Waste Management Symposia, held September 1997, in
Pittsburgh, Pennsylvania, and August 1998, in Boston, Massachusetts, and the Computing in Science and Engineering
Symposium, held May 1997, in Birmingham, Alabama

© Springer Science+Business Media New York 2000
Originally published by Kluwer Academic/Plenum Publishers, New York in 2000
Softcover reprint of the hardcover 1st edition 2000

PREFACE

This book resulted from presentations at the *Emerging Technologies in Hazardous Waste Management* Symposia held in Pittsburgh, Pennsylvania in September 1997, and Boston, Massachusetts in August 1998, and the *Computing in Science and Engineering* symposium held in Birmingham, Alabama in May 1997. Manuscripts submitted from these symposia were subjected to peer review and then revised by the authors. Nineteen manuscripts were selected for inclusion in this volume based upon peer review, scientific merit, the editors' perceptions of lasting value or innovative features, and the general applicability of either the technology itself or the scientific methods and scholarly details provided by the authors.

This volume is a continuation of a theme initiated in 1990. Its predecessors, *Emerging Technologies in Hazardous Waste Management, ACS Symposium Series No. 422 (1990), Emerging Technologies in Hazardous Waste Management II, ACS Symposium Series No. 468 (1991), Emerging Technologies in Hazardous Waste Management III, ACS Symposium Series No. 518 (1993), Emerging Technologies in Hazardous Waste Management IV, ACS Symposium Series No. 554 (1994), Emerging Technologies in Hazardous Waste Management V, ACS Symposium Series No. 607 (1995), Emerging Technologies in Hazardous Waste Management VI, American Academy of Environmental Engineers Publication (1996),* and *Emerging Technologies in Hazardous Waste Management VII, Plenum Press (1997),* are related contributions on hazardous waste management, but each volume is different. By inspection, the reader may quickly recognize the diversity associated with this topic, and the features of its evolution, although no single volume can do justice to the breadth and depth of technologies being developed and applied in practice.

The contributions presented in this volume are divided into three sections. They are: (1) Soil Treatment, (2) Groundwater Treatment, and (3) Radioactive Waste Treatment. We trust that the reader will find them informative and useful.

D. William Tedder
Georgia Institute of Technology
Atlanta, Georgia 30332-0100

Frederick G. Pohland
University of Pittsburgh
Pittsburgh, Pennsylvania 15261-2294

ACKNOWLEDGMENTS

The Industrial and Engineering Chemistry (I&EC) Division of the American Chemical Society organized and sponsored the symposia. ACS Corporate Associates was a major financial cosponsor; the American Institute of Chemical Engineers (AIChE), Environmental Sciences Division, the AIChE Center for Waste Reduction Technologies, and the American Academy of Environmental Engineers were nominal cosponsors. Their generosity was essential to the overall success of the symposia and is gratefully acknowledged.

CONTENTS

GROUNDWATER TREATMENT

RADIOACTIVE WASTE TREATMENT

EMERGING TECHNOLOGIES IN HAZARDOUS WASTE MANAGEMENT 8

An Overview

D. William Tedder[1] and Frederick G. Pohland[2]

[1]School of Chemical Engineering
Georgia Institute of Technology
Atlanta, Georgia 30332-0100
[2]Department of Civil and Environmental Engineering
University of Pittsburgh
Pittsburgh, Pennsylvania 15261-2294

Several long-term trends in technology evolution have become apparent since these symposia first began in 1989. Earlier presenters more frequently discussed treatment methods involving extensive human intervention. Examples of this include soil incineration (e.g., to destroy dioxin at Times Beach) and soil washing techniques that virtually reduce soil to sand and deplete most of its organic and inorganic nutrients. With such harsh treatments, the residues are essentially beach sands with substantially less value, compared to their *in situ* uses prior to contamination. Moreover, the soil incineration option has inherent risks of airborne emissions.[1]

As the symposia have continued, the number of presentations describing extremely harsh and expensive treatment technologies have gradually been supplanted by more subtle and gentler methods. Subsurface engineered barriers, for example, are often very helpful in controlling species migration beyond a designated area, but with less obvious intrusion. Similarly, phytoremediation, the use of existing plant life or mixed cultures of microorganisms, to control hazardous species is continuing to emerge as an important technology. Bioremediation methods, particularly those utilizing *in situ* bacteria, are also of continuing interest as such species can be enlisted to control hazardous chemicals in many instances simply by providing missing nutrients (e.g., phosphate or nitrogen).

The use of plants and bacteria to control organic species is readily obvious. Their use in controlling the movement and concentration of heavy metals is perhaps less obvious, but nonetheless fortuitous. Biological systems frequently exhibit high concentration factors for species, so trickle bed filters or other technologies utilizing suitable

Emerging Technologies in Hazardous Waste Management 8, edited by Tedder and Pohland
Kluwer Academic/Plenum Publishers, New York, 2000.

bacteria are finding increasing numbers of applications, and are often highly cost-effective. In exchange for meeting the requirements of sustaining bacterial life, mankind is able to exploit bacterial properties to control hazards, make the environment cleaner and safer and, by the way, also make it more attractive. Admittedly aesthetic perceptions are highly subjective, but it is suggested that marshlands, even when artificially created, are more attractive than settling ponds or other more obviously man-made devices.

Biological systems and engineered barriers may not be the final answers, particularly when heavy metals are concentrated in them, as these substrates must also be managed. Attractive methods must eventually be used for either recycling such metals to appropriate applications or placing them in appropriate final waste forms for permanent disposal. The latter method continues to be the favored approach for managing many radionuclides. However, biological systems can be a useful first step in concentrating toxicants or in some cases destroying them altogether.

Today, reasonable technical alternatives are available for managing many wastes, but questions still remain. Operational safety and user costs are paramount concerns, but better solutions are being found than those available only 10 years ago. The hazardous waste and environmental management problems resulting from industrialization at the turn of the century are not solved, but substantial progress has been made.

SOIL TREATMENT

In Chapter 2, Hong *et al.* address the use of chelating agents for extracting heavy metals from soil. Chelation is a potentially important method for removing metals (e.g., Pb, Cu, Cd, Zn, Ni and Hg) from contaminated soil. This paper focuses on assessment techniques using predictions that are compared to experimental extraction and recovery results. The prediction of chelator complexing power, selectivity, and eventual chelator recoverability are of particular concern. Assessment techniques are outlined and results are presented for selected species from among 250 alternative chelators.

The ongoing development[2-7] of electrokinetic soil remediation continues to suggest the merit of this technology. In Chapter 3, Inman *et al.* examine the use of novel electrodes and modulated reverse electric fields as a means of dealing with speciation issues that reduce overall efficiency. Modulated reverse electric fields, in conjunction with integrated ion exchange electrodes, are being studied as a means of inducing a more uniform remediation of contaminated soils, and to eliminate non-uniform pH profiles that result from conventional electrokinetics and adversely affect speciation. Their process modifications control soil pH and enhance mass transport of heavy metals between electrodes.

In a related treatment, Rabbi *et al.* discuss the use of electrokinetic injection to enhance *in situ* bioremediation in Chapter 4 (also see Chapter 7). Their objective is to use electrokinetics to maintain higher nutrient concentrations in contaminated soil and thus to promote biological growth and destruction of the contaminants. They demonstrate the potential and versatility for this method to improve treatment at sites where organic contamination cannot be remediated by natural attenuation alone.

The National Research Council of Canada has developed a process for treating highly saline industrial soils contaminated with heavy oils and heavy metals found in Alberta, Canada. Majid and Sparks describe a solvent extraction soil remediation process in Chapter 5. Heavy metal fixation is achieved by incorporating metal binding agents into the soil agglomerates which form during the solvent extraction of organic species. After remediation, the soils remain saline, but soluble salts can then be removed by water per-colation through a fixed bed of dried, agglomerated soil. Their goal is to provide an inte-grated treatment method which will adequately decontaminate soils, but still permit subsequent agricultural use. More aggressive decontamination methods often render the treated soil residues less effective for agricultural applications as mentioned above.

In Chapter 6, Martal *et al.* describe laboratory and field soil-washing experiments with surfactant solutions, and summarize non-aqueous phase liquid (NAPL) recovery mechanisms. They studied solutions consisting of mixtures of anionic surfactants and alcohols, and other solvents in some cases. Their laboratory and field experiments show that residual NAPL recovery can exceed 90% while using less than six pore volumes of concentrated wash solution. Recovery mechanisms depend upon: (1) NAPL type, (2) micellar solution type (alcohol/surfactant or alcohol/surfactant/solvent), (3) ingredient concentrations in the micellar solution, (4) washing direction (upward, downward, or horizontal), (5) injection and pumping patterns, and (6) injection velocities.

In Chapter 7, Maillacheruvu and Alshawabkeh describe experiments to investigate the effects of an electric field on anaerobic microbial activity, a topic related to that discussed by Rabbi *et al.* in Chapter 4. Here the focus is less on the enhancement of nutri-ent transport properties by the electric field, but on effect of the electric field on micro-bial activity. They find that microbial activity, measured as a function of the ability of anaerobic microorganisms to consume readily degradable acetate, generally decreases if the pH and dissolved oxygen concentrations are not controlled. Microbial activity ini-tially decreases under exposure to electric current, but recovers after a period of several hours of exposure. This result suggests that to some extent the culture was able to accli-mate to the current, an important issue if an electric field is used to enhance the trans-port of biological nutrients through contaminated soils as suggested by Rabbi *et al.* in Chapter 4.

GROUNDWATER TREATMENT

Results from studies on Fenton chemistry are reported in Chapter 8 by Tarr and Lindsey. They focus on those chemical mechanisms which affect Fenton oxidation in natural waters. This investigation is an expansion on iron chemistry studies described in earlier volumes of this series.[8-14] Their studies are related to others in this volume (Monsef *et al.* in Chapter 12, Greenberg *et al.* in Chapter 13, and Bower *et al.* in Chapter 14). Tarr and Lindsey find that altered hydroxyl production rates and increased hydroxyl scaveng-ing occur in natural waters. Hydroxyl binding to natural organic matter is also signifi-cant. They conclude that these three factors are key issues in determining the extent of oxidation during pollutant destruction and reaction efficiency. They suggest that the same phenomena may also be important in soil systems.

Javert and Strathmann describe the use of surfactant and cosolvent to enhance the removal of non-aqueous phase contaminants (NAPLs) by modified pump-and-treat methods in Chapter 9. Low NAPL recoveries may result from slow dissolution into groundwater, slow diffusion or desorption, or hydrodynamic isolation. They sum-marize ongoing surfactant and cosolvent studies that are designed to overcome these dif-ficulties and make pump-and-treat technology more generally applicable. This chapter complements earlier studies on NAPLs.[15,16]

Phytoremediation of TNT-contaminated groundwater by a poplar hybrid is described by Thompson *et al.* in Chapter 10. This paper describes a pilot-scale green-house experiment that examined the irrigation of a hybrid poplar tree (*Populus deltoides Xnigra*) with TNT-contaminated groundwater. The results indicate that a poplar tree remediation system may be a feasible solution for low levels of groundwater contamination.

In Chapter 11, Vidic and Pohland describe advances in treatment wall technology. This strategy involves the construction of permanent, semi-permanent or replaceable underground walls across the flow path of a contaminant plume. The main perceived advantage for this technology is reduced operation and maintenance costs. Although considerable design details have already been developed through field- and pilot-scale applications, some critical issues still remain to be resolved. These unresolved issues include establishing tested and proven design procedures, and evaluating interactions (perhaps favorable) with other groundwater remediation technologies.

In Chapter 12, Monsef *et al.* describe the removal of nitroaromatic compounds from water using zero-valent metal reduction and enzyme-based oxidative coupling reactions. Again, iron plays a key role. In this case they observe that zero-valent iron is effective in the reduction of aqueous nitrobenzene to aniline in the absence of oxygen.

A modified Fenton oxidation process, based on a proprietary catalyst, is described by Greenberg *et al.* in Chapter 13. In this case, a chelated iron complex was used to enhance the degradation of organic species at a contaminated site. Laboratory, pilot, and full-scale experiments are described in studies aimed at destroying gasoline and waste oil constitutents. After full-scale treatment, contaminant levels were either not detected, or were reduced to concentrations below New Jersey groundwater standards. The contaminated site was closed after one year.

Advanced oxidation process (AOPs), particularly those utilizing ozone, have been of interest for a number of years. In Chapter 14, Bower *et al.* investigate techniques for making AOP treatment less costly and more effective. They focus on the use of fixed beds containing sands, especially sands with high iron and manganese concentrations. They find an enhancement in phenol degradation rates, possible due to the formation of higher concentrations of hydroxyl radical, at pH 7. Direct ozonation was equally effective at pH 8.9.

RADIOACTIVE WASTE TREATMENT

The development of final waste forms remains as an important topic of study, particularly for radioactive wastes. Waste forms have been a recurrent theme in previous volumes[17–25] of this series. Chapters 15–17 of this volume continue this theme in which waste encapsulation studies utilizing polyethylene, by-product sulfur from refinery operations, and polysiloxane are presented.

The investigation by Adams *et al.* in Chapter 15 considers the effectiveness of using polyethylene for encapsulating depleted uranium trioxide. Using a single-screw extrusion process, they were able to successfully process mixtures with waste loadings up to 90 wt% UO_3. Leach rates increased with waste loadings, but were relatively low. Compressive strengths of samples were nearly constant in samples containing 50–90 wt% UO_3.

In Chapter 16, Kalb *et al.* describe their initial test results investigating the use of by-product sulfur from Kazakhstan to stabilize waste. The by-product sulfur is itself a waste. This stream is primarily elemental sulfur resulting from the refining of petroleum reserves in that country. Kazakhstan also produces hazardous and radioactive wastes;

using by-product sulfur to encapsulate other wastes could therefore "kill two birds with one stone." Kalb et al. focus on the use of a sulfur-polymer cement[21] by reacting the by-product sulfur with organic modifiers (5 wt%). This cement was then used to encapsulate waste, achieving up to 40 wt% loadings. Waste form performance was characterized by measuring compressive strength, water immersion, and accelerated leach testing.

The encapsulation of nitrate salts using polysiloxane is described in Chapter 17. Durik and Miller present results in which compressive strength, metal leaching, and void area measurements of polysiloxane-encapsulated waste were measured as a function of waste loading (27–48 wt%) for three surrogate wastes. Cost comparisons with immobilization using concrete suggest that polysiloxane could be an economical alternative for the U.S. Department of Energy which now manages substantial quantities of nitrate-bearing mixed wastes.

Regardless of the origin or nature of hazardous wastes, management invariably has a common initial need to measure their properties. This can take many forms, but waste assay is inevitable at some point and because the wastes are hazardous, operational risks often result from such analyses. With radioactive wastes, nondestructive assay techniques, in which radwaste drums can be analyzed without actually opening them, have become very important. Such improvements in waste analysis[26,27] are always welcome.

In Chapter 18, Kottle et al. describe the development of an on-line analyzer for vanadous ion. This technique has potential applications in nuclear power plants in conjunction with the removal of corrosion products from heat transfer systems. Its use should reduce personnel exposures during such operations.

The use of magnetic separation for nuclear material detection and surveillance is described by Worl et al. in Chapter 19. In this case, technology is being developed for the capture of submicron actinide particles, or the retrieval of fission products, in order to determine more about the operation of nuclear facilities. Using high-gradient magnetic separation, they are able to separate paramagnetic compounds from those which are diamagnetic, and thus effect some degree of separation and concentration of the paramagnetic species. Their work has potential environmental applications, particularly relating to environmental monitoring of actinide species at very low contamination levels.

Direct chemical oxidation[28] of wastes is an alternative to incineration[29–32] and electrochemical oxidation.[33] In Chapter 20, Balazs et al. describe the application of transition metal catalysts to enhance the ambient temperature destruction of organic wastes with peroxydisulfate. This oxidation technology is potentially applicable to many solid or liquid organics (e.g., chlorosolvents, oils and greases, detergents, contaminated soils and sludges, explosives, chemical and biological warfare agents). Silver ion has the greatest catalytic effect, but reaction rates are reduced somewhat by the presence of chloride ion. Catalysts also affect the production of chlorine gas when treating chlorinated organics, however, this effect is still not well understood.

SUMMARY

Emerging technologies in hazardous waste management are clearly diverse in nature and involve many different disciplines. This diversity is both an advantage and, potentially, a disadvantage. On the one hand, it offers the possibility of favorable collaboration between investigators with different backgrounds and emphases on the problem. This aspect can be advantageous. On the other hand, diversity can easily lead to some degree of confusion and misunderstanding. Thus it can also be a bit of a disadvantage, but such difficulties can be overcome if recognized and properly addressed.

Hazardous wastes still exist and although historical wastes have been reduced, major problems still remain, largely because of significant costs of remediation. Hazardous waste issues can nonetheless be solved if affected agencies, corporations, and political entities are firmly and consistently resolved to do so. Clearly, technical progess has been made since these symposia began in 1989. In many situations several technical alternatives are available to deal with particular problems, but more progress is yet to be made. Additional discovery is needed to further advance the basic science and technology, and to make translation into practice more of a reality. While numerous technical advances are being made, often their implementation is less than desired and needed to at last solve these problems. Thus, the real issue is less one of technology selection, but rather more one of social resolve and determination to see reasonable solutions implemented.

REFERENCES

1. C.D. Stelzer. Another accidental release of dioxin at Times Beach heats up the debate over the incinerator's safety. *Riverfront Times (St. Louis)*, May 15, 1996.
2. E.R. Lindgren, E.D. Mattson, and M.W. Kozak. Electrokinetic remediation of unsaturated soils. In D.W. Tedder and F.G. Pohland, editors, *Emerging Technologies in Hazardous Waste Management IV*, number 554 in ACS Symposium Series, pages 33–50. ACS Books, 1994.
3. E.D. Mattson and E.R. Lindgren. Electrokinetic extraction of chromate from unsaturated soils. In D.W. Tedder and F.G. Pohland, editors, *Emerging Technologies in Hazardous Waste Management V*, number 607 in ACS Symposium Series, pages 10–20. ACS Books, 1995.
4. T.R. Krause and B. Tarman. Preliminary results from the investigation of thermal effects in electrokinetic soil remediation. In D.W. Tedder and F.G. Pohland, editors, *Emerging Technologies in Hazardous Waste Management V*, number 607 in ACS Symposium Series, pages 21–32. ACS Books, 1995.
5. R.F. Thornton and A.P. Shapiro. Modeling and economic analysis of *in situ* remediation of Cr(VI)-contaminated soil by electromigration. In D.W. Tedder and F.G. Pohland, editors, *Emerging Technologies in Hazardous Waste Management V*, number 607 in ACS Symposium Series, pages 33–47. ACS Books, 1995.
6. E.R. Lindgren, R.R. Rao, and B.A. Finlayson. Numerical simulation of electrokinetic phenomena. In D.W. Tedder and F.G. Pohland, editors, *Emerging Technologies in Hazardous Waste Management V*, number 607 in ACS Symposium Series, pages 48–63. ACS Books, 1995.
7. S.B. Martin, Jr., D.J. Dougherty, and H.E. Allen. Electrochemical recovery of EDTA and heavy metals from washing of metal contaminated soil. In D.W. Tedder and F.G. Pohland, editors, *Emerging Technologies in Hazardous Waste Management 7*, pages 159–166. Plenum Press, New York, 1997.
8. R.A. Bull and J.T. McManamon. Supported catalysts in hazardous waste treatment: Destruction of inorganic polutants in wastewater with hydrogen peroxide. In D.W. Tedder and F.G. Pohland, editors, *Emerging Technologies in Hazardous Waste Management*, number 422 in ACS Symposium Series, pages 52–66. ACS Books, 1990.
9. R.A. Larson, K.A. Marley, and M.B. Schlauch. Strategies for photochemical treatment of wastewaters. In D.W. Tedder and F.G. Pohland, editors, *Emerging Technologies in Hazardous Waste Management II*, number 468 in ACS Symposium Series, pages 66–82. ACS Books, 1991.
10. J.J. Pignatello and Y. Sun. Photo-assisted mineralization of herbicide wastes by ferric ion catalyzed hydrogen peroxide. In D.W. Tedder and F.G. Pohland, editors, *Emerging Technologies in Hazardous Waste Management III*, number 518 in ACS Symposium Series, pages 77–84. ACS Books, 1993.
11. C. Sato, S.W. Leung, H. Bell, W.A. Burkett, and R.J. Watts. Decomposition of perchloroethylene and polychlorinated biphenyls with Fenton's reagent. In D.W. Tedder and F.G. Pohland, editors, *Emerging Technologies in Hazardous Waste Management III*, number 518 in ACS Symposium Series, pages 343–357. ACS Books, 1993.
12. J.B. Carberry and S.H. Lee. Enhancement of pentachlorophenol biodegradation by Fenton's reagent partial oxidation. In D.W. Tedder and F.G. Pohland, editors, *Emerging Technologies in Hazardous Waste Management IV*, number 554 in ACS Symposium Series, pages 197–222. ACS Books, 1994.

13. M.D. Gurol, S.S. Lin, and N. Bhat. Granular iron oxide as a catalyst in chemical oxidation of organic contaminants. In D.W. Tedder and F.G. Pohland, editors, *Emerging Technologies in Hazardous Waste Management 7*, pages 9–22. Plenum Press, New York, 1997.

14. G. Kand, J. Jung, K. Park, and D.K. Stevens. Mineralization of hazardous chemicals by Heme reaction. In D.W. Tedder and F.G. Pohland, editors, *Emerging Technologies in Hazardous Waste Management 7*, pages 69–80. Plenum Press, New York, 1997.

15. R.C. Chawla, C. Porzucek, J.N. Cannon, and J.H. Johnson, Jr. Importance of soil-contaminant-surfactant interactions for *in situ* soil washing. In D.W. Tedder and F.G. Pohland, editors, *Emerging Technologies in Hazardous Waste Management II*, number 468 in ACS Symposium Series, pages 316–341. ACS Books, 1991.

16. D.C.M. Augustijn and P.S.C. Rao. Enhanced removal of organic contaminants by solvent flushing. In D.W. Tedder and F.G. Pohland, editors, *Emerging Technologies in Hazardous Waste Management V*, number 607 in ACS Symposium Series, pages 224–236. ACS Books, 1995.

17. C.S. Shieh, I.W. Duedall, E.H. Kalajian, and F.J. Roethal. Energy waste stabilization technology for use in artificial reef construction. In D.W. Tedder and F.G. Pohland, editors, *Emerging Technologies in Hazardous Waste Management*, number 422 in ACS Symposium Series, pages 328–344. ACS Books, 1990.

18. P.L. Bishop. Contaminant leaching from solidified-stabilized wastes: Overview. In D.W. Tedder and F.G. Pohland, editors, *Emerging Technologies in Hazardous Waste Management II*, number 468 in ACS Symposium Series, pages 302–315. ACS Books, 1991.

19. P. Chu, M.T. Rafferty, T.A. Delfino, and R.F. Gitschlag. Comparison of fixation techniques for soil containing arsenic. In D.W. Tedder and F.G. Pohland, editors, *Emerging Technologies in Hazardous Waste Management II*, number 468 in ACS Symposium Series, pages 401–414. ACS Books, 1991.

20. G.R. Darnell, R. Shuman, N. Chau, and E.A. Jennrich. Above grade earth-mounded concrete vault: Structural and radiological performance. In D.W. Tedder and F.G. Pohland, editors, *Emerging Technologies in Hazardous Waste Management II*, number 468 in ACS Symposium Series, pages 415–430. ACS Books, 1991.

21. G.R. Darnell. Sulfur polymer cement as a final waste form for radioactive hazardous wastes. In D.W. Tedder and F.G. Pohland, editors, *Emerging Technologies in Hazardous Waste Management IV*, number 554 in ACS Symposium Series, pages 299–307. ACS Books, 1994.

22. L.J. Staley. Vitrification technologies for the treatment of contaminated soil. In D.W. Tedder and F.G. Pohland, editors, *Emerging Technologies in Hazardous Waste Management V*, number 607 in ACS Symposium Series, pages 102–120. ACS Books, 1995.

23. G.A. Reimann, J.D. Grandy, and G.L. Anderson. Iron-enriched basalt waste forms. In D.W. Tedder and F.G. Pohland, editors, *Emerging Technologies in Hazardous Waste Management V*, number 607 in ACS Symposium Series, pages 121–134. ACS Books, 1995.

24. G.J. Thomas, D.D. Reible, K.T. Valsaraj, L.J. Thibodeaux, and D. Timberlake. Capping of contaminated sediments: Experimental results and validation of mathematical models. In D.W. Tedder and F.G. Pohland, editors, *Emerging Technologies in Hazardous Waste Management VI*, pages 293–310. American Academy of Environmental Engineers, 1996.

25. D.K. Peeler and P.R. Hrma. Compositional range of durable borosilicate simulated waste glasses. In D.W. Tedder and F.G. Pohland, editors, *Emerging Technologies in Hazardous Waste Management VI*, pages 323–338. American Academy of Environmental Engineers, 1996.

26. K.J. Liekhus, M.E. Vaughn, B.A. Jensen, and M.J. Connolly. Method of characterizing VOC concentration in vented waste drums with multiple layers of confinement using limited sampling data. In D.W. Tedder and F.G. Pohland, editors, *Emerging Technologies in Hazardous Waste Management VI*, pages 339–348. American Academy of Environmental Engineers, 1996.

27. R.S. Melarkode, A.C. Bumb, and W.S. Phillips. Use of an analytical transport model to minimize sampling lactions. In D.W. Tedder and F.G. Pohland, editors, *Emerging Technologies in Hazardous Waste Management VI*, pages 349–362. American Academy of Environmental Engineers, 1996.

28. J.R. Smith. Air-nitric acid destructive oxidation of organic wastes. In D.W. Tedder and F.G. Pohland, editors, *Emerging Technologies in Hazardous Waste Management V*, number 607 in ACS Symposium Series, pages 156–162. ACS Books, 1995.

29. T.Y. Xiong, D.K. Fleming, and S.A. Weil. Hazardous material destruction in a self-regeneration combustor-incinerator. In D.W. Tedder and F.G. Pohland, editors, *Emerging Technologies in Hazardous Waste Management II*, number 468 in ACS Symposium Series, pages 12–28. ACS Books, 1991.

30. M. Flytzani-Stephanopoulos, A.F. Sarofim, L. Tognotti, H. Kopsinis, and M. Stoukides. Incineration of contaminated soils in an electrodynamic balance. In D.W. Tedder and F.G. Pohland, editors, *Emerging Technologies in Hazardous Waste Management II*, number 468 in ACS Symposium Series, pages 29–49. ACS Books, 1991.

31. Q.Y. Han, Q.D. Zhuang, J.V.R. Heberlein, and W. Tormanen. Thermal plasma destruction of hazardous waste with simultaneous production of valuable co-products. In D.W. Tedder and F.G. Pohland, editors, *Emerging Technologies in Hazardous Waste Management V*, number 607 in ACS Symposium Series, pages 135–143. ACS Books, 1995.

32. J.L. Graham, B. Dellinger, D. Klosterman, G. Glatzmaier, and G. Nix. Disposal of toxic wastes by using concentrated solar radiation. In D.W. Tedder and F.G. Pohland, editors, *Emerging Technologies in Hazardous Waste Management II*, number 468 in ACS Symposium Series, pages 83–109. ACS Books, 1991.

33. R.G. Hickman, J.C. Farmer, and F.T. Wang. Mediated electrochemical process for hazardous waste incineration. In D.W. Tedder and F.G. Pohland, editors, *Emerging Technologies in Hazardous Waste Management III*, number 518 in ACS Symposium Series, pages 430–451. ACS Books, 1993.

2

CHELATING AGENTS FOR EXTRACTION OF HEAVY METALS FROM SOIL

Complexing Power, Selectivity, And Recoverability

P. K. Andrew Hong*, Chelsea Li, Weimin Jiang, Ting-Chien Chen[†], and Robert W. Peters[‡]

Department of Civil and Environmental Engineering
University of Utah
Salt Lake City, Utah 84112
USA

ABSTRACT

Chelating extraction of heavy metals from contaminated soils is seen as a remediation technique. This work addresses important consideration in the application of chelating agents for soil remediation, namely the complexing power, selectivity, and recoverability of the chelators with respect to heavy metal contaminants. To address these issues, an assessment technique was developed to evaluate the chelators based on complexation equilibria of the chelators toward the target metals including Pb, Cu, Cd, Zn, Ni, and Hg. Predictions in terms of complexing power and selectivity were made using this technique and were compared to experimental extraction results using several chelators including L-5-glutamyl-L-cysteinylglycine (GCG), ethylenediaminetetraacetic acid (EDTA), nitrilotris(methylene)triphosphonic acid (NTTA), and trimethylenedinitrilotetraacetic acid (TMDTA). Experimental results showed an increasing complexing ability in the order of GCG < NTTA < TMDTA < EDTA, and an increasing selectivity toward the metal contaminants in the order of EDTA < TMDTA < GCG < NTTA. These results were largely consistent with equilibrium predictions. Enhanced recovery of metals from a very strong chelator EDTA was demonstrated with the use of metal-competing cationic precipitants (calcium hydroxide and ferric chloride) or anionic sulfide precipitant.

---

* To whom correspondence should be addressed.
[†] Present Address: Department of Environmental Engineering and Health, Tajen Junior College of Pharmacy, En-Pu Hsiang, Pingtung, Taiwan
[‡] Energy Systems Division, Argonne National Laboratory, Argonne, IL 60439-4815, USA

Emerging Technologies in Hazardous Waste Management 8, edited by Tedder and Pohland
Kluwer Academic/Plenum Publishers, New York, 2000.

1. INTRODUCTION

Economic activities in the industrialized countries in the past have resulted in significant contamination of environmental resources including soil and groundwater. These activities include vehicle operation, mining, smelting, metal plating and finishing, battery production and recycling, agricultural and industrial chemical application, and incineration processes. In the U.S., thousands of hazardous waste sites have been placed on the National Priority List for clean-up of heavy metals under Superfund. Metals of particular interest are cadmium, chromium, copper, lead, mercury, nickel, and zinc. Heavy metals contamination has also occurred at military installations with lead, chromium, and cadmium being most prevalent. Military operations that contributed to heavy metals contamination include those in firing ranges, ammunition manufacturing facilities, and open burning pits. Heavy metals are increasingly recognized as major contaminants at military sites, and they now make up five of the six hazardous substances monitored most frequently at Army installations.[1]

Chelating extraction of heavy metals in soil washing operation is seen as an effective remediation technique for contaminated soils. Chelators that have been used for extraction of heavy metals from soils include EDTA, NTA, DTPA, formic, succinic, oxalic, citric, acetic, humic, and fulvic acids, glycine, cysteine, SCMC, ADA, PDA, and others.[2-15] While many studies have shown that chelating agents are effective for metals removal, relatively little work have been published relative to the reclamation and reuse of chelating agents. Chelating extraction of heavy metals with an emphasis on recovering and reusing the chelating agents has only been addressed recently.[15-21] These recent studies have demonstrated the recovery of metals following chelating extraction via (1) the use of electrochemical method to recover strong chelator EDTA, and (2) the elevation of solution pH to recover those chelators of moderate strength.

This paper essays to address three important aspects of chelating extraction for remediation of heavy metals contaminated soils, namely the complexing power, the selectivity, and the recoverability of the chelating agents. Outlined in this paper are assessment techniques for chelators in terms of their complexing ability and selectivity toward target metal contaminants, and the results of evaluation for about 250 chelators employing this technique. Experimental results on extraction and selectivity are presented for several chelators. The recovery of EDTA, a strong chelator with the strength that renders its separation from the extracted metal challenging, was successfully shown in the laboratory.

2. ASSESSMENT TECHNIQUES AND RESULTS

2.1. Chemistry of Metals Extraction Using Chelating Agents

2.1.1. Metal Speciation in Natural Waters. In the presence of ambient natural ligands such as HCO_3^-, CO_3^{2-}, Cl^-, SO_4^{2-}, an aqueous divalent contaminant metal (Me^{II}_{aq}) speciates in various free and complex forms:

$$Me_{aq} = Me^{2+} + Me(OH)_x^{(2-x)} + Me_x(OH)_y^{(2x-y)} + Me(H_nCO_3)_x^{(2-2x+nx)} + MeCl_x^{(2-x)}$$
$$+ Me(SO_4)_x^{(2-x)}$$

The solubilities of contaminant metals are controlled by predominant mineral phases such as $MeO(s)$, $Me(OH)_2(s)$, $MeCO_3(s)$, $Me_x(OH)_y(CO_3)_z(s)$, and $MeSO_4(s)$, depending upon the pH and/or ambient ligands available.

2.1.2. Acid-Base Equilibrium of Chelating Agents. Effective chelating agents typically have multiple coordination sites (i.e., ligand atoms) available for complexation with a metal center. Chelators are often multi-protic acids (H_nL) capable of undergoing acid-base equilibrium reactions in the aqueous phase, e.g.:

$$H_nL = H^+ + H_{n-1}L^- \qquad pK_1$$

and subsequently,

$$H_{n-m}L^{m-} = H^+ + H_{n-(m+1)}L^{-(m+1)} \qquad pK_{m+1}$$

2.1.3. Metals Complexation with Chelating Agents. Each conjugate acid/base of the chelating agent may form a strong complex with the metal, resulting in the formation of various complexes $Me_x(H_{n-m}L)_y^{2x-my}$:

$$xMe^{2+} + yH_{n-m}L^{m-} = Me_x(H_{n-m}L)_y^{2x-my} \qquad pK$$

with the total complexes concentration (MeL_{Tot}) given by MeL_{Tot}:

$$MeL_{Tot} = \Sigma Me_x(H_{n-m}L)_y^{2x-my}$$

Thus, the total metal solubility, Me_{Tot}, is computed by:

$$Me_{Tot} = Me_{aq} + MeL_{Tot}$$

The complexation power of chelating agents toward heavy metals can be determined based on the equilibrium computation procedures formulated above. The strong chelator often demonstrates a total solubility with the chelator (Me_{Tot}) that is orders of magnitude higher than that without the chelator (Me_{aq}).

2.2. Assessments of Complexing Power and Selectivity of Chelating Agents

The metal-ligand complexation equilibrium constant expresses the ligand's affinity toward the target metal (divalent as shown):

$$Me^{2+} + L^{n-} = MeL^{2-n} \qquad pK_{Me}$$

In evaluating chelators, we have examined the equilibrium constants (pK's) of about 250 chelators for complexation with 6 divalent target metals including Pb, Cu, Cd, Zn, Ni, and Hg (Smith and Martell, 1974; 1976; 1982; 1989). The affinity of a chelator toward the target metals, whenever the equilibrium constants are available, was computed as an average pK_{target} value:

$$pK_{target} = (pK_{Pb} + pK_{Cu} + pK_{Cd} + pK_{Zn} + pK_{Ni} + pK_{Hg})/6$$

The affinity of the chelator toward ambient cations such as Ca and Mg was, when available, determined by the average $pK_{ambient}$ value:

$$pK_{ambient} = (pK_{Ca} + pK_{Mg})/2$$

Thus, a selectivity ratio can be defined as SR, where:

$$SR = pK_{targer}/pK_{ambient}$$

According to this formulation, a strong chelator will have large pK_{target} and $pK_{ambient}$ values, whereas a strong and selective chelator toward heavy metals will have a large pK_{target} and a relatively small $pK_{ambient}$, thus a large SR value. Table 1 identifies approximately 40 chelators that have relatively large pK_{target} (>12) and SR (>2) values. This table may serve to guide in the selection of chelators that are both effective and selective toward the 6 target metals.

2.3. Recovery of Moderately Strong and Very Strong Chelating Agents

For a system containing both the ligand and metal at a specified ligand to metal concentration ratio, a degree of complexation (DOC) parameter can be taken as the ratio of the total metal solubility with chelation to the aqueous solubility without chelation, i.e., Me_{Tot}/Me_{aq}.[22] A chelating agent with a large DOC value (>>1) indicates a high degree of complexation of the metal as afforded by the chelating agent and thus a significant enhancement of the solubilization of the metal. Contrarily, a small DOC value (≤ 1) of a chelating agent indicates little enhancement by chelation or the likely occurrence of precipitation of the metal as a mineral at the specified pH of consideration. The effective pH ranges within which that the chelating agents bind adequately with the target metals (Pb, Cu, Cd, Zn, Ni, and Hg) have been reported for about 200 chelating agents.[15,17] For strong chelating agents, the DOC's are high over a broad pH range (e.g. 3 to 12); however, for chelating agents of moderate strength, the DOC's are high at low or moderate pH but low at high pH. This means that a chelator of moderate strength can enhance metal solubilization by complexation at a lower pH but release the complexed metal as hydroxide or other precipitates at an elevated pH condition. This feature can be exploited to separate the chelator from extracted heavy metals, thus allowing the chelator for reuse. Based on this principle, the extraction, recovery, and reuse of several chelators of moderate strength have been demonstrated for Pb, Cu, Cd, and Zn.[15,17-21]

Although the use of strong and selective chelating agents is desirable, the recovery of strong chelating agents such as EDTA for reuse remains a challenge. Beyond the control of pH as a recovery technique, this work further explores the addition of cationic precipitants such as calcium hydroxide and ferric chloride and anionic sulfide precipitant to effect the separation and recovery of spent EDTA. This alternative approach is based on the competition of the added cations (e.g., Ca^{2+} and Fe^{3+}) with the target metal replacing the target metal in the complex, or on the competition of the added anions (e.g., S^{2}) with EDTA for the target metal, in either case resulting in the formation of precipitates of the target metal.

Table 1. Strong and selective chelating agents ($pK_{target} > 12$; SR > 2), as listed in the order of increasing number of carbon atoms in the molecule

Chelating agents	Formula	pK target	SR
N,N-Bis(phosphonomethyl) glycine	$C_4H_{11}O_8NP_2$	12.5	2.0
DL-2-carboxymethyliminodiacetic acid (N-carboxymethylaspartic acid)	$C_6H_9O_6N$	13.0	3.1
N-(2-Mercaptoethyl)iminodiacetic acid	$C_6H_{11}O_4NS$	15.9	3.5
*Ethyphosphinodiacetic acid	$C_6H_{11}O_4P$	25.0	34.2
Ethylenediiminodiacetic acid (EDDA)	$C_6H_{12}O_4N_2$	12.3	3.2
N-(2-Aminoethyl)iminodiacetic acid	$C_6H_{12}O_4N_2$	12.3	2.7
DL-2-carboxy-2'iminodiacetic acid	$C_7H_{11}O_6N$	12.6	3.8
DL-2-(Carboxymethyl)nitrilotriacetic acid	$C_8H_{11}O_8N$	12.7	2.2
Ethylenediiminodipropanedioic acid (EDDM)	$C_8H_{12}O_8N_2$	14.3	2.9
Ethylenedinitrilo-N,N-diacetic-N',N'-bis(methylenephosphonic) acid	$C_8H_{18}O_{10}N_2P_2$	16.2	4.4
Ethylenediiminodibutanedioic acid (EDDS)	$C_{10}H_{16}O_8N_2$	15.0	3.0
Oxybis(ethyleneiminomalonic) acid	$C_{10}H_{16}O_9N_2$	13.7	3.3
N-(2-Hydroxyethyl)ethylenedinitritrilotriacetic acid (HEDTA)	$C_{10}H_{18}O_7N_2$	16.5	2.2
*Ethylenedinitrilo-N,N'-diacethydroxamic-N,N'-diacetic acid	$C_{10}H_{18}O_8N_4$	17.0	2.9
*Oxybis[ethyleneimino(dimethy)methylenephosphonic acid]	$C_{10}H_{26}O_7N_2P_2$	12.0	2.6
N-(2-Hydroxybenzyl)iminodiacetic acid	$C_{11}H_{13}O_5N$	14.5	2.1
Trinethylenedinitrilotetraacetic acid(TMDTA)	$C_{11}H_{18}O_8N_2$	16.6	2.4
DL-1,3-Diamino-2-hydroxypropane-N,N'-dibutanedioic acid	$C_{11}H_{18}O_9N_2$	15.3	4.6
(2-Hydroxytrimethylene)dinitrilotetraacetic acid	$C_{11}H_{18}O_9N_2$	15.6	2.6
*DL-1-Methylethylenedinitrilotetraacetic acid N,N'-diamide	$C_{11}H_{20}O_6N_4$	13.2	2.2
Ethylenediiminodi-2-pentanedioic acid(EDDG)	$C_{12}H_{20}O_8N_2$	12.9	4.6
Meso-(1,2-Dimethylethylene)dinitrilotetraacetic acid	$C_{12}H_{20}O_8N_2$	18.8	2.0
*Ethylenedinitrilo-N,N'-di(3-propanoic)-N,N'-diacetic acid	$C_{12}H_{20}O_8N_2$	14.3	2.1
Tetramethylenedinitrilotetraacetic acid	$C_{12}H_{20}O_8N_2$	15.5	2.6
Thiobis(ethylenenitrilo)tetraacetic acid (TEDTA)	$C_{12}H_{20}O_8N_2S$	16.2	3.0
1,4,7-Triazacyclonoane-1,4,7-triacetic acid	$C_{12}H_{21}O_6N_3$	18.2	2.1
1,4,7,10,13,16-Hexaazacyclooctadecane([18]aneN$_6$)	$C_{12}H_{30}N_6$	20.0	8.3
*DL-(1,3-Dimethyltrimethylene)dinitrilotetraacetic acid	$C_{13}H_{22}O_8N_2$	17.0	2.3
Pentamethylenedinitrilotetraacetic acid	$C_{13}H_{22}O_8N_2$	12.8	2.6
*N-Methylnitrilobis(ethylenenitrilo)tetraacetic acid	$C_{13}H_{23}O_8N_3$	21.1	2.5
Diethylenetrinitrilopentaacetic acid(DTPA)	$C_{14}H_{23}O_{10}N_3$	20.7	2.1
Hexamethylenedinitrilotetraacetic acid	$C_{14}H_{24}O_8N_2$	14.2	3.4
Ethylenebis(oxyethylenenitrilo)tetraacetic acid (EGTA)	$C_{14}H_{24}O_{10}N_2$	16.3	2.0
Oxybis(trimethylenetrilo(tetraacetic acid	$C_{14}H_{24}O_9N_2$	16.5	3.2
N-Benzylethylenedinitrilo-N,N',N'-triacetic acid	$C_{15}H_{20}O_6N_2$	16.8	2.5
Ethylanabis-N,N'-(2,6-dicarboxy)piperidine	$C_{16}H_{24}O_8N_2$	16.1	2.4
*Octamethylenedinitrilotetraacetic acid	$C_{16}H_{28}O_8N_2$	14.3	3.0
*Ethylenebis[(N-methylnitrilo)ethylenenitrilo]tetraacetic acid	$C_{16}H_{30}O_8N_4$	27.7	4.0
N'-(Allyloxyethyl)diethylenetriaminetetraacetic acid	$C_{17}H_{29}O_9N_3$	17.3	17.3
1,4,7,10-Tetraazacyclotridecane-1,4,7,10-tetraacetic acid (TRITA)	$C_{17}H_{30}O_8N_4$	20.6	2.1
Ethylenediiminobis[(2-hydroxy phenyl)acetic acid] (EHPG)	$C_{18}H_{20}O_6N_2$	18.7	2.5
*N,N'-Bis(2-pyridylmethyl)ethylenedinitrilo-N,N'-diacetic acid	$C_{18}H_{22}O_4N_4$	16.7	2.5
Triethylenetetranitrilohexaacetic acid (TTHA)	$C_{18}H_{30}O_{12}N_4$	20.1	2.2
1,4,8,11-Tetraazacyclotetradecane-1,4,8,11-tetraacetic acid (TETA)	$C_{18}H_{32}O_8N_4$	19.3	3.7
21,24-Dimethyl-4,7,13,16-tetraoxa-1,10,21,24-tetraazabicyclo-[8,8,8]hexacosane	$C_{20}H_{42}O_4N_4$	12.7	3.7

3. EXPERIMENTAL METHODS

Deionized water ($18\,M\Omega$-cm) obtained from a Milli-Q system (Millipore) was used throughout. Chelating agents EDTA (Sigma), GCG (Aldrich), NTTA (Aldrich), and TMDTA (Fluka) were used as received. Soils were taken from a contaminated site in Salt Lake City, air-dried for a month, then passed through a 2 mm sieve. Soil properties were characterized by Utah State University Soil Testing Laboratory: loamy sand; sand, 78%; silt, 14%; clay, 8%; very coarse, 7.5%; coarse, 25.5; medium, 20.1%; fine, 30.1; very fine,

0.1%; organic matter, 2.67%; organic carbon, 1.55%; Cu, 440 mg/Kg; Pb, 1,900 mg/Kg; Ni, 23.3 mg/Kg; Zn, 1,800 mg/Kg; and As, 685 mg/Kg. Typical experiments were conducted in 125 mL glass Erlenmeyer flasks using a batch solution volume (V) of 100 mL. All flasks were sealed with stoppers to reduce exchange with the atmosphere during experiments. All pH adjustments were performed manually by addition of either a 5 M HNO_3 or NaOH solution. pH measurements were with an Orion model SA 720 pH meter. Stock metal solutions (1,000 mg/L) were prepared according to ASTM Methods. A gyratory shaker table (New Brunswick Scientific Co., Model G-2) provided agitation during extraction procedures. The soil was kept in suspension by operating the shaker table at 260 rpm. All experiments were conducted at the room temperature of $23 \pm 1\,°C$. Total dissolved metal concentrations (Me_T) were measured in aliquots withdrawn from the reaction mixtures and filtered through a 0.45 μm filter (Gelman Sciences sterile aerodisc), then acidified with nitric acid. Metal analyses were by atomic absorption (AA) spectrometry (Perkin Elmer Model 280) using ASTM methods.[23] Standard procedures were followed when available.[23-26] All equilibrium constants used in calculation were obtained from Smith and Martell.[27]

Extraction experiments were conducted with 5% soil slurries (i.e., 5 g of soil in 100 mL of chelator solution). Concentrations of chelators were 10–50 mM. To commence extraction, the soil was added to the chelator solutions and the pH was initially adjusted to about pH 7.5 if necessary. The mixture was continuously maintained in suspension by a shaker table. Either 4 or 24 hours was allowed for extraction. The tests of chelator recovery were performed by adding various amounts of precipitant solids (cationic or anionic) into solutions of metal-ligand complexes. The mixtures were continually agitated for 1 or 2 hours to promote precipitation, then centrifuged, filtered, and the concentrations of metals in the filtrate analyzed.

4. RESULTS AND DISCUSSION

Table 2 shows the results of metals extraction from the contaminated soil sample using chelators at different concentrations. The extraction was carried out using a 100 mL of 5% soil slurry at pH 7.4 to 8.4 for 24 hours. The extraction results were compared to metal loadings characterized using a 10% HNO_3 extraction for 4 hours. As shown, the extraction levels with all chelators increased with increasing chelator concentration, with 50 mM of EDTA extraction of Pb (the major contaminant) approaching 90% of that achieved by 10% HNO_3. The extraction effectiveness among the chelators appears to follow this decreasing order: EDTA > TMDTA > NTTA > GCG. This observation is consistent with assessment results based on comparisons of the pK_{target}, which predict the chelators' complexing power in exactly the same order (i.e., pK_{target}'s = 18.3, 16.6, 14.1, and 10.9 for EDTA, TMDTA, NTTA, and GCG, respectively). The amounts of Ca extracted by these chelators appear to follow this increasing order: NTTA ~ GCG < TMDTA < EDTA. A comparison of SR's of the chelators predicts the selectivity in this decreasing order: GCG > TMDTA > NTTA ~ EDTA (i.e., SR's = 2.87, 2.44, 1.92, and 1.90 for GCG, TMDTA, NTTA, and EDTA, respectively). According to the relative amounts of Pb and Ca extracted, the trend in selectivity of the chelators toward target metals is consistent with the predictions, with an exception for NTTA which seems to show, relative to other chelators, a greater preference for target metals than predicted. It should be noted that the pK's of NTTA vary greatly among the different metals with pK's of 17.4, 11.6, 11.1, and 16.4 for Cu, Cd, Ni, and Hg, respectively, and pK's not being available for Pb and Hg. However,

Table 2. Metals extraction from soil using different chelator concentrations. Conditions were 100 mL of 5% soil slurry, extraction pH 7.4–8.4, 24 hrs flask shaking. Metal loadings (10% HNO_3 for 4 hrs) were Pb = 2,320 g/Kg, Cu = 246 mg/Kg, Ni = 27.0 mg/Kg, Zn = 391 mg/Kg, Fe = 9,850 mg/Kg, and Ca = 10,100 mg/Kg

Chelator	Metal (mg/Kg)	Chelator Concentration (mM)		
		50	20	10
1. EDTA	Pb	2,080	1,970	1,570
	Cu	109	93.4	68.4
	Ni	9	5.4	3.8
	Zn	231	194	154
	Fe	588	621	408
	Ca	11,500	8,990	5,400
2. NTTA	Pb	1,690	1,330	1,080
	Cu	83.6	74.6	58.8
	Ni	10.8	8.6	5.2
	Zn	272	232	167
	Fe	969	797	984
	Ca	4,140	1,790	1,570
3. TMDTA	Pb	1,820	1,500	1,410
	Cu	87	63.4	55.8
	Ni	8.6	4.8	3.2
	Zn	189	167	157
	Fe	346	248	232
	Ca	9,400	7,330	5,090
4. GCG	Pb	1,170	835	162
	Cu	45.4	30.8	162
	Ni	5.4	3.8	3.2
	Zn	141	64.8	53.4
	Fe	138	53.4	74.4
	Ca	4,890	1,870	1,350

the high selectivity of NTTA is not surprising as it contains three P atoms as the ligand atoms which highly prefer the target metals of higher polarizability.

Table 3 shows the results of metals extraction from the contaminated soil sample during consecutive runs. The extraction was carried out using a 100 mL of 5% soil slurry at pH 7.4 to 8.1 for 4 hours during each run. The results show the same trends in the chelators' effectiveness (i.e., EDTA > TMDTA > NTTA) and selectivity (i.e., NTTA > TMDTA > EDTA). These results further show that during 3 consecutive washings with 20 mM of the chelators, Pb removals ranging from 85% to nearly 100%, as compared to removals using 10% HNO_3 for 4 hours, can be achieved.

It should be noted that the introduction of an average pK (pK_{target}) was intended to provide an estimate of the general complexing ability of a chelator toward six most prevalent metal contaminants, and that this average was computed using metals for which the complexation equilibrium constants are available. There are inherent errors in this approach, e.g., a singularly large or small pK for one metal may over-estimate or under-estimate complexation for other metals, and be consequential in the estimation of the SR values as well. An alternative, likely more accurate, approach is to tailor to a specific contaminated soil by calculating the pK_{target} of the chelator based only on the major contaminant metals actually occurring in that soil.

Table 3. Metals extraction from soil during consecutive runs.
Conditions were 20 mM chelator, 100 mL of 5% soil slurry,
extraction pH 7.4–8.1, 4 hrs flask shaking. Metal loadings
(10% HNO_3 for 4 hrs) were Pb = 2,320 mg/Kg, Cu = 246 mg/Kg,
Ni = 27.0 mg/Kg, Zn = 391 mg/Kg, Fe = 9,850 mg/Kg, and
Ca = 10,100 mg/Kg

Chelator	Metal (mg/Kg)	Consecutive Extraction Run			Runs Total
		1	3	5	
1. EDTA	Pb	1,690	431	262	2,380
	Cu	61.8	7.60	20.4	89.8
	Ni	12.2	8.8	5.6	26.6
	Zn	101	30.2	37.4	169
	Fe	227	117	548	892
	Ca	6,270	1,360	1,030	8,660
2. NTTA	Pb	1,160	442	369	1,970
	Cu	38.4	6.4	19.4	64.2
	Ni	20.8	5.8	5.8	32.4
	Zn	93.6	22	33	149
	Fe	351	209	721	1,280
	Ca	1,640	1,420	1,280	4,340
3. TMDTA	Pb	1,300	569	392	2,260
	Cu	49.6	17.2	23	89.8
	Ni	9	5.6	5.2	19.8
	Zn	68.4	33	23.4	125
	Fe	116	154	453	723
	Ca	3,980	2,290	1,670	7,940

The recovery of metals and the chelator from metal-EDTA complexes was studied using different cationic (Ca^{2+}, Fe^{3+}) and anionic (S^{2-}, PO_4^{3-}, CO_3^{2-}) precipitants. Table 4 shows results of separating Pb from EDTA complexes employing different amounts of cationic precipitants. The separation was promoted by contacting the Pb-EDTA complex solution with added $Ca(OH)_2$ and $FeCl_3$ soilds for 2 hours with pH adjusted to 10.0–10.5. The results indicated that the recovery of Pb was far from completion (i.e., 7–50%) when neither or only one of the precipitants was added. However, the recovery was nigh complete when both precipitants were added. The enhanced separation of Pb was attributed to Ca^{2+} cation competing with Pb^{2+} for EDTA and the subsequent coagulation/co-precipitation of oxyhyroxides of Fe^{3+} and Pb^{2+}, resulting in the significant removal of Pb from the solution and the Ca-EDTA complex remaining in the solution.

Based on the data presented in Table 4, the appropriate amounts of addition for the effective recovery of Pb (>97%) appeared to be 60–100 mg/L of $Ca(OH)_2$ and 200–600 mg/L of $FeCl_3$ for the tested conditions. In the interest of generating a lesser amount of sludge to be further handled, the addition of 100 ppm of $Ca(OH)_2(s)$ together with 200 ppm of $FeCl_3(s)$ would be appropriate for the tested Pb loading in the solution. On the mole basis, this translates to be approximately 2 times excess of Ca^{2+} over Pb^{2+}, whereas the amount of excess $FeCl_3$ is secondary because its major role appears to be in the coprecipitation of lead and iron oxyhydroxide.

The separation of metals from their EDTA complexes was also studied by adding various amounts of anionic precipitant sulfide, phosphate, or carbonate, respectively, to

Table 4. Separation of lead from EDTA as a function of
Ca and Fe dosages. Conditions were Initial: $pH_0 < 2.0$,
$[Pb]_0 = 300\,ppm$, $[EDTA] = 1\,mM$. At separation:
pH = 10.0–10.5, time = 2 hrs

Solid Added		Measured amounts of metals remaining in solution			
$Ca(OH)_2$ (mg/L)	$FeCl_3$ (mg/L)	Pb (mg/L)	Ca (mg/L)	Fe (mg/L)	Pb Recovery* (%)
0	0	280	0.3	<1	6.70
0	600	143	8.8	1.1	52.3
100	0	150	450	<1	50.0
100	200	8.7	390	<1	97.0
100	300	4	385	<1	98.7
100	400	3.2	360	<1	98.9
100	500	0.95	340	<1	99.8
100	600	0.95	300	<1	99.8
100	700	0.6	290	<1	99.8
19	600	29	27	1.2	90.3
37	600	14	39	0.2	95.3
56	600	6	120	0.2	98.0
74	600	6	184	0.1	98.0
93	600	2.8	212	0.1	99.1
110	600	2.3	404	0.3	99.2
150	600	1.3	—	0.1	99.6
190	600	2	—	0.2	99.3
240	600	2.8	—	1.3	99.1
300	600	3	—	0.4	99.0
350	600	2.3	—	0.6	99.2
410	600	1.3	—	0.2	99.6

*Calculation based on decrease of Pb concentration in solution from an
initial level of 300 ppm.
— Not measured.

EDTA complexes solution containing Pb, Cu, and Cd. Table 5 shows the recovery of Pb, Cu, and Cd from their EDTA complexes at different sulfide precipitant concentrations at an adjusted solution pH of 10.5. As shown, the ease in recovering the metals follows a decreasing order Cu > Cd > Pb. This trend of recovery is largely determined by the solubility of the metal in distilled or natural water at a given pH, as well as the stability (or solubility) of the minerals formed in the presence of the natural and added ligands such as sulfide, phosphate, or carbonate.

The results of Table 5 show that sulfide is an effective precipitant for the metals. Under the studied conditions, sulfide at about 3 times mole excess (10 mM) of the total metal (3 mM) was sufficient for the effective recovery of Pb, Cu, and Cd (to below detection level or 0.1 ppm). Additions of phosphate and carbonate were also carried out separately under conditions identical to the case of sulfide. While phosphate showed marginal effects on the recovery of Pb, its use in the recovery of Cu and Cd was dismal even at 20 mM. Carbonate offered little aid in the recovery of the metals even at 20 mM. The enhanced separation of metals with sulfide was attributed to the relatively successful competition of sulfide with EDTA for the metals, resulting in the formation of metal sulfide precipitates and free EDTA remaining in the solution.

Table 5. Recovery of lead, copper, and cadmium from EDTA using an anionic sulfide precipitant. Conditions were pH = 10.5, $[Pb]_0 = [Cu]_0 = [Cd]_0 = 1 \, mM$, $[EDTA] = 5 \, mM$, time = 1 hr

Sulfide Precipitant (mM)	Measured amounts of metals remaining in solution		
	Pb (ppm)	Cu (ppm)	Cd (ppm)
0	154	52.8	97.6
3	117	8.4	33
6	51.2	nd	0.3
10	nd	nd	0.1
20	nd	nd	0.1

nd = not detected.

5. CONCLUSIONS

An assessment technique was formulated to evaluate the complexing power and selectivity of chelators toward six metal contaminants. Using this technique, pK_{target} and SR parameters were computed for about 250 chelators that resulted in 40 of them being identified as strong and selective. Several chelators (EDTA, NTTA, TMDTA, and GCG) were experimentally tested for extraction and selectivity for lead in a contaminated soil. The experimental results largely supported assessment results. Therefore, the two parameters, pK_{target} and SR are recommended for rapid evaluation of potential chelators regarding their ability and selectivity, respectively, for extraction of target metals from contaminated soils. A chelator showing a high pK_{target} (>12 or higher) that is computed based on actual major contaminant metals of the soil would be particularly effective, whereas a chelator having a high SR value (>2) computed likewise would be more selective. This work also demonstrates a solution to a problem concomitant with a strong chelator, i.e., the recovery and reuse of it. The recovery of a strong chelator such as EDTA is achievable by the addition of a calcium precipitant (as $Ca(OH)_2(s)$) at a calcium-to-metal ratio of 2, along with sufficient $FeCl_3$ for coprecipitation and coagulation of the metal. Effective separation of the metal from a strong chelator is also achievable by an addition of anionic sulfide precipitant (as NaS(s)) at a sulfide-to-metal mole ratio of 3.

ACKNOWLEDGMENT

We thank the Great Plains/Rocky Mountain Hazardous Substance Research Center for project funding.

REFERENCES

1. Bricka, R., and C. Williford, L. Jones (1994) "Heavy metal soil contamination at U.S. Army installations: Proposed research and strategy for technology Development," Technical Report IRRP-94-1, U.S. Army Engineer Waterways Experiment Station, Vicksburg, MS.
2. Brewster, M.D., R.W. Peters, M.P. Henry, T.L. Patton, and L.E. Martino (1995). "Physical/Chemical Treatment of Heavy Metals-Contaminated Soils", Paper presented at the 2nd International Symposium on Waste Processing and Recycling in Mineral and Metallurgical Industries", Vancouver, British Columbia, Canada, August 19–23.

3. Bulman, R.A., A.J. Wedgwood, and G. Szabo (1992). "Investigations into the Chemical Forms of ^{239}Pu in a West Cumbrian Saltmarsh Soil Radiolabelled by an Environmental Process", *Sci. Total Environ.*, *114*(1–3):215–226.

4. Burckhard, S.R., A.P. Schwab, and M.K. Banks (1995). "The Effects of Organic Acids on the Leaching of Heavy Metals from Mine Tailings", *J. Haz. Mater.*, *41*(1–2):135–145.

5. Davis, A.P. and I. Singh, (1995). "Washing of Zinc(II) from Contaminated Soil Column", *J. Environ. Engrg. (ASCE)*, *121*(2):174–185.

6. Doepker, R.D. (1991). "Enhanced Metal Mobilization through Leachants Containing Acetate Ion", pp. 365–381 in *Emerging Technologies in Hazardous Waste Management II, ACS Sympos. Series 468*, D.W. Tedder and F.G. Pohland, Eds., American Chemical Society, Washington, D.C.

7. Elliott, H.A., G.A. Brown, G.A. Shields, and J.H. Lynn (1989). "Restoration of Pb-Polluted Soils by EDTA Extraction", pp. 64–67 in *7th Internat. Conf. on Heavy Metals in the Environ.*, *Vol. II*, J.-P. Vernet, Ed., Geneva, Switzerland (1989).

8. Ellis, W.D., T.R. Fogg, and A.N. Tafuri (1986). "Treatment of Soils Contaminated with Heavy Metals", pp.201–207 in *Land Disposal, Remedial Action, Incineration and Treatment of Hazardous Waste*, 12th Ann. Res. Sympos., EPA 600/9–86/022, Cincinnati, OH.

9. Hessling, J.L., M.P. Esposito, R.P. Traver, and R.H. Snow (1989). "Results of Bench-Scale Research Efforts to Wash Contaminated Soils at Battery-Recycling Facilities", pp. 497–514 in *Metals Speciation, Separation, and Recovery, Vol. II*, J.W. Patterson and R. Passino, Eds., Lewis Publishers, Inc., Chelsea, MI.

10. Kocher, W.M. (1995). "The Use of Soil Washing Processes for the Reclamation and Reuse of Foundry Waste Sands", *Proc. 27th Mid-Atlantic Indus. Waste Conf.*, 27:450–459.

11. Peters, R.W. and L. Shem (1992), "Use of Chelating Agents for Remediation of Heavy Metal Contaminated Soil", *ACS Sympos. Series 509 on Environmental Remediation: Removing Organic and Metal Ion Pollutants*, G.F. Vandegrift, D.T. Reed, and I.R. Tasker, Eds., *509*:70–84, American Chemical Society, Washington, D.C.

12. Peters, R.W. (1995). *Feasibility/Treatability Studies for Removal of Heavy Metals from Training Range Soils at the Grafenwöhr Training Area, Germany*, ANL/ESD/TM-81, Argonne National Laboratory, Argonne, IL.

13. Sistani, K.R., D.A. Mays, R.W. Taylor, and C. Buford (1995). "Evaluation of Four Chemical Extractants for Metal Determinations in Wetland Soils", *Commun. Soil Sci. Plant Anal.*, *26*(13–14):2167–2180.

14. Tsang, K.W., P.R. Dugan, and R.M. Pfister (1992). "Mobilization of Bi, Cd, Pb, Th, and U Ions from Contaminated Soils and the Influence of Bacteria on the Process", pp. 78–93 in *Emerging Technologies in Hazardous Waste Management IV*, D.W. Tedder and F.G. Pohland, Eds., ACS Sympos. Series No. 554, American Chemical Society, Washington, D.C.

15. Chen, T.-C., E. Macauley, and A. Hong (1995). "Selection and Test of Effective Chelators for Removal of Heavy Metals from Contaminated Soils", *Canadian Journal of Civil Engineering*, 22:1185–1197.

16. Allen, H.E. and P.-H. Chen (1993). "Remediation of Metal Contaminated Soil by EDTA Incorporating Electrochemical Recovery of Metal and EDTA", *Environ. Prog.*, *12*(4):284–293.

17. Hong, A., T.-C. Chen, and R. Okey (1995a). "Chelating Extraction of Copper from Soil with S-Carboxymethylcysteine", *Water Environment Research*, 67:971–978.

18. Hong, A., T.-C. Chen, and R. Okey (1995b). "Chelating Extraction of Zinc from Soil with N-(2-acetamido)iminodiacetic Acid", *ACS Symposium Series 607*, Chapter 17, 210–223.

19. Macauley, E. and A. Hong (1995). "Chelation Extraction of Lead from Soil Using Pyridine-2,6-dicarboxylic Acid", *J. Haz. Mater.*, *40*(3):257–270.

20. Chen, T.-C. and A. Hong (1995). "Chelating Extraction of Lead and Copper from an Authentic Contaminated Soil Using N-(2-Acetamido)iminodiacetic Acid and S-Carboxymethyl-L-cysteine", *J. Haz. Mater.*, *41*(1–2):147–160.

21. Hong, A. and T.-C. Chen (1996). "Extractive Recovery of Cadmium from Soil Using Pyridine-2,6-dicarboxylic Acid", *Water, Air, And Soil Pollution*, 86:335–346.

22. Stumm W. and J.J. Morgan (1996). Aquatic Chemistry. John Wiley & Sons, Inc., New York.

23. ASTM Standards, (Annual), American Society for Testing and Materials, Philadelphia, PA.

24. American Public Health Association, Greenberg, A.E., L.S. Clesceri, and A.D. Eaton, Eds., (1992). *Standard Methods for the Examination of Water and Wastewater*, 18th Ed., American Public Health Association, Washington, D.C.

25. U.S.EPA. (1979) Methods for the Chemical Analysis of Water and Wastes. EPA600/4-79-020.

26. U.S.EPA. (1986) Testing Methods for Evaluating Solid Waste (SW-846). Volume IA, IB, IC, and II, Third Edition, U.S.EPA 955-001-00000-1.

27. Smith, R.E. and A.E. Martell, Critical Stability Constants. Vol. 1, Vol. 4, 5, and 6, New York: Plenum Press, 1974, 1976, 1982, 1989.

ABBREVIATIONS AND STRUCTURES OF CHELATING AGENTS

ADA   N-(2-Acetamido)iminodiacetic acid, $C_6H_{10}N_2O_5$
$H_2NCOCH_2N(CH_2COOH)_2$

DTPA   Diethylenetrinitrilopentaacetic acid, $C_{14}H_{23}O_{10}N_3$
$(HOOCCH_2)_2NCH_2CH_2N(CH_2COOH)CH_2CH_2NC(CH_2COOH)_2$

EDTA   Ethylenediaminetetraacetic acid, $C_{10}H_{16}O_8N_2$
$(HOOCH_2C)_2NCH_2CH_2N(CH_2COOH)_2$

GCG   L-5-Glutamyl-L-cysteinylglycine (Glutathione), $C_{10}H_{17}O_6N_3S$
$H_2N(HOOC)CHCH_2CH_2C(O)NHCH(CH_2SH)C(O)NHCH_2COOH$

NTA   Nitrilotriacetic acid, $C_6H_9O_6N$
$HOOCCH_2N(CH_2COOH)_2$

NTTA   Nitrilotris(methylene)triphosphonic acid, $C_3H_{12}O_9NP_3$
$N[CH_2P(O)(OH)_2]_3$

PDA   Pyridine-2,6-dicarboxylic acid, $C_7H_5O_4N$
$C_5H_3N(COOH)_2$

SCMC   S-Carboxymethylcysteine, $C_5H_9O_4NS$
$HOOCCH_2SCH_2CHNH_2COOH$

TMDTA   Trimethylenedinitrilotetraacetic acid, $C_{11}H_{18}O_8N_2$
$(HOOCH_2C)_2NCH_2CH_2CH_2N(CH_2COOH)_2$

3

ELECTROKINETIC SOIL REMEDIATION USING NOVEL ELECTRODES AND MODULATED REVERSE ELECTRIC FIELDS

Maria E. Inman, E. Jennings Taylor, Debbra L. Myers, and Chengdong Zhou

Faraday Technology, Inc.
315 Huls Drive, Clayton
Ohio 45315

1. ABSTRACT

Soil contamination is a widespread problem in the United States with more than 4,500 sites currently on the EPA's national priorities list for clean up. The remediation of *heavy metal* contamination is particularly urgent due to the high levels of contamination and toxicity. This paper describes the investigation into a novel, two-fold approach to electrokinetic remediation of soils contaminated with heavy metals. *Modulated reverse electric fields*, in conjunction with *integrated ion exchange (IIX^{TM}) electrodes*, are being used to (a) induce a more uniform and enhanced remediation of heavy metal contaminated soils, and (b) eliminate the non-uniform pH profile encountered in conventional *electrokinetics*, by acidification of the soil surrounding the cathode. Concentration of the heavy metal contaminants from the soil into the IIX^{TM} electrode eliminates the need for soil excavation and further treatment. The heavy metals absorbed into the IIX^{TM} electrodes can be recovered, resulting in regeneration of the electrodes, which can then be reused in the soil treatment process.

2. BACKGROUND

The main contaminants in Superfund sites are heavy metal cations, organics and inorganic complexing agents or anions. The remediation of the heavy metal ion contamination is particularly urgent due to their high levels of contamination and their toxicity. Cleanup has typically been achieved by soil excavation followed by incineration or chemical treatment of the soil. However, this process is extremely costly; it has been estimated to cost $2,400/yd^3 for cleanup by soil excavation.[1] In-situ soil remediation techniques are therefore gaining ground as desired technologies for Superfund site cleanup.

Current address: Mitsubishi Silicon America, South Campus, 3990 Fairview Industrial Dr. SE, Salem, OR 97302

Emerging Technologies in Hazardous Waste Management 8, edited by Tedder and Pohland
Kluwer Academic/Plenum Publishers, New York, 2000.

In electrokinetic soil remediation, a current or voltage is applied between two electrodes buried in the soil, and the ionic contaminants are driven towards the electrodes under the influence of the electric field, resulting in localization of the contaminants.[2,3] This localization of both anionic and cationic metal contaminants around the anode and cathode, respectively, means less soil to excavate, thereby significantly reducing the costs associated with soil removal. Additionally, the only species introduced into the soil during treatment are non-hazardous protons and hydroxyl ions. One of the technical barriers to commercial implementation of electrokinetics is due to the side reactions of water electrolysis at the electrodes, under the influence of the applied electric field, which result in generation of hydroxyl ions at the cathode:

$$4H_2O + 4e^- \rightarrow 4OH^- + 2H_2 \tag{1}$$

and protons at the anode:

$$2H_2O \rightarrow 4H^+ + O_2 + 4e^- \tag{2}$$

This leads to an increased pH at the cathode and lowered pH at the anode with the resulting non-uniform pH profile depicted in Fig. 1. Protons generated at the anode move toward the cathode and the hydroxyl ions generated at the cathode move toward the anode. Metal cations combine with hydroxyl ions to precipitate as a non-recoverable metal hydroxide, shown as a large concentration of Pb between the anode and cathode in Fig. 1. Protons and hydroxyl ions meet, and combine to form water, resulting in a drop in soil conductivity and leading to increased power consumption and soil heating. Heating the soil a few degrees causes evaporation of the pore fluid which results in additional precipitation of contaminants, a further drop in soil conductivity, and poorer contaminant mobility. This non-uniform pH profile, or pH front, must be eliminated for commercial utilization of electrokinetics for site remediation to be realized.

Our novel process utilizes integrated ion exchange (IIX™) electrodes to adsorb the heavy metals directly from the soil, eliminating the requirement for soil excavation and further treatment, as well as providing a means of soil pH control. The process also utilizes modulated reverse electric fields (MREF) to further control the soil pH and enhance mass transport of metal contaminants.

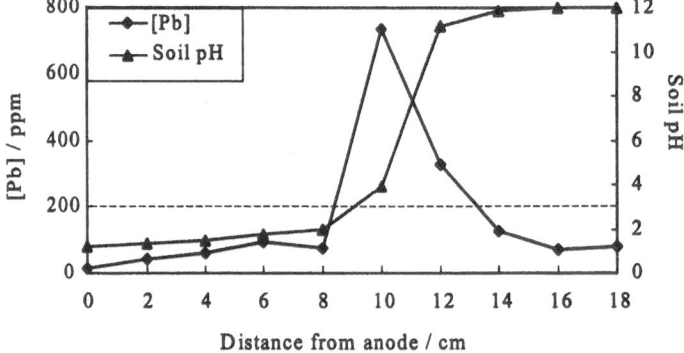

Figure 1. pH profile and resultant [Pb] profile for conventional electrokinetic soil treatment using a graphite cathode. DC = 20 V; $pH_{initial} = 7$; $[Pb]_{initial} = 200$ ppm.

3. TECHNICAL APPROACH

3.1. Integrated Ion Exchange (IIX^{TM}) Cathodes

The integrated ion exchange (IIX^{TM}) cathode is shown schematically in Fig. 2. Graphite particles are intimately mixed with ion-exchange resin beads, and contained in a polypropylene mesh bag. The graphite:resin ratio may be dependent upon the soil contamination level and the desired regeneration cycle, or graphite may not be required at all. This is the subject of investigation. A graphite rod provides electrical contact with the cathode. Ion-exchange resin exchanges metal cations for protons via a chemical reaction:

$$nR\text{-}H + M^{n+} \Leftrightarrow Rn\text{-}M + nH^+ \tag{3}$$

The H$^+$ ions generated by the ion exchange reaction neutralize the OH$^-$ ions produced at the cathode as a result of water electrolysis, allowing excellent control of the soil pH around the cathode. Once the ion-exchange resin is saturated with contaminant ions, the IIX^{TM} cathode can be removed from the ground and regenerated using an acid solution, then placed back in the soil, and the treatment continued. The metal removed from the soil can then be recovered using electrowinning techniques or waste treatment processes. If the IIX^{TM} cathode is left in the soil after the resin has reached its full capacity, H$^+$ ions will no longer be released into the soil, and the OH$^-$ ions will not be neutralized, resulting in an undesirable increase in soil pH. Obviously, there is an optimum treatment time for regeneration of the cathode, to remove the maximum amount of Pb while maintaining an acidic soil pH. The regeneration cycle time is currently being investigated in this program, and the results will be reported at some later date.

3.2. Modulated Reverse Electric Fields

A modulated reverse electric field (MREF) is shown schematically in Fig. 3. The waveform consists of a forward cathodic modulation, followed by an off-time, followed

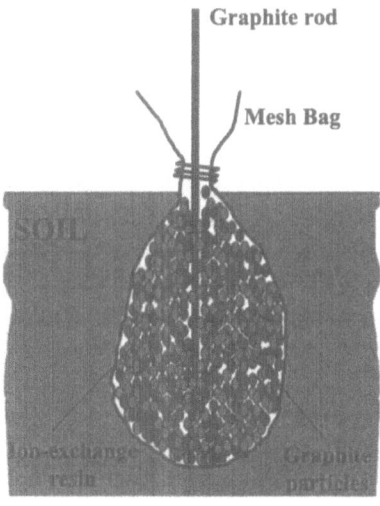

Figure 2. Schematic diagram of a IIX^{TM} cathode.

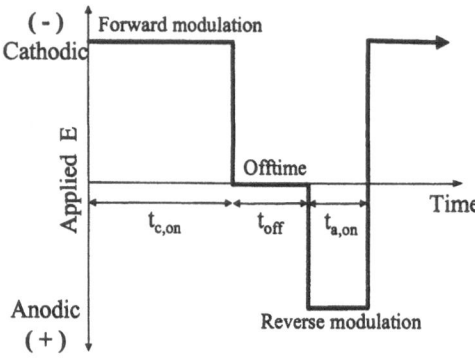

Figure 3. Schematic diagram of a modulated reverse electric field (MREF) waveform.

by a reverse cathodic modulation. The electrochemical reactions occurring at the cathode during the forward cycle are:

$$M^{n+} + ne^- \rightarrow M \qquad (4)$$

$$4H_2O + 4e^- \rightarrow 4OH^- + 2H_2 \qquad (5)$$

The reactions occurring at the same electrode during the reverse modulation (which is strictly speaking an anode during this part of the cycle):

$$M \rightarrow M^{n+} + ne^- \qquad (6)$$

$$H_2 \rightarrow 2H^+ + 2e^- \qquad (7)$$

During the forward modulation, M^{n+} is adsorbed onto the ion-exchange resin, and may also be plated out on the graphite particles that are part of the cathode. Therefore, the M^{n+} concentration at the cathode surface will be depleted, compared to the bulk concentration in the soil. An offtime after the forward modulation allows further chemical ion-exchange to take place, and H^+ to be desorbed from the resin for neutralization of the OH^- ions generated during the forward modulation. Additionally, the offtime allows the diffusion layer around the electrode to be replenished with cations from the bulk soil, by diffusion down the concentration gradient. If the chemical ion-exchange reaction is slow, the off-time needs to be large, to allow sufficient H^+ to be desorbed, to enhance neutralization of the OH^- generated during the forward modulation. During the reverse modulation, the oxidation reaction is dominated by the H^+ generation reaction, due to the high concentration of metal cations in the soil. Therefore, there is an excess of protons in the soil that are driven away from the cathode during the reverse modulation, resulting in an acidic soil pH around the cathode. Additionally, adjustment of MREF parameters during the reverse cycle, can favor reaction (7) to generate H^+ ions.

3.3. Mass Transfer in MREF

Unlike DC electrolysis, mass transfer characteristics of MREF electrolysis are a time dependent process. MREF causes concentration fluctuations near the electrode surface to reduce the effective Nernst diffusion layer thickness. Consequently, very high instantaneous limiting current densities can be obtained with MREF electrolysis as compared to DC electrolysis. Modeling work by Chin[4] has indicated that limiting current densities obtained under the pulse reverse current (PRC) conditions of low duty cycle and high frequency, can be two to three orders of magnitude greater than the DC limiting current density. Vilambi and Chin[5,6] confirmed the earlier modeling work with experimental studies for a copper sulfate bath for selected pulse periods and duty cycles in PRC electrolysis, reporting diffusion controlled current densities as high as several hundred A/cm^2, while the corresponding values for DC electrolysis were less than $1 A/cm^2$. Therefore, electrokinetics utilizing MREF can produce enhanced movement of ionic species through the soil, as compared to the DC process.

4. EXPERIMENTAL PROCEDURE

The soil remediation experiments were performed in rectangular plexiglas cells measuring $20 cm \times 2 cm \times 4 cm$. White quartz sand (Aldrich Chemicals, Wisconsin) was used to simulate soil. The sand was contaminated with 200 ppm Pb, from a 1,015 ppm Pb AA standard solution in 1 wt% HNO_3 (Aldrich Chemicals, Wisconsin). The sand pH was adjusted to 7 using 1.01 M NaOH solution (Aldrich Chemicals, Wisconsin). Bromothymol Blue was added to the sand to show color changes with changes in sand pH over the course of the experiment. The initial sand moisture content was approximately 20 wt%. The anode and cathode were placed approximately 1 cm from opposite ends of the cell. The contaminated sand was packed firmly into each cell, and covered with parafilm to limit evaporation. The cells were hooked up in parallel to either the DC or pulse reverse rectifier for treatment. At the completion of the experiments, the sand in each cell was divided up into 10 lots, spaced evenly along the length of the cell, and analyzed for the pH, moisture content, and Pb concentration.

The IIX™ cathode was constructed from graphite particles and WACMPH ion exchange resin, at a weight ratio of 1:2. The WACMPH ion exchange resin (ResinTech Inc., Cherry Hill, New Jersey) was chosen from a variety of ion exchange resins, for the near neutral pH sand conditions and the Pb contaminant used in testing. This is a significant advantage of this technology, in that the cathode makeup can be chosen for the sand pH and the specific contaminants in the ground to maximize performance.

Testing was performed under both DC conditions, at 20 V, and using modulated reverse electric field conditions. The MREF parameters were:

Forward duty cycle, $D_f = 50\%$
Forward peak voltage, $V_f = 35 V$
Reverse duty cycle, $D_r = 10\%$
Reverse peak voltage, $V_r = 25 V$
Frequency $= 100 Hz$
Average voltage $= D_f V_f - D_r V_r = 15 V$

Please note that although the peak forward and reverse voltages used in the MREF waveform are higher than the DC voltage, the average MREF voltage is less than that of the DC case.

5. RESULTS AND DISCUSSION

The initial results described below will illustrate that, compared to a conventional electrokinetic process utilizing simple graphite electrodes and applied DC, our novel approach maintained an acidic sand pH throughout the contaminated test cell, and allowed greater removal of Pb^{2+} cations from the sand.

5.1. Maintenance of Acidic Sand pH

Figure 4 compares the effect on the sand pH of using a simple graphite cathode under DC control (20 V), an IIX™ cathode under DC control (20 V), and an IIX™ cathode under MREF control (15 V avg), after 15 days of treatment.

Using DC control and a graphite cathode, there was an iso-electric pH front formed approximately 8 cm from the cathode. Under DC control, but using an IIX™ cathode, there was a significant decrease in the sand pH, remaining acidic throughout the entire test cell. The use of an IIX™ cathode in conjunction with a modulated reverse electric field, further enhanced pH control around the cathode, causing an additional small decrease in the pH close to the cathode.

5.2. Enhancement of the Pb Removal Rate

Figure 5 compares the effect on the Pb concentration profile of using a simple graphite cathode under DC control (20 V) and an IIX™ cathode under MREF control (15 V avg), after 15 days of treatment, and an IIX™ cathode under DC control (20 V), after 10 days of treatment (Pb concentration data for the 15 day experiment shown in Fig. 4 was not available).

The results for the graphite electrode with a DC field show a prominent peak in the concentration of Pb at the same location as the iso-electric pH front shown in Fig. 4. This indicates substantial precipitation of the Pb ions as they encounter the region of alkaline pH close to the cathode. The experiment using an IIX™ cathode in conjunction with a DC field showed a low Pb concentration up to about 4 cm from the cathode,

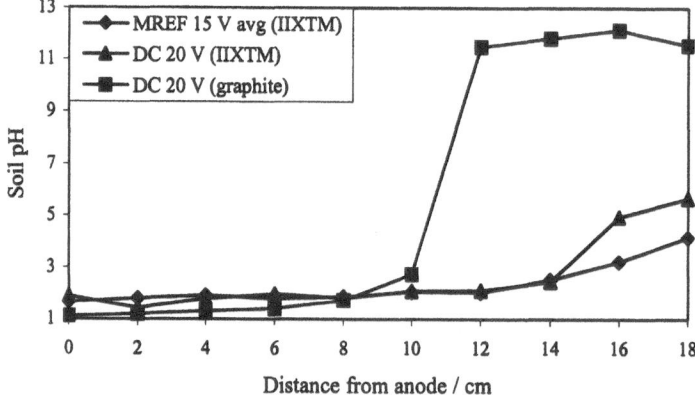

Figure 4. Sand pH as a function of distance from the anode for conventional and novel electrokinetics processes. [Pb]$_{initial}$ = 200 ppm; pH$_{initial}$ = 7.

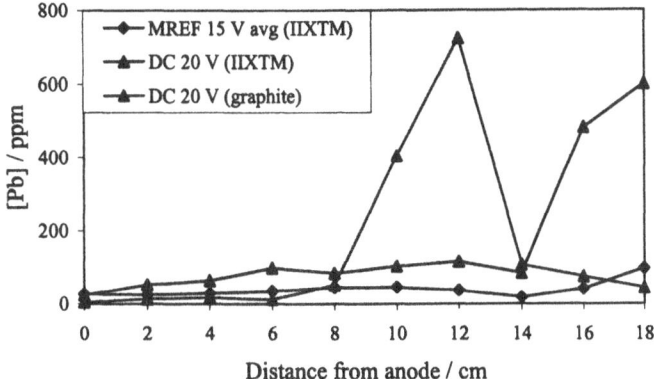

Figure 5. Sand [Pb] as a function of distance from the anode for conventional and novel electrokinetics processes. $[Pb]_{initial} = 200\,ppm$; $pH_{initial} = 7$.

increasing from that point to the cathode. This results represents an improvement over the results obtained using a graphite cathode. Furthermore, the results would probably have been improved further if the experiment had been continued for a full 15 days. The experiment using an IIX™ cathode with a MREF waveform showed a low value for the lead concentration at all points throughout the cell, suggesting that a substantial fraction of the lead originally present had been collected in the cathode.

Table 1 shows that the % removal of Pb from the sand under varying treatment methods and times. The total amount of Pb originally in the sand was 78 mg. This data shows that the amount of Pb removed from the sand increased with increasing treatment time, for both graphite electrodes and IIX™ cathodes. The IIX™ cathode outperformed the graphite cathode, for a DC process, removing up to 45% of the Pb after 619 hours. By comparison, the graphite cathode removed only 12% after 500 hours. The results also show that by applying an MREF waveform to the IIX™ cathode, the % of Pb removed from the sand increased to 59%, after only 360 hours. It should be noted that the parameters of the MREF waveform have not yet been optimized. It is expected that the % removal, and the removal rate, will be increased once the optimum MREF parameter values have been determined. This work is currently ongoing and will be presented in a future publication.

Table 1. Percent removal of Pb from contaminated sand using conventional and IIX™ cathodes, under a DC or Modulated Reverse Electric Field electrokinetics process

Waveform	Cathode	Treatment Time / hours	Original Pb in sand/mg	Final Pb in cathode/mg	% Removal
DC	Graphite	163	78	3	3.8%
DC	Graphite	504	78	9	11.5%
DC	IIX™	163	78	21	26.9%
DC	IIX™	360	77	25	32.5%
DC	IIX™	619	77	35	45.5%
MREF	IIX™	360	78	46	59.0%

Note: IIX™ cathode = 1:2 graphite:WACMPH ion exchange resin; DC = 20 V; MREF = V_f = 35 V, D_f = 50%, V_r = 25 V; D_r = 10%, Frequency = 100 Hz; Initial pH = 7; Initial [Pb] = 200 ppm.

6. CONCLUSIONS

The advantages of using IIX™ cathodes and modulated reverse electric fields over conventional processes are summarized below:

1. Use of modulated reverse electric fields and IIX™ cathodes prevents formation of the pH front by producing H^+ ions at the cathode to neutralize the OH^- ions resulting from electrolysis reactions.
2. Use of modulated reverse electric fields and IIX™ cathodes allowed accelerated in-situ removal of Pb from the sand, as compared to the conventional DC process.
3. IIX™ cathodes can be regenerated ex-situ, allowing recovery of heavy metals without the need for soil excavation and further treatment.
4. Multiple heavy metal contaminants can be removed from the soil simultaneously.
5. Selection of the appropriate resin will allow the treatment process to be tailored to the metal cations in the contaminant site.

Future work will concentrate on optimization of the MREF waveform, determination of the regeneration time for an IIX™ cathode, and transfer of this technology to actual site testing.

ACKNOWLEDGMENTS

This research was supported by a SBIR Phase II grant from the U.S. Environmental Protection Agency (Contract Number: 68D60051).

REFERENCES

1. Department of Energy (1996), Subsurface Contaminants Focus Area, DOE/EM-0296.
2. Acar Y.B. and Alshawabkeh A. (1993), Principles of Electrokinetic Remediation, Environmental Science and Technology, 27, 13, p. 2638.
3. Acar Y.B., Gale R.J., Alshawabkeh A. Marks R.E., Puppla S., Bricka M., and Parker R. (1995), Electrokinetic Remediation: Basics and Technology Status, Journal of Hazardous Materials, 40, 3, pp. 117–137.
4. Chin D.-T. (1983), Mass Transfer and Current-Potential Relations in Pulse Electrolysis, J. Electrochem. Soc., 130, p. 1657.
5. Vilambi N.R.K. and Chin D.-T. (1987), AESF Project 68: Study of Selective Pulse Plating, 74th AESF Annual Technical Conference Proceedings for Sur/Fin '87, American Electroplaters and Surface Finishers Society Meeting, Chicago.
6. Vilambi N.R.K. and Chin D.-T. (1988), Selective Pulse Plating of Copper from an Acid Copper Sulfate Bath, J. Plating and Surface Finishing, 75, pp. 67–74.

4

ELECTROKINETIC INJECTION OF NUTRIENTS IN LAYERED CLAY/SAND MEDIA FOR *BIOREMEDIATION* APPLICATIONS

M. F. Rabbi[1], R. J. Gale[2,3], E. Oszu-Acar[3], G. Breitenbeck[4],
J. H. Pardue[1], A. Jackson[1], R. K. Seals[1], and D. D. Adrian[1]

[1]Department Of Civil and Environmental Engr.
Louisiana State University
Baton Rouge, Louisiana 70803
[2]Department Of Chemistry
Louisiana State University
Baton Rouge, Louisiana 70803
[3]Electrokinetics Inc.
11552 Cedar Park Avenue
Baton Rouge, Louisiana 70809
[4]Department of Agronomy
Louisiana Sate University
Baton Rouge, Louisiana 70803

1. ABSTRACT

Electrokinetic processing has been used to inject representative cationic and anionic nutrients into soil layers, which have widely different hydraulic and electrical conductivities. In this case, using parallel clay and sand layers and at a practical current density, ammonium ion concentrations reached 300–35x0 mg/L in the sand and 250–300 mg/L in the clay layer, while the respective values for sulfate ion were 900–1,000 mg/L and 1,000–1,200 mg/L. This demonstrates the potential for bio-electrokinetic enhancement by nutrient injection for those sites which cannot be bioremediated by *natural attenuation* alone. Although considerable electroosmosis occurred at the neutral pH of the system, the anions could be injected upstream about one meter by 1,028 hours of processing. Both the power consumption costs and the process rates are practicable for the injection of such process additives.

2. INTRODUCTION

In situ biotreatment is an important technology for the remediation of toxic organics encountered in the National Priority List. However, natural attentuation may be inadequate in many instances since the indigenous microbes, capable of the appropriate

Emerging Technologies in Hazardous Waste Management 8, edited by Tedder and Pohland
Kluwer Academic/Plenum Publishers, New York, 2000.

degradation processes, may have insufficient *nutrients* and *cometabolites* to be effective. Recent surveys have concluded that the ineffective transport of remediation additives has been the primary cause for system inefficiencies and failures e.g.[1,2] Difficulties encountered for the effective introduction of process additives and nutrients into *biologically active zones* (BAZs) include the existence of preferential flow paths, heterogeneities, adsorption or chemical reactions in the soil, and biofouling. Some of these problems can be ameliorated with electrokinetic injection since the introduced additive concentrations and the direction of the plumes can be more carefully controlled in the electric field.[3-7] Although it has been clearly demonstrated that charged nutrient species may be introduced into porous soil media by electrokinetic processing to accelerate or enhance indigenous bioremediation,[6,8] it was not clear what might occur at a site which has heterogeneities. The purpose of this study has been to test the simultaneous injection of ammonium and sulfate ions into a fine *sand* and *clay layered bed*. These soils were chosen because of the large differences in their *hydraulic permeabilities* and *electrical conductivies*. The electrical conductivity of a kaolinite bed was $124\,\mu S/cm$,[2] approximately an order of magnitude higher than that of the fine sand bed, whereas its hydraulic conductivity was $2 \times 10^{-7}\,cm/s$, about three orders of magnitude lower than that of the fine-grained sand bed.[6]

In previous, separate injection experiments, into a fine-grained sand bed and a clay bed,[6,8] it was demonstrated that the uniformity of transport of the injected ions under electrical fields depends on the initial pore fluid conductivity and the boundary conditions. Typically, sands may have electrical conductivities of $10\,\mu S/cm$, where practical values for clays are $10-500\,\mu S/cm$. The cations and anions transport mechanisms were found to be different due to electroosmosis. Generally, electroosmotic flow is from the anode to the cathode but an asymmetrical ion transport mechanism occurs, in which the injected cations migrate from the anode to the cathode by elec-tromigration accelerated by *electroosmosis*, whereas the injected anions migrate from the cathode to the anode by *electromigration* opposed by electroosmosis. As a result, initial pore fluid cations are flushed out from the system, which renders the injected cations more uniformly transported than the anions in the media. The transport rates of the injected ions under the electrical field are not only functions of the ionic mobilities but also depend on the concentrations and charges of the injected ions relative to all other competing ions. The transportability (or transference number) of the injected ions is high when that ion becomes the predominate ion in the medium.

Media heterogeneity can be in a direction perpendicular to the electrical field, or in the same direction. In the case of a perpendicular heterogeneity, such as a clay layer placed perpendicular to the electric field and sandwiched in between more permeable layers, or vice versa, the results of the earlier studies[6,8] show that the transport rate in the clay layer would be expected to be slightly lower. However, it might still be possible to employ the electrokinetic injection technique efficiently to introduce and migrate nutrients and process additives across the medium since the transport rates will be of the same order of magnitude. When the heterogeneity is in the same direction with the electrical field, there is the possibility that the layer with the higher electrical conductivity (clay in this case) would transmit most of the current, hindering the transport of the injected species across the medium with the lower ionic conductivity (the fine-grained sand layer).

3. EXPERIMENTAL

The apparatus and method of controlling the pH at the catholyte and anolyte electrode wells have been described in detail elsewhere.[6,8] Ammonium hydroxide, added from a pH servo system, was added to the anolyte to prevent the pH from decreasing due to the formation of hydrogen ions from water electrolysis. Similarly, sulfuric acid was added into the catholyte to maintain the pH 7. The cell was constructed of 5 mm thick Acrylic of length 80 cm, thickness 5.1 cm, and height 80 cm. The distance between C rod electrodes of diameter 4.8 cm was 110 cm. The quantity of fluid in each electrode chamber was 5.6 L and the cell held about 31.4 L of soil. Fine-grained sand and kaolinite clay were used to create a layered bed. The clay was placed first in the cell at a water content of 45% to a thickness 47 cm. The lowest two lines of sampling ports remained within this layer. The fine-grained sand layer was placed on top of the clay layer at a thickness of 27 cm. The two rows of sampling ports were about 1 and 11 cm above the clay/fine-grained sand interface and the water table was 15 cm above this interface. All ports were sampled to determine the ion transport across the layered clay/sand medium. The electroosmotic flow was determined with an outflow hole at the catholyte solution level.

4. RESULTS AND DISCUSSION

4.1. Geotechnical Parameters

The electrical gradients in the clay and the fine-grained sand layers were 1.6 and 2.5 V/cm, respectively, under an average current density of 151 μA/cm.[2] It is necessary to consider the different cross sectional areas of the two layers if average electrical conductivities are to be calculated; that of the clay layer was 80–90 μS/cm and that of the fine-grained sand layer 10–12 μS/cm. The anode and cathode liquid were flushed after 596 hours of processing. Immediately after that time the potential difference across both layers increased. The voltage gradients developed during processing in both layers are presented in Fig. 1 and decreased to become approximately constant after 600–1,000 hours. The voltage gradient dropped to 0.4 V/cm in both layers after 1,028 hours of processing. The calculated electrical conductivities in the sand layer increased to 500 mS/cm during the first 596 hours of processing. After flushing the anode and cathode compartments, the conductivity decreased to a value 300 mS/cm. The equivalent conductivity values (or apparent conductivity, which disregards the localized changes across the specimen) for both layers are presented in Fig. 2. Depending upon the local pH and ionic strength, the electrical conductivity varies from one position to another.

During 1,028 hours of processing, 35 L or around 2.5 pore volumes of water had transported from the anode to the cathode. This electroosmotic flow is significantly higher than the tests with kaolinite only but the causes for this are unknown. Similarly to the previous experiments, the electroosmotic permeability increased with time from an initial value 1.5×10^{-5} cm^2/Vs to final value 9.5×10^{-5} cm^2/Vs (Fig. 3). The reasons for the increase in K_e values with time is perhaps because the soil pH rises during these experiments, increasing the electrokinetic zeta potential. Figure 4 displays the changes in electroosmotic water transport efficiency (k_i) with time. The k_i values varied within a range of 0.14 to 0.18 cm^3/As during 1,028 hours of processing.

4.2. Ammonium and Sulfate Profiles

The ammonium ion concentrations across the two layers are shown in Figs. 5 and 6. Firstly, ammonium ions were efficiently transported across both of the layers. The transport rate was faster across the fine-grained sand layer than across the clay layer

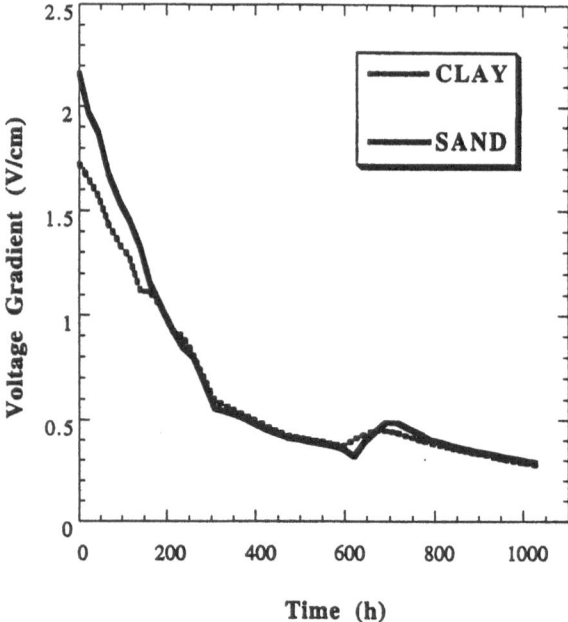

Figure 1. Voltage gradients developed across sand and clay layers.

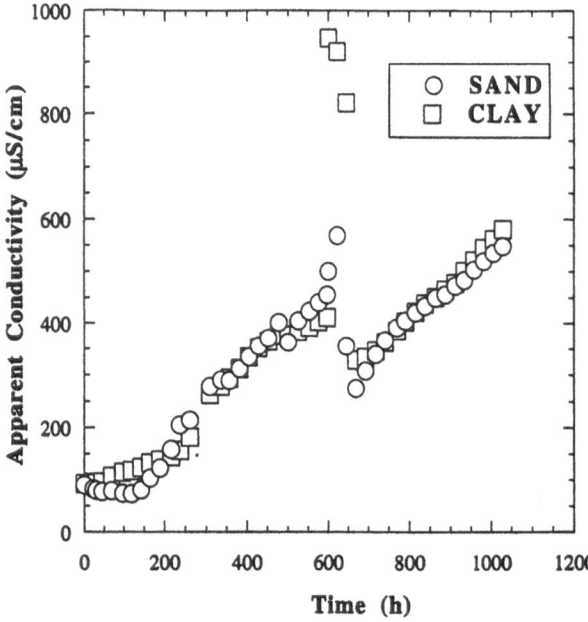

Figure 2. Variation in the apparent conductivities in the sand and clay.

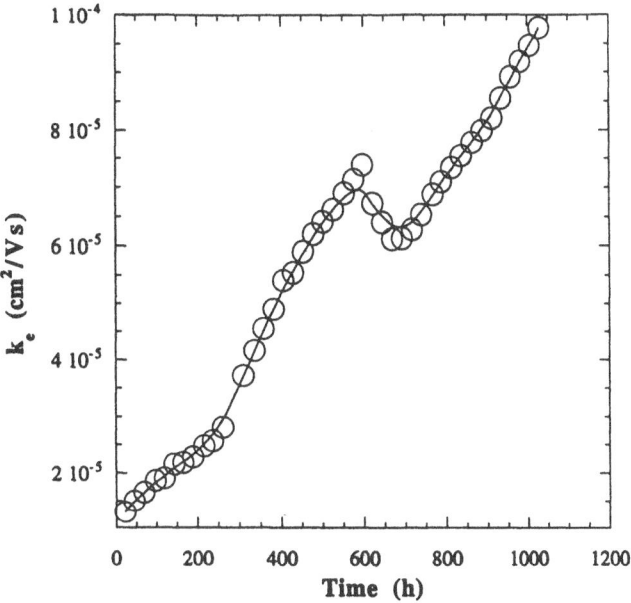

Figure 3. Variation in the coefficient of electroosmotic permeability.

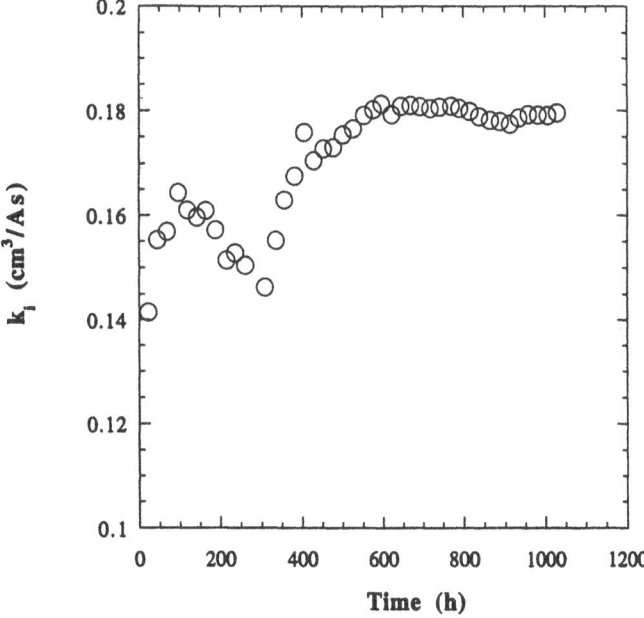

Figure 4. Variation in the electroosmotic water transport efficiency.

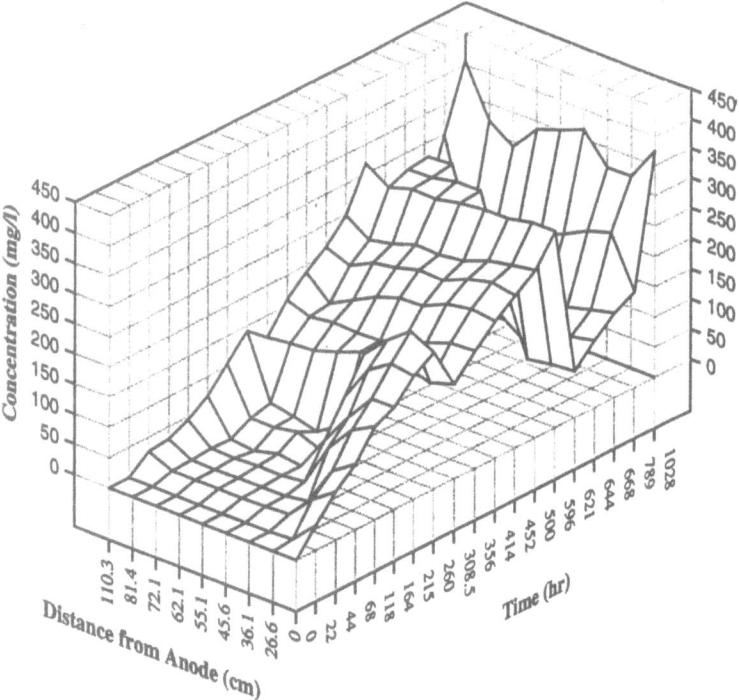

Figure 5. Ammonium ion concentrations across the sand layer.

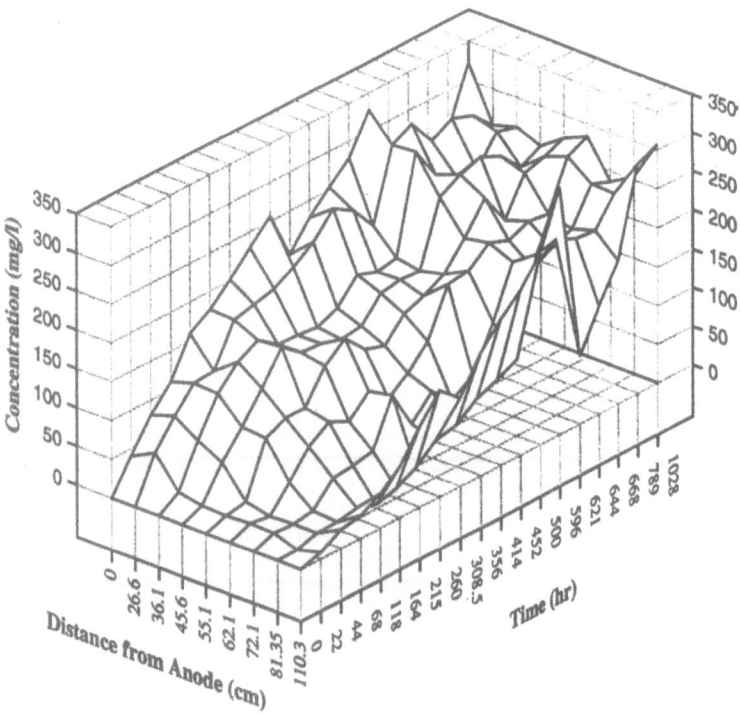

Figure 6. Ammonium ion concentrations across the clay layer.

within the first 200 hours due both to its higher electrical gradient and lower conductivity. The effective transport rates for the sand and clay beds were 18 and 13 cm/day, respectively. Subsequently, the transport rates decreased. After 596 hours of processing, the concentration of ammonium ion across the fine-grained sand bed was slightly higher than that in the clay layer, both around 250 mg/L. After 1,028 hours of processing, the final concentration of ammonium ions varied within a range in the sandlayer of 300–350 mg/L and in the clay layer of 250–300 mg/L.

Sulfate transport across the layers is also interesting (Figs. 7, 8). In the clay layer, there was no transport until the boundary concentration at the anode increased above the initial sulfate concentration. After 596 hours processing, the sulfate ions saturated both the fine-grained sand and clay layer at about 700–800 mg/L. After flushing, the sulfate ions profiles in both layers were similar and final concentrations of sulfate in the sand and clay layer were in the ranges 900–1,000 mg/L and 1,200–1,000 mg/L, respectively. The final distributions of ammonium and sulfate ions at different rows in the clay (two rows) and sand (two rows) are presented in Figs. 9 and 10.

5. CONCLUSIONS

The technique of neutralizing the water electrolysis reactions at the anode and/or cathode of an electrokinetic process can be used to inject simple ionic nutrients uniformly

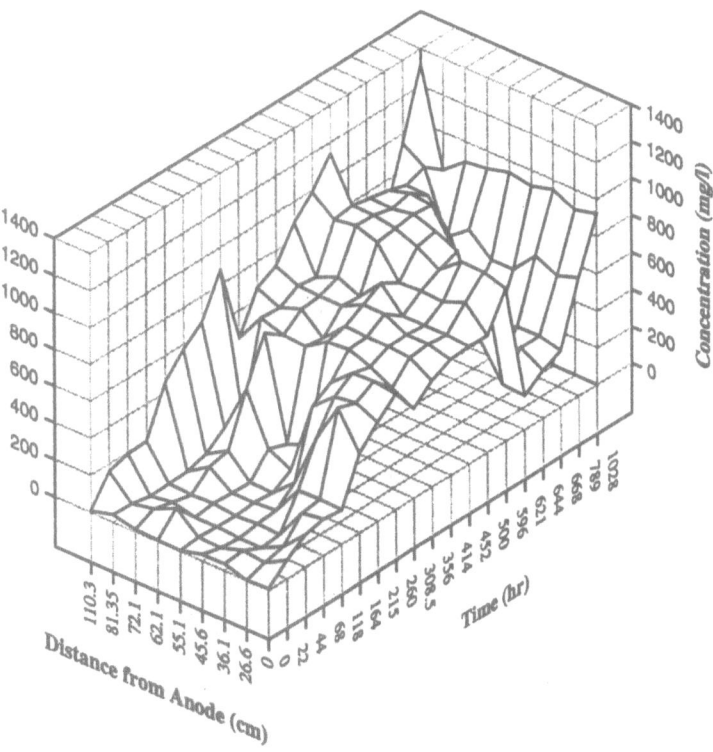

Figure 7. Sulfate ion profile across the sand layer.

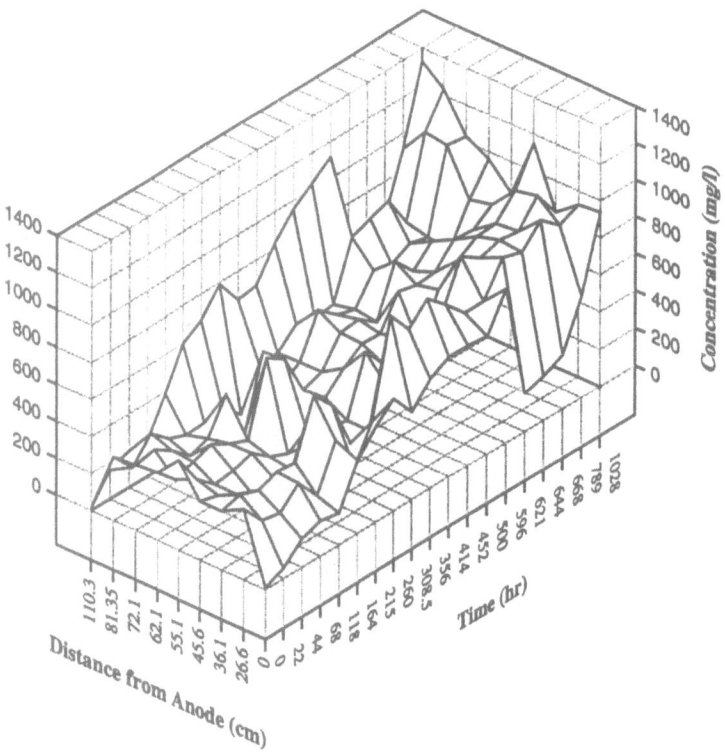

Figure 8. Sulfate ion profile across the clay layer.

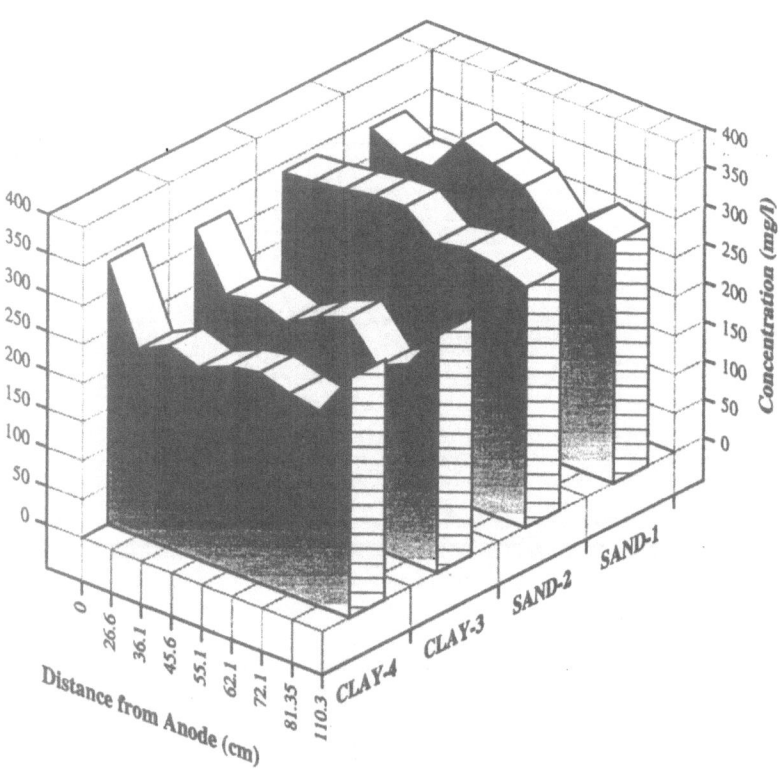

Figure 9. Final ammonium ion concentrations across the layered bed.

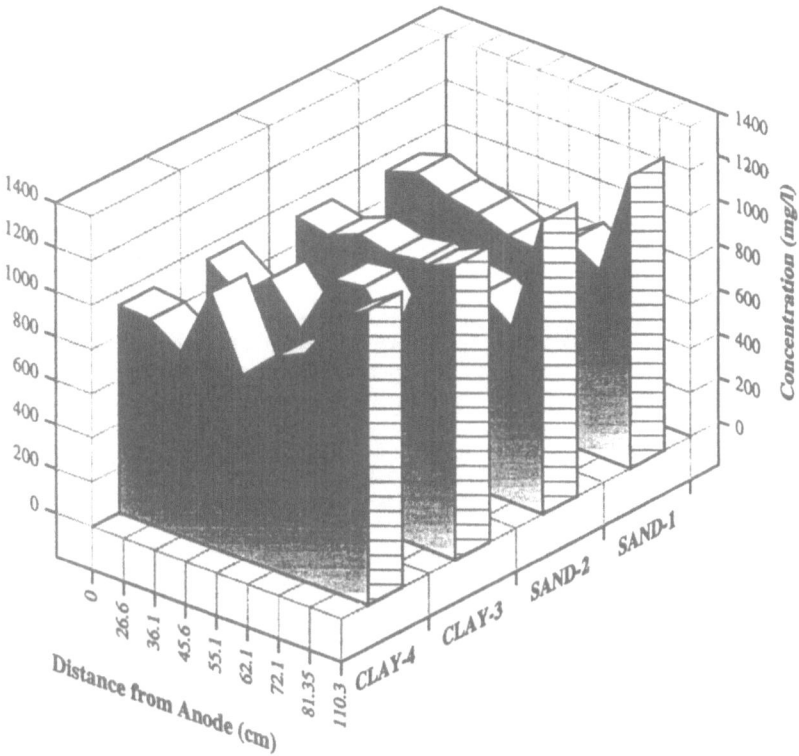

Figure 10. Final sulfate ion concentrations across the layered bed.

into layered soil media for *in-situ* bioremediation. The results of these experiments demonstrate the potential and feasibility to inject these species into layered soils under electrical fields to enhance bioremediation. Transport rates on the order of 10–20 cm/day were achieved for ammonium and sulfate ions in a fine sand bed, a clay bed, and a layered bed. Despite differences in initial conductances, the disimilar layers each load to about the same extent with the additives. This is important in the field since the toxic organic contaminants often accumulate in clays clods, or at clay interfaces, through preferential binding. The electrical gradients of 1 V/cm, or less, constituted the predominant driving force for transport under current densities of 15–150 μA/cm.[2] At a current density of 500 μA/cm,[2] the upper conductance range provides a 1 V/cm electric gradient, i.e., 4 meters electrode placement requires 400 V DC. Corresponding power consumption is 12 kWhr/m[3] for these processes and the cost for four months of operation at $0.03–0.05/kWhr is estimated as $39–50/m[3] treated soil. It is noted that the major site implementation costs will be higher due to installation, maintenance expenses, etc. No evidence for measurable losses of the injected ionic species due to microbial processes is seen in these tests, perhaps because of a lack of organic substrate. In cases where organic cometabolites need to be injected to enhance the degradation of a toxic pollutant, it is important to assess their local rates of bio-consumption by indigenous microbes. A follow-up study[9] at LSU and Electrokinetics Inc., sponsored by the NRMRL USEPA program, is currently attempting to assess the biodegradation of TCE in clay using electrokinetically injected benzoic acid additive. A numerical computer simulation of ion injection from neutralized electrode wells has also been completed.[10]

ACKNOWLEDGMENTS

This work has been supported at Electrokinetics Inc. and LSU through the SBIR Program of the DOE and the National Risk Management Research Laboratory (NRMRL), USEPA. Support from the LEQSF Industrial Ties Program at Louisiana State University is also gratefully recognized.

REFERENCES

1. Zappi, M., Gunnison, D., Pennington, J., Cullinane, J., Teeter, C., Brannon, C.L., and Myers, T., Tech. Report No. IRRP-93-3, US Army Corps of Engineers, Waterways Experiment Station, August 1993.
2. Cookson, J.T., *Bioremediation Engineering: Design and Application*, McGraw Hill, New York, NY, 1995.
3. Marks, R.E., Acar, Y.B., and Gale, R.J., USP 5,458,747, 1995.
4. Marks, R.E., Acar, Y.B., and Gale, R.J., In: *Remediation of Hazardous Waste Contaminated Soils*, Wise, D.L. and Trantolo, D.J. (eds.), Marcel Dekker, 1994, pp. 406–435.
5. Acar, Y.B., Rabbi, M.F., Ozsu-Acar, E.E., Gale, R.J., and Alshawabkeh, A.N., Chemtech, April 1996, pp. 40–44.
6. Acar, Y.B., Rabbi, M.F., and Ozsu-Acar, E.E., J. Geotech. & Geoenviron. Engr. **1997** 123(3):239–249.
7. Loo, W.W. and Chilingar, G.V., Remediation Management, Second Quarter 1997, pp. 40–41.
8. Rabbi, M.F., PhD Thesis, Louisiana State University, 1997.
9. Pardue, J.H., Jackson, W.A., Gale, R.J., Rabbi, M.F., Clark, B., and Ozsu-Acar, E.E., in *Physical, Chemical and Thermal Technologies, Remediation of Chlorinated and Recalcitrant Compounds*, Wickramanayake, G.B. and Hinchee, R.E. (eds.), Batelle Press, Columbus, Ohio, 1998, pp. 461–465.
10. Zheng, W., PhD Thesis, Louisiana State University, 1997.

5

INTEGRATED REMEDIATION PROCESS FOR A HIGH SALINITY INDUSTRIAL SOIL SAMPLE CONTAMINATED WITH HEAVY OIL AND METALS

Abdul Majid[1] and Bryan D. Sparks[2]

[1]Institute for Chemical Process & Environmental Technology
National Research Council of Canada
Montreal Road Campus
Ottawa, Ontario, K1A 0R6, Canada
[2]Institute for Chemical Process & Environmental Technology
National Research Council of Canada
Montreal Road Campus
Ottawa, Ontario, K1A 0R6, Canada

1. ABSTRACT

A highly saline industrial soil sample contaminated with heavy oils and several heavy metals, was tested for remediation using NRC's Solvent Extraction Soil Remediation (SESR) process. The sample was provided courtesy of Newalta Corporation, a soil remediation company, based in Calgary, Alberta. Hydrocarbon contaminants were removed by applying both single and multistage extraction, using toluene as the solvent. Heavy metal fixation was achieved by incorporating metal binding agents into the soil agglomerates formed during the solvent extraction of organic contaminants.

The extracted solids were evaluated for their heavy metal leaching potential using the US-EPA's toxicity test method 1310A and Toxicity Characteristics Leaching Procedure method 1311. Long term stability of the treated solids in terms of metal leaching was tested by the US-EPA's multiple extraction procedure method 1320. The effect of metal binding agents on the extraction efficiency of SESR was also investigated.

After remediation by the SESR process the contaminated soil sample remained saline. Leaching of soluble salts from the dried agglomerates was carried out by water percolation through a fixed bed of dried, agglomerated soil.

2. INTRODUCTION

Contamination of soil by hazardous organic pollutants and/or heavy metals is a serious environmental problem facing the global community. Over time, these pollutants,

Emerging Technologies in Hazardous Waste Management 8, edited by Tedder and Pohland
Kluwer Academic/Plenum Publishers, New York, 2000.

39

trapped in the soil matrix, leach through inadequate holding facilities and migrate deep into the earth, finally making their way to groundwater aquifers. Once contaminated, these aquifers carry the toxins through the ecological system, bringing them into the food chain.

Typically, both heavy metals and hydrocarbon contaminants are associated with the finer particle fractions of soils and sediments. Most techniques for the removal of metals involve contacting the soil with an aqueous solution. Similarly, some processes remove hydrocarbon contaminants by aqueous washing methods. For fine textured, high clay soils these techniques tend to produce intractable sludges having poor solid-liquid separation characteristics. Other soil cleaning technologies, such as thermal desorption, are also poorly suited for treating fine textured top soils because of adverse effects on the associated humic matter or soil mineralogy. In such cases the treated soil may have to be landfilled or used as subsoil because of impaired soil fertility.

At the National Research Council of Canada, liquid phase agglomeration techniques, in combination with solvent extraction have been successfully used for the remediation of fine textured, organic contaminated soils. Co-agglomeration of fines with coarse particles is achieved by the addition of water to a vigorously mixed slurry of soil in the selected solvent. Water acts as a solids bridging liquid and dense soil agglomerates are formed under appropriate conditions. A simplified diagram, illustrating the conditions required for agglomeration, is shown on Fig. 1. The bridging liquid remains in the agglomerate pores where interfacial tension provides the forces necessary to hold the particles together in the aggregates. Judicious selection of agglomerate size, by controlling water content, greatly improves the efficiency of solids-solvent separation. The latter is a prime requirement for effective process operation. Typically, only a few extraction steps are required. This combination of solvent extraction and soil remediation is known as the SESR process.

As an extension of this work, metal binding amendments are incorporated during the mixing stages for solvent extraction of the oily contaminants and agglomeration of the soil. The combined process thus allows concurrent removal of organics and fixation of heavy metals. A flow diagram of this integrated remediation process is shown in Fig. 2. Soil treatment by this approach should be more economical, in terms of material handling and equipment costs, than other processes, which require separate extraction and leaching steps for treatment of organic contaminants and heavy metals.

Another advantage of the SESR process is the ease of salt leaching from the dried, extracted soil agglomerates. Agglomerates produced using liquid phase agglomeration techniques are more stable than similar sized, natural aggregates of the same soil. During drying of agglomerated soil, water-soluble salts, dissolved in the agglomerate pore water, effloresce. This behavior facilitates their removal from the agglomerate surfaces by leaching with water.[1]

In this work we present the results for tests in which the SESR process is used to treat a highly saline industrial soil sample contaminated with both heavy oils and heavy metals. We investigated concurrent removal of the hydrocarbon, by solvent extratction, and fixation of heavy metals by incorporation of metal binding agents into the agglomerates. Phosphate rock and coal combustion fly ashes were used as amendments for this purpose. The degree of heavy metal fixation was determined by the US EPA's Toxicity Characteristics Leaching Procedure (TLCP) on the dried agglomerates. The leaching of soluble salts by percolation of water through a packed bed of agglomerated soil was also carried out.

- During agitation, particles suspended in a liquid will bridge together to form compact spherical pellets if a second liquid is present which is immiscible with the suspending liquid and which preferentially wets the particles.

- Liquid Phase Agglomeration is a novel form of wet pelletizing that depends on the wettability of particles.

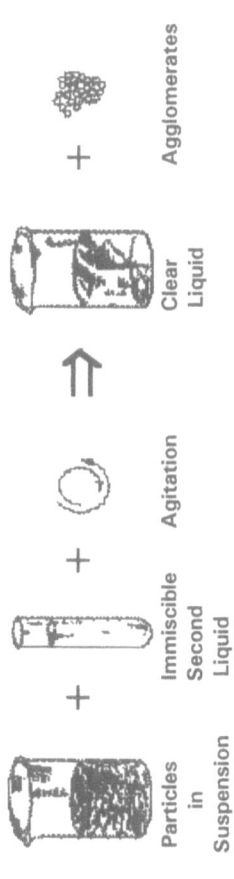

Figure 1. Simplified Schematic for Liquid Phase Agglomeration Procedure.

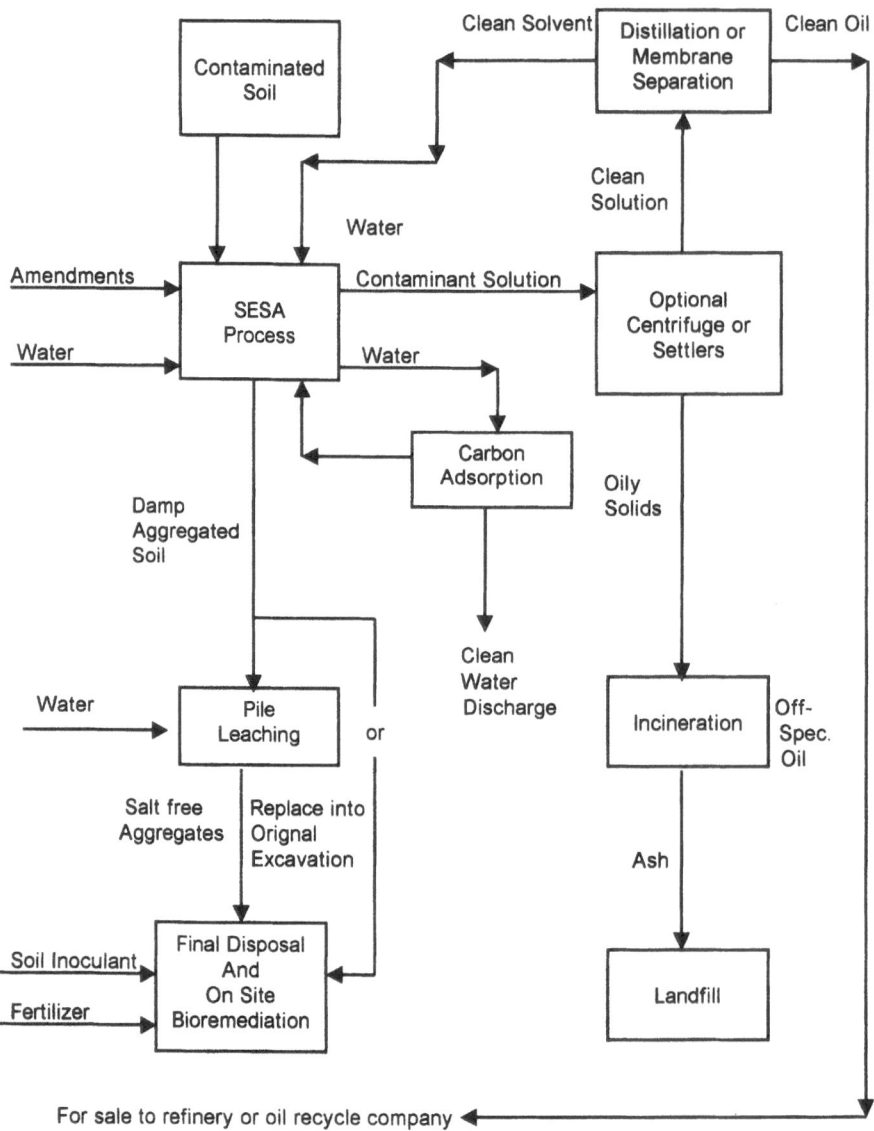

Figure 2. Flow Diagram for the Integrated Soil Remediation Process.

3. EXPERIMENTAL METHODS

3.1. Materials

The soil was a mixture of tank bottoms, frac sand and spill material containing a range of organic and inorganic contaminants. It was provided courtesy of Newalta Corporation, Calgary, Alberta. The composition and properties of the contaminated sample are given in Table 1.

Table 1. Properties of Newalta Soil Sample

Parameter	Value
Oil, w/w% of wet sample[1]	12.9 ± 0.3
Solids, w/w% of wet sample[1]	82.2 ± 1.0
Water, w/w% of wet sample[1]	6.2 ± 1.1
Total of Heavy oil, Solids and Water, w/w% of wet sample[1]:	101.3 ± 1.5

Size Distribution

−325 mesh (−45 μm) solids:	12.2% of total dry solids
100/325 mesh (150/45 μm) solids:	18.6% of total dry solids

Electrical conductivity, mS/cm

OMEE Method[2],	7.4
Alberta Environment method[3],	22.2

Heavy metals:

Metal	Metal Concentrations (mg/g of solids)		OMEE Guidelines*	
	Newalta	ICPET	Industrial Soil	Agric, Prklnd, Res
B	2.87	ND	2.0[4]	1.5[4]
Cd	<0.90	ND	12.0	3.0
Cr	<4.0	85	750.0	750.0
Cu	134.0	ND	225.0	150.0
Pb	198.0	1,480	1,000.0	200.0
Hg	<0.02	ND	10.0	10.0
Ni	20.2	ND	150.0	150.0
V	<2.00	ND	200.0	200.0
Zn	137.2	104	600.0	600.0

*Ontario Ministry of Environment and Energy (Soil, drinking water and air quality criteria: ACES Report to OMEE, 1994).
[1]Determined by Dean & Stark Soxhlet Extraction Method with toluene.
[2]Ontario Ministry of Environment Method with water to soilds ratio of 2:1, OMEE guidelines call for a value of 0.7 mS/cm for agricultural/residential/parkland soil and 1.4 mS/cm for commercial/industrial use.
[3]This method requires the formation of a paste; Alberta guidelines: 2.0 mS/cm for agricultural/residential/parkland and 4.0 mS/cm for commercial/industrial use.
[4]available boron based on hot water extraction.

3.2. Additives

Agricultural peat moss was used as a source of peat for this study; it was ground to about 150 μm particle size, using a Brinkmann Centrifugal Grinding Mill ZM-1. Distilled water containing suspended peat had a pH of 4.0 ± 0.1. Ignition in a muffle furnace at 400 ± 10 °C gave an ash content of 2.3 w/w% (dry basis). No detectable amounts of heavy metals were found in the TCLP leachate from peat.

Flue gas scrubber sludge, obtained courtesy of Joel Colmar of Western Ash Company, located in Arizona, was used as a potential source of gypsum. The sample was dried at 110 °C before use. X-ray analysis of this sample showed gypsum ($CaSO_4 \cdot 2H_2O$) to be the major component while calcite ($CaCO_3$) and quartz (SiO_2) were minor constituents. Western Ash also supplied other samples of coal combustion by-products. X-ray analyses of these materials showed CaO (25–40%) and MgO (5–15%) to be the active ingredients.

The sample of lignin was obtained courtesy of Vince Di Tullio of Albany International (Canada) Inc. A sample of phosphate rock was obtained from White Springs Agricultural Chemicals, Inc. Florida. According to the product specification sheet the major components were fluorapatite and quartz.

All solvents used were of HPLC grade.

3.3. Estimation of Oil Content of Contaminated Solids

The amount of oil contained in contaminated solids was estimated by extraction with toluene, using the Soxhlet-Dean and Stark method[2]; extraction was carried out for 20 hours. The quantitation of the oil component was carried out using a spectrophotometric method[3] based on the linear relationship between the absorbance at 530 nm and the concentration of oil in solution.

For calibration purpose oil samples were obtained from a methylene chloride extract of the soil. Non-filterable solids were removed from the oil extract by centrifugation. The solvent was removed at 40 °C in a Brinkmann rotary evaporator under reduced pressure. The amount of residual solvent in the oil was quantitatively measured using proton NMR.[4] A correction for residual solvent content was applied to the amount of oil used to preparae standard solutions. For spectrophotometric measurements, absorbances at 530 nm were determined for toluene solutions of the oil in the concentration range of 0.01–0.4 w/w%. Plots of the percent oil vs. absorbance produced a straight line passing through the origin.

3.4. Agglomeration Procedure

In a typical test a polypropylene Waring Blendor Jar (500 mL), equipped with Teflon washers and a plastic cover was first weighed. Contaminated solids (100 g), and solvent (250 mL) were then weighed into the tared jar. The contents were agitated at high shear for 1 minute. The solution was carefully drained through a preweighed glass filter paper and the filtrate collected in a tared, 250 mL glass measuring flask. After filtration the amounts of solid and solution retained by the filter paper were determined gravimetrically. The filter paper was dried at 110 °C and the amount of solid on the filter paper measured; the weight of solution associated with the filter paper was estimated by difference. The flask containing the extract was also reweighed to obtain the amount of solution collected. The extract was analyzed by the spectrophotometric method. The amount of extracted solids plus residual contaminant and solvent was obtained from the weight of jar plus contents, after draining the solution.

Metal fixation reagent, additional water and fresh solvent (250 mL) were added to the Blendor jar containing the extracted solids from the primary extraction. The contents were agitated at high shear for 1 minute followed by 5–10 minutes at low shear until the slurry became clear as discrete agglomerates formed. The solution was drained and agglomerates surface washed three times with fresh toluene. All washings and the solution were combined and analyzed spectrophotometrically to determine the amount of oil extracted in stage 2. The washed agglomerates were transferred to a weighed thimble for the determination of residual contaminant by extraction using the Soxhlet-Dean and Stark method. Mass balance calculations were made from the above information. The average mass balance, i.e., the average sum of all component determinations from a series of tests was 99.2 ± 0.7.

3.4.1. Multi-stage Extraction. A multi-stage extraction procedure was used to test the ultimate contaminant removal efficiency of the extraction process. The agglomerates obtained from each stage were dispersed in fresh solvent. High shear agitation was applied for one minute in order to break the agglomerates. This was followed by low shear agitation for 5–10 minutes. The process was repeated four more times with fresh solvent. Five separate tests were carried out in order to obtain an adequate amount of agglomerated soil after each extraction stage. The agglomerates were washed three times with fresh solvent before transferring into the extraction thimble for the determination of residual contaminant.

3.4.2. Mixing Efficiency. The effect of more efficient agglomerate breakdown on contaminant removal during staged extraction was determined by crushing the agglomerated soil in a mortar and pestle between stages and then reagglomerating the solids under normal mixing conditions.

3.5. Electrical Conductivity of the Extracted Solids

The electrical conductivity of aqueous suspensions of extracted, dry solids was measured using the Ontario Ministry of Environment and Energy (OMEE) revised procedure.[5] The revised method requires a water to soil ratio of 2:1; in the previous method only sufficient water is used to form a paste with the amount varying, depending on soil type. Under these conditions it was often difficult to separate enough water for the conductivity measurements. The use of additional water in the revised procedure requires that the existing guidelines of 2.0 mS/cm (agricultural/residential/parkland) and 4.0 mS/cm (commercial/industrial soils) be reduced to 0.7 mS/cm and 1.4 mS/cm respectively.

3.6. Leaching of Soluble Salts

After agglomeration the soil remained saline because all of the connate and process water, with dissolved salts, remains in the agglomerate pores. However, drying, to recover residual solvent, causes migration of the water-soluble components to the agglomerate surfaces as water is evaporated. Once at the surface the salts are readily removed by leaching a fixed bed of agglomerates with water.

An amount of water corresponding to a water:soil ratio of 2:1 was added to the bed of agglomerates. The leachate flow rate was controlled using an adjustable clamp attached to the leachate outlet tube. Conductivity of the leachate was measured using a Horiba model ES-12 conductivity meter. Collected leachate was recycled until its conductivity did not change more than 0.1 mS/cm between successive washes. At this point fresh water was added and the procedure continued until the conductivity of the leachate was lower than 0.7 mS/cm. At this stage the residual conductivity of the leached agglomerates themselves was measured according to the OMEE revised method.

3.7. Heavy Metal Leaching Tests

Heavy metal leaching potential of the treated solids was determined by the US-EPA's toxicity test method 1310A and Toxicity Characteristics Leaching Procedure method 1311 as described in the US Federal Register[6] and specified in SW-846.[7] Long term stability of the treated solids in terms of metal leaching was tested using the agency's multiple extraction procedure method 1320.

3.8. Analysis of Metals

Heavy metals were analyzed by Inductively Coupled Plasma Spectroscopic Analysis (ICP).

3.9. Carbon Analysis of Dry Solids

Total carbon was determined using a Leco CR12 carbon analyser. Organic carbon was determined after decomposing carbonate carbon with dilute hydrochloric acid; carbonate carbon was then determined by difference.

3.10. Surface Analysis

X-ray photoelectron spectroscopy (XPS) was performed with a monochromatic Kα source of x-rays. Samples were lightly pressed into a shallow powder sample holder, evacuated to less than 10^{-6} torr pressure and then introduced into the analysis chamber of a Kratos Axis XPS. A survey spectrum was run first followed by analysis of two typical areas at high resolution.

4. RESULTS AND DISCUSSION

The SESR process has been tested on a highly saline industrial soil sample contaminated with heavy oil and several heavy metals. Table 2 summarizes the results for a single stage extraction using toluene as a solvent; over 90% of the contaminant is removed in a single stage application of the SESR extraction process.

4.1. The Effect of Additives (Metal Binding Agents)

Several metal binding agents, e.g., flue gas scrubber sludge, sodium and calcium phosphates, peat, lignin and coal combustion ashes were incorporated individually into the agglomerates during processing in order to test their ability to fix leachable heavy metals. The effect of these additives on the efficiency of hydrocarbon contaminant removal by the SESR process is illustrated by the bar chart shown on Fig. 3.

Whereas, phosphates, peat, lignin and coal combustion ashes did not affect the basic extraction efficiency of the SESR process, flue gas scrubber sludge appears to have a beneficial effect on oil removal. In this case the amount of contaminant removed exceeded the amount recovered using the traditional Soxhlet extraction procedure.

4.2. Residual Contaminant

The amount of residual contaminant in the treated solids was measured for mass balance purposes and to assess contaminant removal efficiency. Figure 4 plots the residual contaminant data for single stage extraction (Table 2) as a function of total contaminant removed. If the data for blank tests and the tests with ash "C" are excluded then the remaining data can be fitted to a second order equation with a correlation coefficient

Table 2. Summary Results for Single Stage Extraction of Heavy oil from Newalta Sample

| Test No. | Additive* | Feed recovery (%) | Amount of non-settling solids[†] | Amount of Oil extracted[‡] | | Residual Oil[††] ppm |
				w/w% of wet solids	% of total**	
1	—	98.9 ± 0.4	2.4 ± 1.5	12.18 ± 0.245	94.4 ± 1.9	4,867 ± 723
2	Na$_3$PO$_4$ (1)	98.6 ± 1.1	2.6 ± 0.2	11.84 ± 0.348	91.8 ± 2.7	7,259 ± 577
3	FBA (5)	99.4 ± 0.2	6.7 ± 1.3	12.15 ± 0.335	94.2 ± 2.6	6,387 ± 368
4	Lignin (1)	99.0 ± 0.8	3.6 ± 0.1	12.37 ± 0.103	95.9 ± 1.3	6,745 ± 365
5	Peat (2)	99.5 ± 0.1	4.2 ± 0.7	12.68 ± 0.258	98.3 ± 2.0	5,713 ± 171
6	Western Ash "C"(5)	99.3 ± 0.2	3.0 ± 1.2	12.01 ± 0.297	93.1 ± 2.3	4,491 ± 209
7	Scrubber sludge (5)	99.5 ± 0.2	3.5 ± 0.4	13.38 ± 0.490	103.7 ± 3.8	4,252 ± 191
8	Phosphate Rock (1)	99.3 ± 0.2	2.7 ± 0.5	11.96 ± 0.335	92.7 ± 2.6	7,056 ± 23

*Amount as w/w% of wet solids in parenthesis.
[†]Unagglomerated fine solids reporting with solution (w/w% of wet solids).
[‡]Average of five tests.
**SOXHLET extractable amount (12.9 ± 0.2).
[††]Average amount of residual oil remaining on solids after extraction by SESR; determined by Soxhlet extraction using toluene. Values are in ppm, calculated on dry solids basis.
OMEE guidelines for sites contaminated with heavy hydrocarbons:
Surface soil/subsurface (potable groundwater condition): 1,000 ppm
Surface soil (non-potable ground water situation case 1: residential/parkland: 1,000 ppm
Surface soil (non-potable ground water situation case 2: Industrial/commercial: 5,000 ppm
Subsurface soil (non-potable ground water situation case 1: residential/parkland: 5,000 ppm
Subsurface soil (non-potable ground water situation case 2: Industrial/commercial: 10,000 ppm

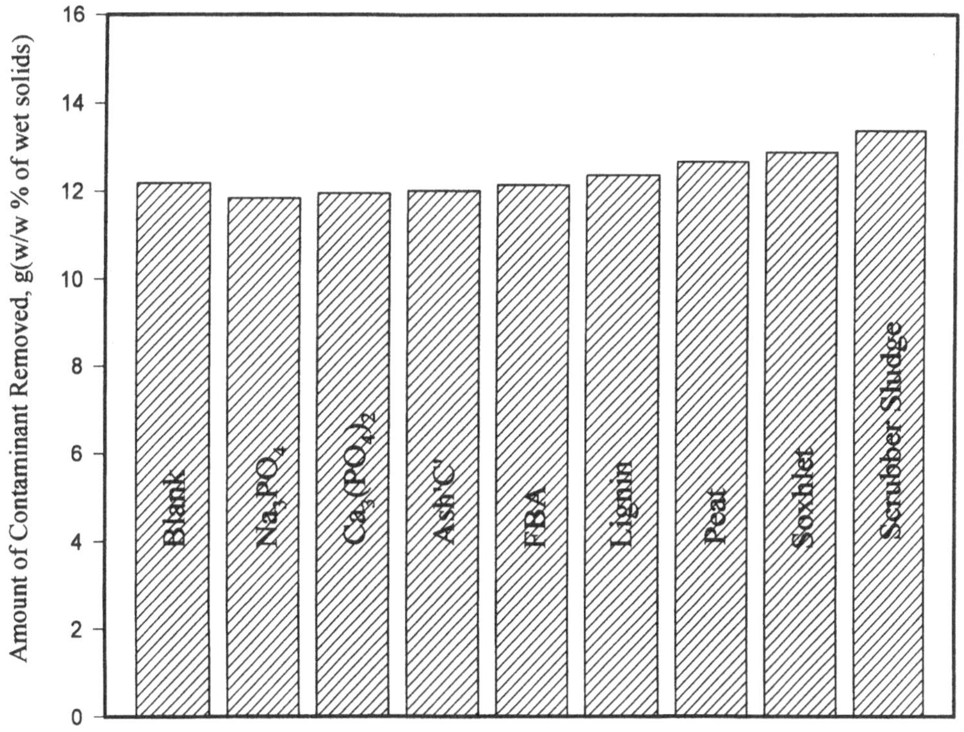

Figure 3. The Effect of Additives on the Contaminant Removal Efficiency of SESR.

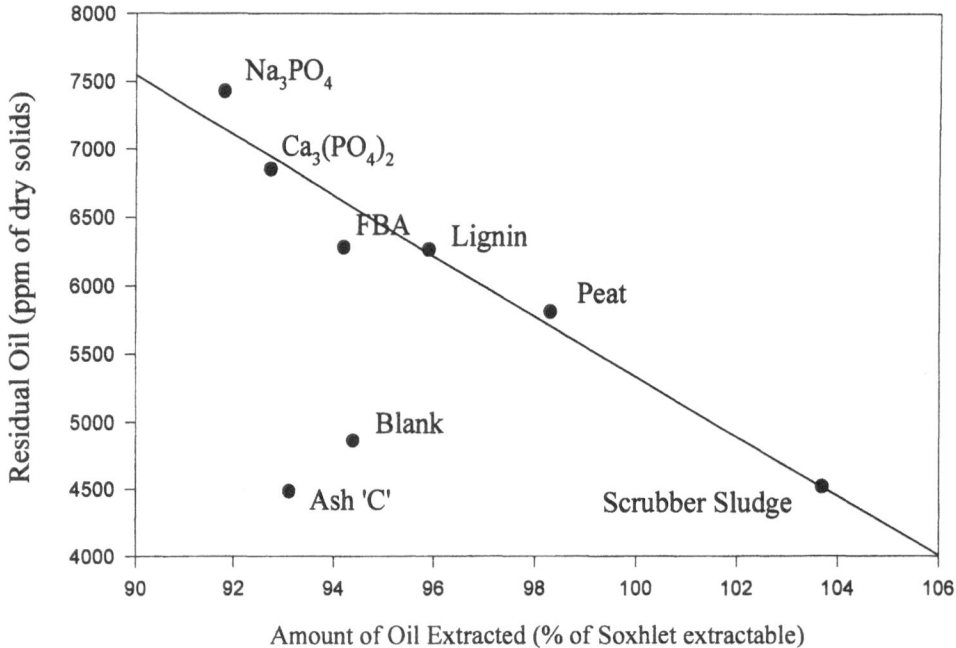

Figure 4. Extraction Efficiency of Single Stage SESR Process.

of 0.957. This suggests that a part of the contaminant is strongly adsorbed on the solids and is not removed during Soxhlet extraction. This is consistent with the higher recovery of the contaminant in the presence of flue gas scrubber sludge. The results also suggest that the additive Ash "C" acts as an adsorbent for the contaminant.

The residual amount of contaminant, after a single extraction stage, was within the OMEE guidelines of 5000 ppm for sub-surface soil disposal. However, for use as surface soil the guidelines for residual contaminant require levels of 1000 ppm or less. This level cannot be achieved in a single stage process.

4.3. Multistage Extraction by SESR

Multistage extraction was evaluated to determine the process's capability to achieve residual contaminant levels within the strict guidelines for residential and parkland use. The results are listed in Table 3; they show that after five stages of extraction, with conventional mixing, the residual contaminant levels are quite close to the guidelines of 1000 ppm oil. However, better results were achieved by mechanically breaking down agglomerates between stages and then reforming them during each conventional mixing step. This procedure improved contact between solvent and contaminant entrapped within the agglomerates during their rapid formation. This approach has the potential to reduce the number of cycles required to achieve the regulated levels of residual contaminant in the treated solids.

4.4. Heavy Metals Fixation

The solids were contaminated with heavy metal salts of As, Cd, Cr, Pb, Hg and Zn. US-EPA' TCLP tests indicated that the leachable amounts of mercury (1.25 mg/L), and

Table 3. Multistage Extraction of Contaminant

Test #	Description	# of cycles*	Residual Contaminant (ppm)†
1	No additive	1	4,867 ± 723 (3)
2	No additive	2	2,624 ± 284 (2)
3	No additive	3	2,408 ± 26 (2)
4	No additive	4	1,266 ± 40 (2)
5	No additive	5	1,060 ± 39 (2)
6	1% Scrubber Sludge	1	3,092 ± 61 (3)
7	5% Scrubber Sludge	1	4,252 ± 191 (3)
8	5% Scrubber Sludge	3	2,284 ± 7 (2)
9	5% Scrubber Sludge	4	1,429 ± 51 (2)
10	5% Scrubber Sludge	5	1,217 ± 8 (2)
11	1% $Ca(OH)_2$	1	2,568 ± 771 (4)
12	1% $Ca(OH)_2$	2	2,334 ± 151 (3)
13	1% $Ca(OH)_2$	3	1,693 ± 276 (3)
14	1% $Ca(OH)_2$	4	1,091 ± 39 (3)
15	1% $Ca(OH)_2$	5	1,031 ± 18 (3)
16	1% $Ca(OH)_2$	3‡	1,372 ± 194 (3)
17	1% $Ca(OH)_2$	4‡	839 ± 69 (3)
18	1% $Ca(OH)_2$	5‡	687 ± 204 (3)
19	1% Scrubber Sludge	5‡	880 ± 55 (3)

*For single cycle, the extracted solids were surface washed three time. For multiple cycles, one washing was carried out between cycles and three at the end.
†Dry solids basis, Figures in parenthesis represents number of tests carried out.
‡Agglomerates produced in each stage were transferred into a mortar and pestle; crushed and then transferred back into the Blendor jar for reagglomeration.

lead (29.3 ± 5.0) exceeded OMEE's guidelines of 0.2 mg/L and 5.0 mg/L respectively. Tests were carried out to fix these heavy metals by incorporating, metal binding materials into the soil agglomerates during solvent extraction of organic contaminants. The extracted dried soil samples were evaluated for their heavy metal leaching potential using US-EPA's toxicity test method 1310A and Toxicity Characteristics Leaching Procedure method 1311. Long term stability of the treated solids, in terms of metal leaching, was tested using US-EPA's multiple extraction procedure, method 1320. This method is designed to simulate leaching from repetitive precipitation of acid rain on an improperly designed sanitary landfill. The multiple extractions determine the highest concentration of each constituent that is likely to leach in a natural environment.

Analysis of the leachates obtained using test methods 1310A and 1320 showed that the concentrations of targeted heavy metals were below the detection limit (see Table 4). However, use of the more severe TCLP method showed that only phosphates, flue gas scrubber sludge and coal combustion ash "C" were completely effective, see Table 4. For peat, lignin and FBA fixation was complete for mercury but only partial for lead.

4.5. Leaching of Soluble Salts

The electrical conductivity data, shown in Table 1, demonstrates the high salinity of the contaminated soil. The electrical conductivity of soil is a measure of the total concentration of soluble salts in the soil pore water. High salinity can have a large osmotic influence on plant growth, as well as on soil organisms. OMEE guidelines suggest that the electrical conductivity of soil, measured according to a fixed 2:1 water:soil ratio, should not exceed 0.7 mS/cm for agricultural, residential and parkland use and 1.4 mS/cm for industrial and commercial use. An electrical conductivity value of 0.7 mS/cm

Table 4. Heavy Metals Leachability Results from US-EPA TCLP tests at 3.0 ± 0.1*

Metal	Feed	EPA & OMEE regulatory limit	Additives: 5% Ash"C", 1% phosphates, 5% Scrubber sludge	Additive: 2% Peat	Additive: 5% FBA	Additive: 1% Lignin
As	2.7 ± 0.1	5	<2.0[†]	<2.0	<2.0	<2.0
Cd	0.4	1	<0.3	<0.3	<0.3	<0.3
Cr	<0.05	5	<0.1	<0.1	<0.1	<0.1
Pb	29.3 ± 5.0	5	<2.0	24.0	12.8	26.0
Hg	1.25	0.2	<1.0	<1.0	<1.0	<1.0
Zn	1.6	250	<0.05	<0.05	<0.05	<0.05

Extractable Heavy Metals (mgs/L of Extract)

*Leachates obtained using test methods 1310A and 1320 did not contain any of the metals tested.
[†]< represents method detection limit.

corresponds to the boundary between what may "result in a slightly stunted condition" and "slight to severe burning" in most plants.[5]

After agglomeration there was essentially no change in the salinity of dried agglomerates. However, we have demonstrated that during drying, dissolved salts, e.g., brine in the water that fills the agglomerate pores, are subject to efflorescence effects which cause them to concentrate at the surfaces of the agglomerates.[1] These salts are then readily removed in a fixed bed, water leaching system. Because attrition of agglomerates is relatively low the treated soil still retains its desirable size distribution after leaching.

The leaching of brine contaminated agglomerates from this study was carried out by percolation of water through a packed bed of agglomerated soil. The results are listed in Table 5. The leaching of salt was very rapid from agglomerated soil compared to the case for Soxhlet extracted, but unagglomerated, soil solids. The agglomerated blank soil and the samples containing phosphate gave the best results, requiring approximately

Table 5. Leaching of Soluble Salts from Extracted Solids**

Test No.	Additive*	Total volume of H₂O used for Leaching, mL[†]	H2O:Pore Volume Ratio	Flow Rate mL/Min	Total Time, hrs	OMEE Method[‡]	Final Leachate	Leached Solids
1[a]	—	1,265	59.7	3	20.0	7.4	0.30	0.53
2[b]	—	600	31.8	10	4.0	7.3	0.36	0.31
3	Na₃PO₄ (1)	600	25.2	10	3.0	6.7	0.48	0.65
4	Ca₃(PO₄)₂ (1)	600	25.2	10	2.5	6.7	0.53	0.65
5	Scrubber sludge (5)	900	39.6	10	8.0	6.2	0.40	0.40
6	FBA (5)	1,100	47.9	10	5.7	7.4	0.36	2.00
7	FBA (10)	1,600	70.4	3	26.0	13.2	0.28	4.00

Electrical Conductivity, mS/cm (columns: OMEE Method[‡], Final Leachate, Leached Solids)

*Amount as w/w% of wet solids in parenthesis.
[†]Amount of solids used in all tests: 50 g, The leachate was recycled until the conductivity of the subsequent leachate did not change more than 0.1 mS/cm.
[‡]H2O: Solids Ratio: 2:1.
[a]SOXHLET extracted blank solids.
[b]SESR blank solids.
**OMEE Guidelines for Electrical Conductivity of Soil:
  a) Agricultural, Residential and Parkland Land use Categories: 0.7 mS/cm.
  b) Industrial and Commercial Land use Categories: 1.4 mS/cm.

half the amount of water for leaching compared with the Soxhlet extracted solids. Similarly, leaching times were also much reduced for the agglomerated samples. Several of the leached soil samples met OMEE guidelines for agricultural, residential and parkland use.

The leaching of soluble salts from agglomerates containing FBA as an additive was very slow. Also, the electrical conductivity of the leached solids was higher than OMEE guidelines for both residential and industrial soil uses. We believe that the slow leaching of lime, which is a major constituent of FBA, from the agglomerate matrix, causes this result. This observation suggests that this method is not suitable for evaluating the conductivity of samples containing lime.

4.6. Surface Analysis of Solids

Surfaces of cleaned solids and non-settled solids from the solvent phase, after separation of agglomerated soil, were characterized using X-ray photoelectron spectroscopy (XPS). The results are given in Table 6. XPS provides information on the topmost 10 nm layer of the particles. The fact that Si, Al, Fe, Na, Ca and Mg were seen in the XPS spectra for extracted solids suggests that the solids have only a thin or patchy layer of organics on the surface. The organic layer consists mostly of carbon atoms with small amounts of O and S. The high-resolution spectra of the carbon peak suggests the presence of some functionalities such a C–OH, C=O, HO–C=O, C–SH.

XPS spectra of the ashed samples suggest that most of the oxygen detected in these samples is associated with the mineral matter. The lower amount of carbon in the HCl treated samples, compared with the blank samples, occurs because of the acid decomposition of carbonate. These results also suggest that the carbonate content of the non-settling solids is much higher than in the bulk solids. Whereas, the treatment of bulk solids with HCl results in the reduction, or complete dissolution of Fe, Na, Ca and Mg, an enrichment of these elements was observed in the residue from non-settling solids after acid treatment.

Table 6. Concentrations (atomic %) of elements by XPS

	Soxhlet extracted solids			SESR extracted with Scrubber sludge			Non-settling solids		
	Blank	Ashed	HCl treated	Blank	Ashed	HCl treated	Blank	Ashed	HCl treated
Total C	30.3	4.1	22.2	27.8	2.9	19.5	43.8	2.5	11.6
C-286.1ev*	17.7	4.1	14.1	16.5	2.9	13.5	29.2	2.5	7.0
C-286.1ev[†]	7.9	—	6.2	7.6	—	4.6	9.7	—	3.0
C-288.4ev[‡]	2.7	—	1.9	2.2	—	1.5	2.9	—	1.6
O	50.5	66.1	54.6	51.2	73.1	53.3	43.2	69.4	64.7
S	1.8	4.3	—	4.3	5.1	—	2.0	7.9	3.6
Si	8.6	12.0	20.2	7.2	8.1	25.1	3.6	4.5	5.7
Al	2.5	4.8	2.4	2.6	2.5	1.3	1.8	2.3	2.9
Fe	2.4	4.4	0.5	1.4	2.4	—	2.4	6.8	7.9
Na	0.5	0.7	—	0.5	0.3	—	0.8	0.4	1.0
Ca	2.7	3.6	—	4.8	4.9	—	1.2	5.9	2.6
Mg	0.3	—	—	0.3	0.6	—	trace	0.4	trace

*Saturated carbon.
[†]C–OH, C–O–C, C-NH$_2$, C-SH.
[‡]C=O.

The lower carbon content of the solids extracted in the presence of flue gas scrubber sludge compared with the solids extracted by the Soxhlet method supports the extraction results found in the SESR tests. The results suggest a beneficial effect of added flue gas scrubber sludge on oil extraction efficiency.

5. CONCLUSIONS

A highly saline, industrial soil sample, contaminated with a heavy oil and several heavy metals, was successfully remediated using a modified SESR process. During processing, concurrent removal of the hydrocarbon, by solvent extraction, and fixation of heavy metals, by incorporating metal fixation agents into the soil agglomerates formed, was achieved. Phosphate rock and coal combustion fly ashes were particularly effective for metal fixation. Over 90% of the oily contaminant was removed in the first stage of the SESR process. At this point the amount of residual contaminant in the treated soil was within OMEE guidelines of <5000 ppm for the subsurface soil category. A five stage extraction process, without optimum mixing, reduced the residual oil contaminant to levels close to the OMEE guidelines for residential and parkland use. However, better results were achieved by improving the efficiency of mixing to provide more breakdown and reforming of agglomerates. An improvement in mixing has the potential to reduce the number of cycles required to achieve the regulated levels of residual contaminant in the treated solids.

Additives used for the fixation of lead and mercury did not, in general, affect the basic extraction efficiency of the SESR process. However, flue gas scrubber sludge had a beneficial effect on oil removal; in this case the amount of contaminant removed exceeded the amount recovered by the Soxhlet extraction procedure. The effect was confirmed by measurement of surface carbon concentration by XPS analysis of the solids treated in different ways.

The additives tested for the fixation of heavy metals, were effective in rendering these metals non-leachable under the conditions of acid rain. However, when a more rigorous leaching procedure, such as EPA's TCLP, was used only phosphates, flue gas scrubber sludge and coal combustion ash "C" were found to be completely effective. Peat, lignin and FBA completely fixed mercury but were only partially effective for lead.

Dried agglomerates were leached with water in a fixed bed system to remove residual brine in order to meet OMEE Guidelines for soil conductivity. The required conductivity was reached with less water and in a shorter time for agglomerated solids compared to unagglomerated soil extracted using conventional means.

ACKNOWLEDGMENT

The authors are grateful to, Gerald Pleizier and V. Boyko for technical help.

REFERENCES

1. Meadus, F.W., Sparks, B.D., and Majid, A. "Solvent Extraction Using a Soil Agglomeration Approach", Emerging Technologies in Hazardous Waste Management VI, 1994, 161–176, D. William Tedder and Frederick G. Pohland Editors, American Chemical Society, Washington DC.

2. Syncrude analytical methods for oil sand and heavy oil processing, J.T. Bulmer and J. Star Editors. Alberta oil Sands Information Centre, Edmonton, AB, pp. 46–51 (1979).
3. Patel, M.S., "Rapid and Convenient Laboratory Method for Extraction and Subsequent Spectrophotometric Determination of Heavy Oil Content of Bituminous Sands"; Anal. Chem. **46**, 794(1974).
4. Majid, A. and Sparks, B.D., "Total Analysis of Mineral Wastes Containing Heavy Oil, Solvent, Water and Solids", AOSTRA J. Res. **1**, 21–29(1984).
5. Ontario Ministry of Environment and Energy: Standards Development Branch, "Rationale for the Development and Application of Generic Soil, Groundwater and Sediment Criteria for Use at Contaminated Sites in Ontario", May, 1996.
6. US EPA Federal Register. **51**, (142), March 1990, Office of Solid Waste, Washington DC.
7. US EPA—"Test Methods for Evaluation of Solid Waste: Physical/Chemical Methods", SW-846, 3rd Ed., Office of Solid Waste and Emergency Response, Washington DC, Nov. 1986.

ABBREVIATIONS

SESR   Solvent Extraction Soil Remediation
OMEE   Ontario Ministry of the Environment and Energy
EPA   Environmental Protection Agency
TCLP   Toxicity Characteristics Leaching Procedure
ICP   Inductively Coupled Plasma
ICPET   Institute for Chemical Process and Environmental Technology
ND   Not Detected
MS/cm   Conductivity in mlliSiemens/cm

6

LABORATORY AND FIELD SOIL WASHING EXPERIMENTS WITH SURFACTANT SOLUTIONS

NAPL Recovery Mechanisms

Richard Martel[1], Pierre J. Gélinas[2], René Lefebvre[1], Alain Hébert[2], Stefan Foy[2], Laurent Saumure[2], Annie Roy[1], and Nathalie Roy[1]

[1]*In situ* Research Group
INRS-Géoressources, 2535 boulevard Laurier
C.P. 7500, Sainte-Foy, Quebec
Canada G1V 4C7
[2]*In situ* Research Group
Département de géologie et de génie géologique
Université Laval
Sainte-Foy, Quebec, Canada G1K 7P4

1. ABSTRACT

During the last six years, surfactant solutions have been developed to dissolve or mobilize different NAPL types. These solutions were made with anionic surfactants and alcohols, as well as solvents in some cases. Laboratory and field tests were performed using these solutions to recover residual NAPL in sediments and etched-glass micromodels. Surfactant washing experiments were done on: (1) small sand columns (110 g); (2) a large sand column (65 kg); (3) an etched glass micromodel and; (4) a test plot in an aquifer contaminated with a complex mixture of DNAPL. Washing experiments were applied to the recovery of diesel, Aroclor 1248, and a complex DNAPL mixture. Lab and field experiments show that it is possible to recover more than 90% of residual NAPL with less than 6 pore volumes of a concentrated washing solution. Recovery mechanisms depend on: (1) NAPL type, (2) micellar solution type (alcohol/surfactant or alcohol/surfactant/solvent), (3) ingredient concentrations in the micellar solution, (4) washing direction (upward, downward, horizontal), (5) injection/pumping pattern, and (6) injection velocity.

Emerging Technologies in Hazardous Waste Management 8, edited by Tedder and Pohland
Kluwer Academic/Plenum Publishers, New York, 2000.

2. INTRODUCTION

Since 1990, our research group has been developing an *in situ* technology to recover NAPL in contaminated aquifers. This method distinguishes itself from conventional pump- and-treat by the use of a very small volume of washing solution composed of alcohol/surfactant or alcohol/surfactant/solvent. The principle of the method is to form a NAPL-in-water microemulsion by the use of a concentrated washing solution. The developed technology can be applied at depth in the water-saturated zone or under buildings. The technology is also well adapted for zones highly contaminated with long chain organic compounds which have low aqueous solubility and low volatility.

Tests were conducted at different scales to develop the technology (Fig. 1): in the laboratory at the test tube scale for washing solution selection by phase diagrams (Martel *et al.*, 1993 and Martel and Gélinas, 1996); in small and large sand columns for washing solution optimization; (Martel and Gélinas, 1996 and Martel *et al.*, 1998a and 1998b) in etched glass micromodels to simulate NAPL trapping and recovery mechanisms in fractured media (Tittley, 1994); in sand boxes to reproduce permeability constrasts of stratified media in 2-D (Martel, K. E. *et al.*, 1998); and finally at the field scale for a demonstration of the technology (Martel *et al.*, 1998c). This paper discusses NAPL recovery mechanisms observed during injection of micellar solutions in contaminated sand columns and etched glass micromodels and during a field test in a contaminated aquifer.

3. METHODOLOGY

Two phases are observed when water containing surfactant is mixed with an equal amount of NAPL (Fig. 2): a NAPL phase and an aqueous phase which contains part of the NAPL. The objective is to maximize the amount of NAPL in the aqueous phase

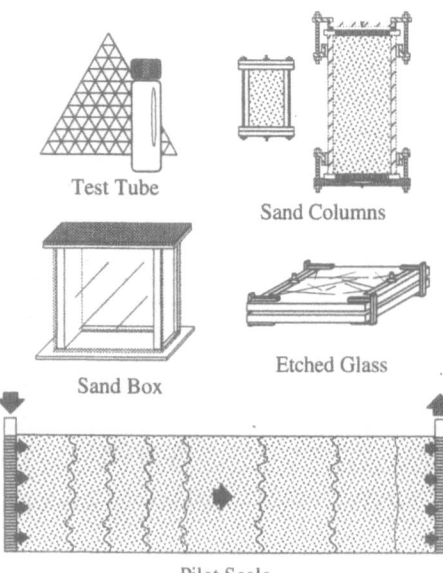

Test Tube

Sand Columns

Sand Box

Etched Glass

Pilot Scale

Figure 1. Scales of study for the development of *in situ* soil washing with surfactant solution.

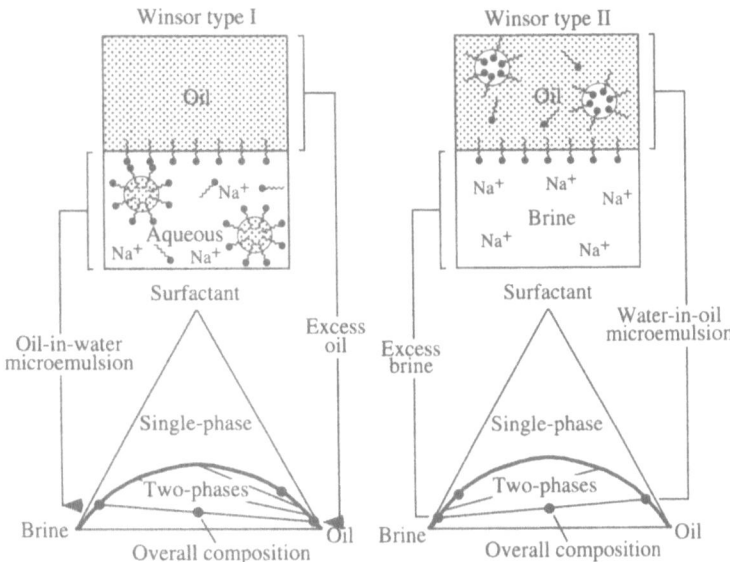

Figure 2. Phase behavior in winsor systems (adapted from lake, 1989).

(NAPL-in-water microemulsion zone) by changing salinity. In the porous media, NAPL can then be transported in the aqueous phase and is displaced by solubilization. This is miscible displacement of NAPL in a Winsor type I system.

Increasing the water salinity transforms the two phases: part of the water is dissolved in the NAPL phase (water-in-NAPL microemulsion) and no NAPL is found in the aqueous phase. In the porous media, the NAPL can be displaced by mobilization because of the decrease in interfacial tension (around 0.1 mN/m) between the aqueous and the NAPL phases. This is immiscible displacement of NAPL in a Winsor type II system. At intermediate salinity, a Winsor type III system is observed with an ultra low interfacial tension. This system is difficult to obtain and to maintain in the porous media. Because of ultra low interfacial tension between the three phases, this system is not recommended for DNAPL recovery.

The same changes in phase behavior are observed when alcohol is added to the water-NAPL system instead of salt. The methodology used to select surfactant solutions for a maximum NAPL-in-water microemulsion zone in phase diagrams is explained in detail elsewhere (Martel *et al.*, 1993; Martel and Gélinas 1996). With this technique, washing solutions were developed to dissolve diesel and Aroclor 1248.

For heavy and viscous hydrocarbons, it could be advantageous to add solvents in the washing solution (Martel *et al.*, 1998a). Different types of emulsions can be observed by changing the proportion of solvents in the surfactant/alcohol solutions (Fig. 3). With this approach, washing solutions were developed for the *in situ* recovery of a complex DNAPL mixtures including chlorinated solvents found in some Quebec contaminated sites.

Laboratory and field tests were performed using these solutions to recover residual NAPL in sediments and etched-glass micromodels. Surfactant washing experiments were done on (1) small sand columns (110 g) contaminated with diesel, Aroclor 1248 and a complex DNAPL mixture, (2) a large sand column (65 kg) contaminated with a complex

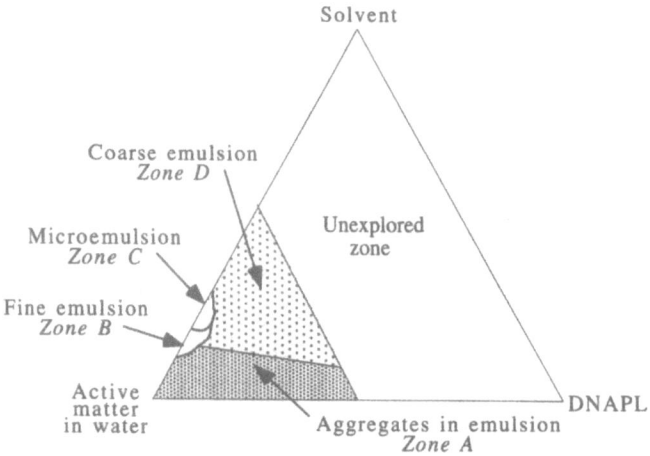

Figure 3. DNAPL extraction zones in phase diagrams with the selected washing solution containing solvents with surfactant and alcohol (active matter).

DNAPL mixture, (3) an etched glass micromodel containing residual diesel and (4) a test plot in an aquifer contaminated with a complex DNAPL mixture. The methodology for small sand column tests is explained in Martel and Gélinas, (1996), for a large sand column test in Martel *et al.*, (1998b), for etched glass experiments in Tittley, (1994), and for field test in Martel *et al.*, (1998c).

4. RESULTS

4.1. DNAPL Recovery in Small and Large Sand Column

The washing solution that shows a DNAPL-in-water microemulsion in the phase diagram (*Zone C*), has the best performance in sand column. With this solution (9.21 wt% n-butanol, 9.21 wt% Hostapur SAS (Hoechst GmbH), 13.16 wt% toluene, 13.16 wt% d-limonene), close to 100% of residual DNAPL was recovered with less than 3 pore volumes of washing solution injected downward at 8 °C in Mercier sand. The same solution was injected in a large sand column containing 65 kg of Mercier sand with a residual DNAPL saturation of 0.19. In this test, close to 90% of the residual was recovered after the injection of only 0.8 pore volume (PV) of washing solution (Fig. 4). Of the DNAPL recovered, 34% was mobilized as a pure product, 51% was mobilized in a water-in-NAPL microemulsion (Winsor type II) (35% as a NAPL denser than water and 16% as a NAPL lighter than water) and 4% was solubilized as DNAPL-in-water microemulsion (Winsor type I).

More than 70% of the DNAPL was recovered in the oil bank ahead of the washing solution (Fig. 5). The maximum DNAPL production is observed at 0.8 PV where 55% of the effluent is composed of DNAPL. At 1.2 pore volume, ingredients of the washing solution are at optimal concentration and the oil recovery, within a DNAPL-in-water microemulsion, decreases rapidly as the small volume of washing solution is diluted by the following rinsing solution (0.27 PV of 10 wt% n-butanol, 10 wt% Hostapur SAS

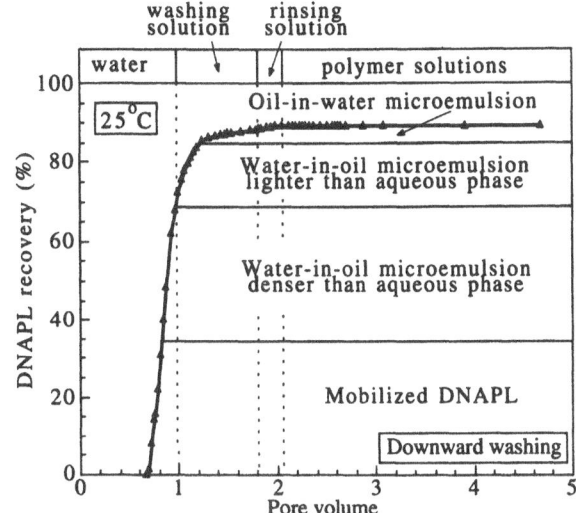

Figure 4. Cumulative DNAPL recovery in the 65 kg sand column effluent.

(Hoechst GmbH) and 620 mg/L xanthan gum). Less DNAPL is then available in the porous media.

The reduction in interfacial tension created by surfactants in the washing solution increases DNAPL mobilization and its proportion in the effluent. The oil bank is formed by an increase in oil saturation resulting in a higher relative oil permeability that favors DNAPL flow in the sand column (Fig. 6). The oil bank is maintained by alcohols and solvents transfer from the washing solution to the oil phase as shown by the reduction in viscosity of the oil phase.

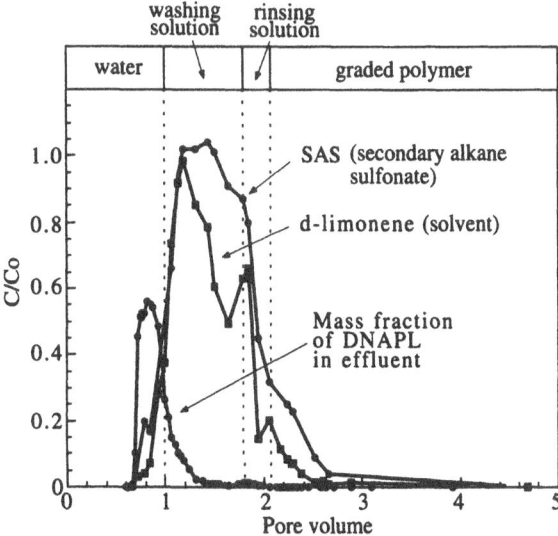

Figure 5. Mass fraction of DNAPL produced and relative concentration of ingredients in the 65 kg sand column effluent.

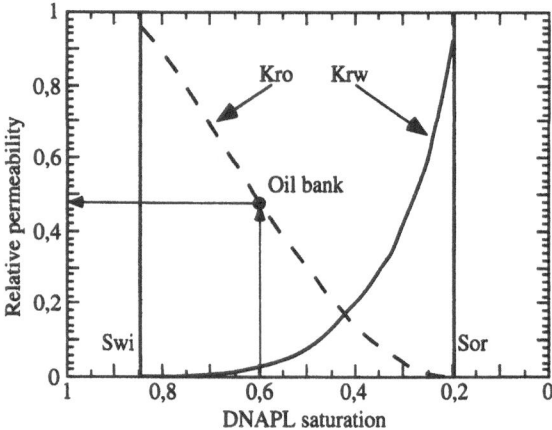

Figure 6. Relative permeability of oil (Kro) and water (Krw) in the 65 kg sand column as a function of DNAPL saturation.

4.2. DNAPL Recovery in the Field

In the fall of 1994 a field test was designed and carried out in a sand pit which received several millions liters of oil wastes during the 60's. The experimental plot of 4.3 m by 4.3 m was equiped with a five-spot recovery pattern consisting of one central injection well and four corner recovery wells, together with 12 multilevel observation wells. The recovery wells are located 3 m from the injection well. The site is characterized by a 2 m thick silty sand unit underlained by 30 m of silty clay which act as an impervious base. The washing experiment was made in the water saturated silty sand unit located between 1 m and 2 m depth. A detailed description of this test is published by Martel *et al.*, (1998c). The washing proceeded in several steps (Fig. 7):

1) Water flooding with 1.1 PV and polymer flooding with 0.6 PV to reach steady state flow in the cell and to decrease non-residual oil saturation;
2) Injection of 0.9 PV of surfactant solution (11.3 wt% n-butanol, 11.3 wt% Hostapur SAS (Hoechst GmbH), 4.5 wt% toluene, 18.0 wt% d-limonene) to wash the sediments;

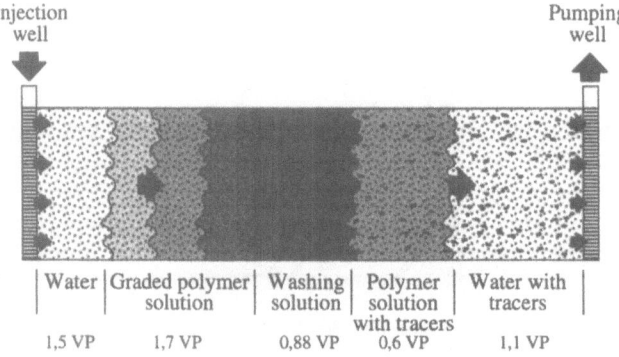

Figure 7. Injection sequence for field application at the thouin sand pit.

3) Polymer flooding with 1.7 PV and water flooding with 1.5 PV to push the washing solution out and rinse the sediments; and

4) Injection of acclimated autochthon bacteria and nutrients to increase biodegradation of the remaining DNAPL and ingredients of the washing solution.

This test is one of the very few environmental field applications of polymer solution used to help NAPL recovery. The mass balance of DNAPL in the cell before and after the washing experiment suggest that more than 86% of the initial DNAPL mass was removed in the cell. Also, more than 1,200 fluid samples were recovered in the monitoring and recovery wells. The proportion of DNAPL recovered by the different mechanisms was defined from the fluids collected in monitoring wells located at 0.65 m and 1.4 m from the injection well. At 0.65 m (Fig. 8) close to 80% of the DNAPL which circulated in the porous media was mobilized and 20% was solubilized. At 1.4 m (Fig. 9) the DNAPL is mainly solubilized. Also, the sharp shape of recovery curves indicates a stable displacement front obtained by the use of the polymer solution. The relative importance of mobilization and solubilization can be explained by the variation in fluid velocity in the sediment for radial injection geometry. For this geometry, the velocity profile decreases with the distance from the injector (Fig. 10). At 1.4 m, the fluid velocity slows down, the viscous forces are decreased which reduces also DNAPL mobilization. However, contact time between the washing solution and the trapped DNAPL is increased which favors its solubilization.

4.3. Aroclor 1248 Recovery in Small Sand Column

Leaks of pure Aroclor 1248 (A1248) from a heat exchanger penetrated the cracked concrete floor of the hot oil furnace room (HOFR) of an industrial facility in Ontario. A1248 infiltrated into the underlying silty sand unit down to a silty clay layer 2.5 meters below. Twenty-one (21) boreholes were drilled in the HOFR (Fig. 11) and the

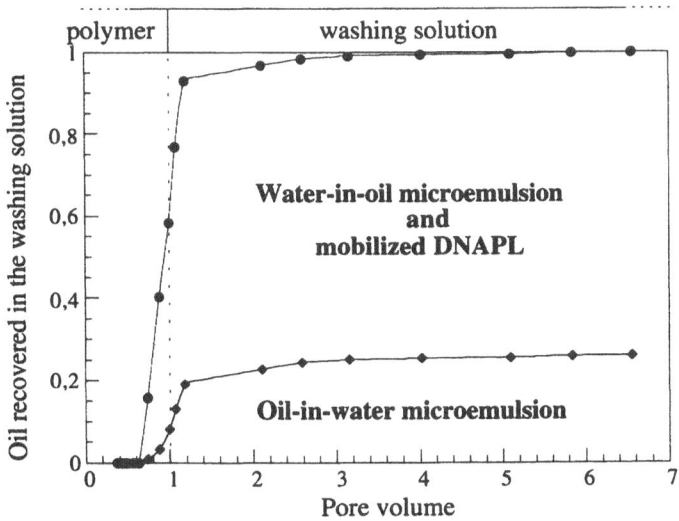

Figure 8. Proportion of DNAPL recovered by mobilization and solubilization at well WP2, 0.65 m form the injector.

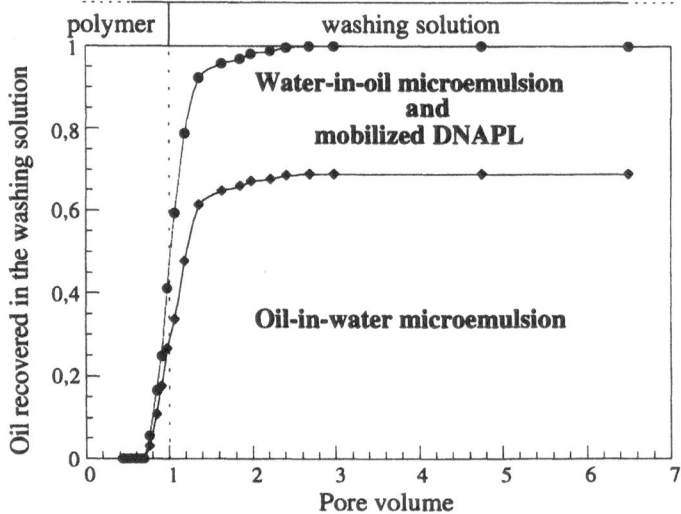

Figure 9. Proportion of DNAPL recovered by mobilization and solubilization at well 11A2, 1.4 m from the injector.

surrounding area to define the stratigraphy and the extent of PCB contamination in soils. Average total organic carbon in the silty sand unit is 9% and *in situ* permeability tests in observation wells give an average hydraulic conductivity of 9.6×10^{-6} m/s. Average A1248 concentrations are 5,000 mg/kg and vary from 70 to 37,000 mg/kg that is above the federal and provincial criteria of 50 mg/kg. Most of the conventional restoration techniques (biodegradation, air sparging, soil venting, pump and treat . . .) cannot be used efficiently to recover PCB in water saturated sediments. New approaches must thus be developed and tested on a small scale prior to their full scale application.

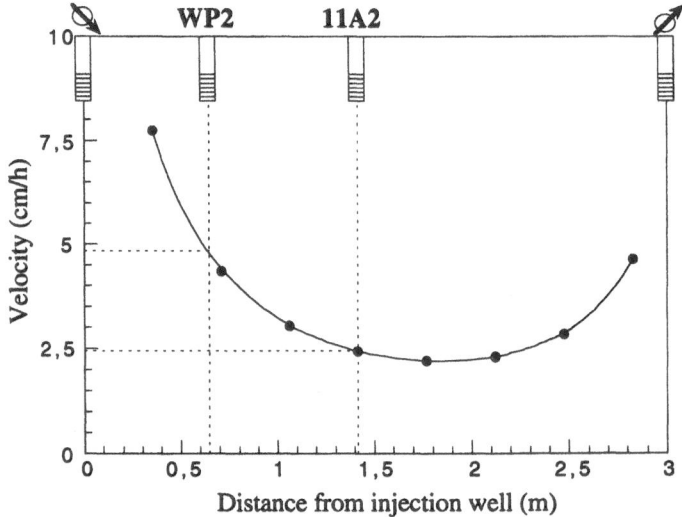

Figure 10. Fluid velocity profile in sediments between the injector and a recovery well.

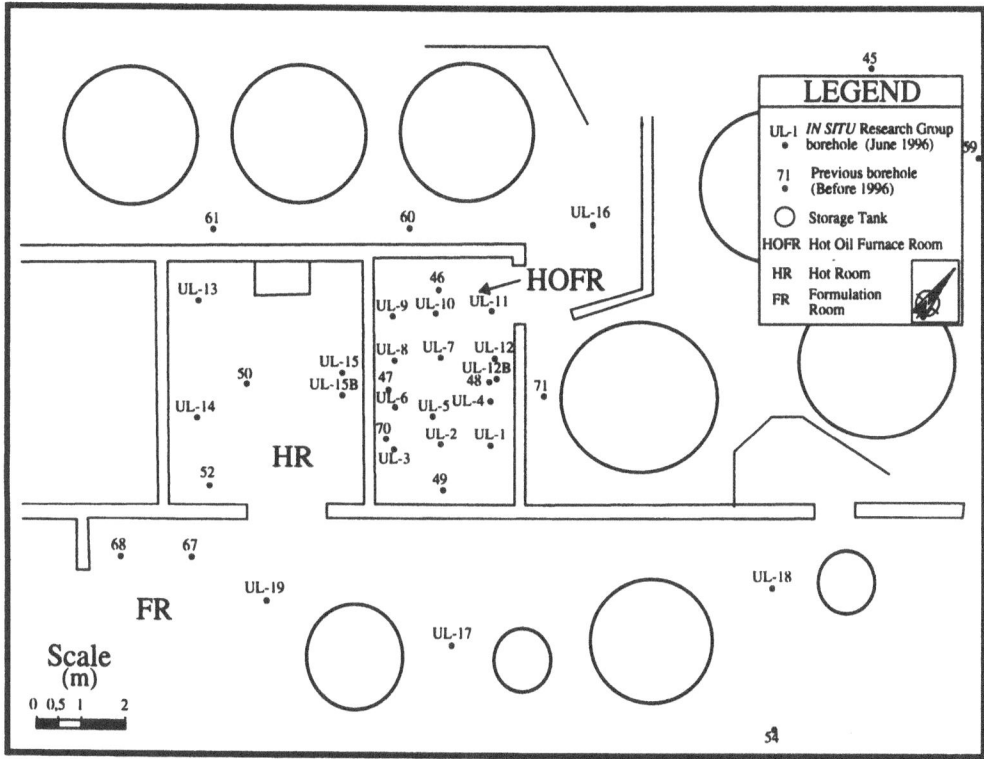

Figure 11. Characterization of the hofr area contaminated by aroclor 1248.

In situ mobilization of A1248 is very difficult because of its low saturation in soils and its high viscosity (450 cP) at soil temperature under the building (20 °C) (Fig. 12). A washing solution must be developed to dissolve A1248 *in situ*. Fifteen (15) surfactants and seven (7) alcohols were screened to test their potential for A1242 dissolution in phase diagrams (Fig. 12). Pure Aroclor 1242 was used because A1248 can no longer be found commercially or for scientific experiments. Twenty sand column tests carried out at 25 °C with the site sand show that up to 99% of the initial A1248 concentration (6,000 mg/kg) can be recovered after the injection of less than 8 pore volumes of washing solution (Fig. 13). In that case, A1248 was only solubilized even if the downward injection favors its mobilization.

4.4. Diesel Recovery in Small Sand Column and Etched Glass Micromodel

Other oil recovery mechanisms were observed with a washing solution injected (3.0 wt% n-butanol, 4.5 wt% n-pentanol, 7.5 wt% Hostapur SAS (Hoechst GmbH)) in sand column contaminated with diesel at residual saturation. In these tests, 100% of the diesel is recovered with less than 5 pore volumes of washing solution (Fig. 14). For this type of oil, which is less viscous than A1248, oil recovery mechanisms depend on injection direction. An upward injection, i.e. the same direction as buoyancy forces, displaces by mobilization an additional quantity of diesel compared to a downward injection where the diesel is recovered only by solubilization.

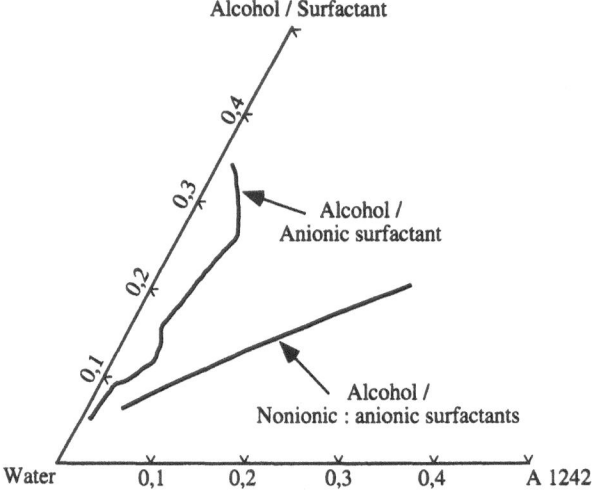

Figure 12. Pseudo ternary phase diagram with two surfactant solutions selected for A1242 dissolution.

In etched glass experiments oil recovery mechanisms where found to depend on injection velocity. The micromodel was built according to the technique described in McKellar and Wardlaw (1982). Two glass plates of 14 cm × 14 cm × 0.6 cm are etched with a mirror image fracture network and assembled with a spacing of 5 um. Five fracture apertures were selected between 180 and 580 um with an etching depth of 100 um. Water and diesel was injected into the micromodel and residual diesel was recovered with surfactant solutions (1.43 wt% n-butanol, 3.57 wt% n-pentanol, 5.0 wt% Hostapur SAS (Hoechst GmbH)). A detailed description of the micromodel and the washing experiments can be found in Tittley, (1994). For washing solution injection velocity in the micromodel equivalent to a groundwater velocity of 300 m/y (Fig. 15), diesel is miscibly displaced whereas 3,000 m/y recovery is obtained by mobilization.

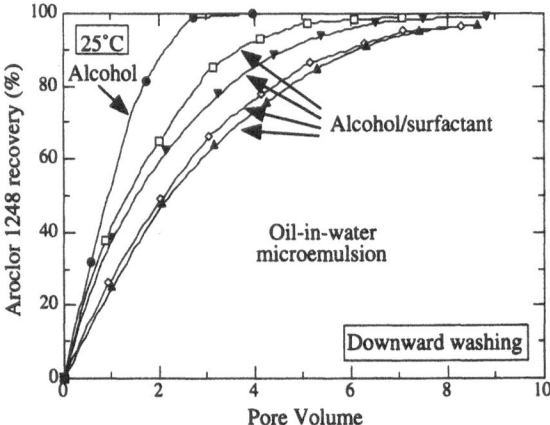

Figure 13. Cumulative A1248 solubilization in the 110 g sand column as a function of pore volume of washing solution injected.

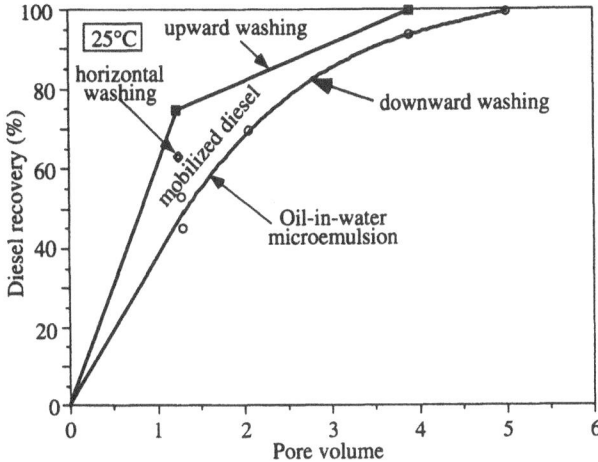

Figure 14. Cumulative diesel solubilized and mobilized in 110 g sand columns as a function of washing direction and the number of pore vulumes of a washing solution.

5. CONCLUSION

Lab and field experiments, summarized in Table 1, show that it is possible to recover *in situ* more than 90% of residual NAPL with the injection of less than 6 pore volumes of a concentrated washing solution. Recovery mechanisms depend on: (1) NAPL type, (2) micellar solution type (alcohol/surfactant or alcohol/surfactant/solvent), (3) ingredient concentrations in the micellar solution, (4) washing direction (upward, downward, horizontal), (5) injection/pumping pattern, and (6) injection velocity (contact time).

It is necessary to understand these NAPL recovery mechanisms to predict the behavior of recovery systems and to design these systems. For example, the NAPL recovery processes can be modified near an injection well because the recovery mechanisms

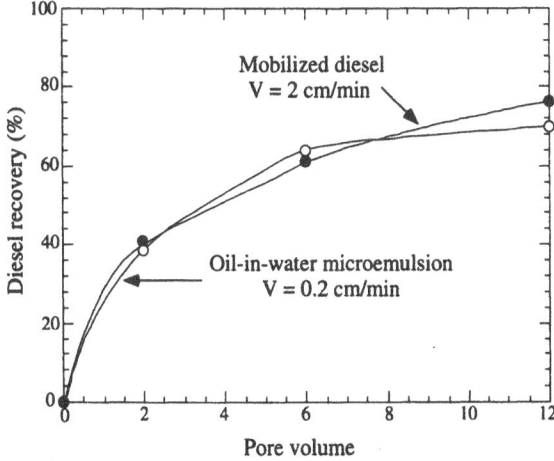

Figure 15. Cumulative diesel solubilized and mobilized in an etched glass micromodel as a function of washing solution velocity.

Table 1. Summary of soil washing experiments

Contaminant	Washing Solution	Experiment Type	Flow Direction	% Recovery @ nb. PV	Observations on Recovery Mechanisms
Mercier DNAPL	Alcohol/anionic surfactant/solvents	Mercier sand in small column	Downward	100% @ 3 PV	Suspected dominant solubilization and some mobilization
Mercier DNAPL	Alcohol/anionic surfactant/solvents	Mercier sand in large column	Downward	89% @ 0.8 PV	Observed a sequence of 34% mobilization as pure product, 51% mobilization as water-in-NAPL micoemulsion, 4% solubilization
Thouin DNAPL	Alcohol/anionic surfactant/solvents	Field test	Horizontal in a radial 5-spot pattern	86% @ 0.9 PV	Recovery mechanisms depend on fluid velocity: mobilization close to injector even with a system designed for solubilization
Aroclor 1248 PCB	Alcohol anionic surfactant	Test site sand in small column	Downward	99% @ 7 PV	Strictly solubilization
Aroclor 1248 PCB	Alcohol	Test site sand in small column	Downward	99% @ 4 PV	Strictly solubilization
Diesel	Alcohols/anionic surfactant	Ottawa sand in small column	Downward and upward	100% @ 5 PV 100% @ 4 PV	Strictly solubilization when flow is downward. Mobilization is added to solubilization when flow is upward.
Diesel	Alcohols/anionic surfactant	Etched glass micromodel	Horizontal	75% @ 12 PV	Complete mobilization at high fluid velocity. Only solubilization at low fluid velocity

are a function of fluid velocity. This may have important implications during DNAPL recovery where vertical mobilization of the DNAPL may occur deeper in the aquifer even if the system was developed to dissolve the DNAPL.

ACKNOWLEDGMENTS

This project was funded by the DESERT Program (Development and Demonstration of Site Remediation Technologies) of the Quebec Ministry of Environment and Wildlife and Environment Canada, by DOMTAR Inc. and the National Research Council of Canada through a University-Industry grant.

REFERENCES

Lake, L.W., 1989. *Enhanced Oil Recovery*. Prentice Hall, Englewood Cliffs, NJ. 550 pp.

Martel, R., Gélinas, P.J. Desnoyers, J.E. and Masson, A. 1993. Phase diagrams to optimize surfactant solutions for oil and DNAPL recovery in aquifers. Ground Water. v. 31, no. 5, pp. 789–800.

Martel, R. and P.J. Gélinas. 1996. Surfactant solutions developed for NAPL recovery in contaminated aquifers. Ground Water. v. 34, no. 1, pp. 143–154.

Martel, R., P.J. Gélinas, and J.E. Desnoyers. 1998a. Aquifer washing by micellar solutions: 1-Optimization of alcohol/surfactant/solvent solutions. Journal of Contaminant Hydrology. V. 29, pp. 319–346.

Martel, R., R. Lefebvre, and P.J. Gélinas. 1998b. Aquifer washing by micellar solutions: 2-DNAPL recovery mechanisms for an optimized alcohol/surfactant/solvent solution. Journal of Contaminant Hydrology. V. 30, pp. 1–31.

Martel, R., P.J. Gélinas, and L. Saumure. 1998c. Aquifer washing by micellar solutions: 3-Field test at the Thouin sand pit (L'Assomption, Quebec, Canada). Journal of Contaminant Hydrology. V. 30, pp. 33–48.

Martel, K.E., R. Martel, R. Lefebvre, and P.J. Gélinas. 1998. Polymer solutions used to improve *in situ* NAPL recovery in polluted aquifers: A laboratory evaluation. Ground Water Monitoring and Remediation. Summer pp. 103–113.

McKellar, M. and N.C. Wardlaw. 1982. A method of making two-dimensional glass micromodels of pore systems, The Journal of Canadian Petroleum Technology, July–August pp. 39–41.

Roy, N., R. Martel, and P.J. Gélinas. 1995. Laboratory assessment of surfactant biodegradation in aquifers under denitrifying conditions. In: 5th Annual Symposium on Groundwater and Soil Remediation. Toronto, Ontario, October 2–5, 1995.

Tittley, M. 1994. Study of Diesel and Surfactant Solution Behavior in a Fractured Bedrock Aquifer, Using Glass Micromodels. M.Sc. Thesis. Department of Geological Engineering, Laval University, Quebec, Canada, 103 pp.

7

ANAEROBIC MICROBIAL ACTIVITY UNDER ELECTRIC FIELDS

Krishnanand Maillacheruvu[1] and Akram N. Alshawabkeh[2]

[1]Assistant Professor
Department of Civil Engineering and Construction
Bradley University
Peoria, Illinois
[2]Assistant Professor
Department of Civil and Environmental Engineering
Northeastern University
Boston, Massachusetts

ABSTRACT

The impact of electric fields on some environmental conditions and on the activity of anaerobic microorganisms was investigated in this study. Experiments were conducted in plexiglass chambers with coated titanium mesh electrodes mounted on either end. Electric fields of 1.5 V/cm through 6 V/cm were applied. Anaerobic cultures (primarily a consortium with methane producing bacteria) obtained from a mesophilic digester from the Redhook Wastewater Treatment Plant, Brooklyn, were used in these experiments. Results indicated that the pH decreased to as low as 4.5 even under completely mixed conditions in fed-batch reactors, if pH-control measures were not in place. Microbial activity, measured as a function of the ability of the anaerobic microorganisms to consume readily degradable acetate, generally decreased as expected if the pH and dissolved oxygen were not controlled. Using sodium sulfite slightly in excess of the stoichiometric amount and sodium bicarbonate, both pH variation as well as oxygen generation were adequately controlled. Microbial activity initially decreased under exposure to electric current but recovered after a period of several hours of exposure, suggesting a degree of acclimation of the anaerobic culture to the electric current.

INTRODUCTION

Economical restoration of contaminated soils to environmentally acceptable conditions is an important challenge facing the scientific and technical community. Many

Emerging Technologies in Hazardous Waste Management 8, edited by Tedder and Pohland
Kluwer Academic/Plenum Publishers, New York, 2000.

current in situ soil remediation technologies as well as pump-and-treat and vacuum extraction technologies depend on hydraulic and air flow for effective remediation of soils. These technologies are not as effective in the clean-up of lower hydraulic conductivity (less than 10^{-5} cm/sec), such as fine sands, silts and clays. Soil heterogeneities will also limit the feasibility of remediation by hydraulic and/or airflow. In situ bioremediation is an attractive and often cost-effective option to remediate contaminated soils and groundwater. However, successful implementation of in situ bioremediation is dependent upon presence, or effective injection, of electron acceptors and nutrients into the porous medium. Effective introduction and transport of electron acceptors and nutrients is hindered by low soil permeabilities, preferential flow paths (channeling), soil heterogeneity, biological utilization, and chemical reactions in the soil. Several surveys have concluded that this ineffective transport of remediation additives is the primary cause of system failure for some in situ bioremediation efforts (Zappi *et al.*, 1993; NRC 1993). Complications of site geohydrology, additive transport and associated reactions, coupled with the observed inefficiency in the field, have usually been approached by gross overinjection. Excessive dosing combined with the shortcomings of the hydraulically driven transport processes can result in elevated remedial costs, potential for gas binding, and excessive biological growth (biofouling). A technology for uniform introduction of electron acceptors and nutrients has been the principal bottleneck in the successful field implementation of in situ bioremediation (Zappi *et al.*, 1993).

Electric fields can be used to overcome injection problems associated with aquifer heterogeneities and low permeabilities. Migration of charged additives (including electron acceptors and nutrients) under electric fields can result in uniform and efficient introduction of these additives to stimulate in situ bioremediation in the contaminated zone. Recent research has shown that ion migration through soil pore fluid under applied direct current can be substantial. Acar and Alshawabkeh (1993) demonstrated effective ion transport on the order of 5 to 10 cm/day under a 1 volt/cm electric field in kaolinite soil samples with hydraulic conductivity in the order of 10^{-7} cm/sec. However, application of electric current in saturated soils may result in significant changes in the environmental conditions including pH, temperature, and geochemistry. These environmental changes will potentially impact microbial activity and thus the efficiency of bioremediation of target contaminants. Several factors, including electrolysis reactions at the electrodes, transport processes under electric fields, and microbial adhesion, transport, and growth under electric fields may affect the success of enhancing bioremediation by injection of electron acceptors and nutrients under electric fields. A schematic of possible field implementation of enhancing bioremediation by electric fields is presented in Fig. 1. This paper will briefly review of the impacts of electric fields on pH and geochemistry, and present the results of a study investigating the impact of electric field intensity on the behavior of anaerobic microorganisms under electric fields using completely mixed bioreactor systems to eliminate soil heterogenity effects and other complicating factors identified above.

BACKGROUND

The use of electric fields has emerged as an innovative method for in situ restoration of contaminated hazardous waste sites. This technique uses low-level direct current on the order of few Amps per square meter of cross-sectional area between the electrodes or an electric potential difference on the order of a few volts per cm across electrodes placed in

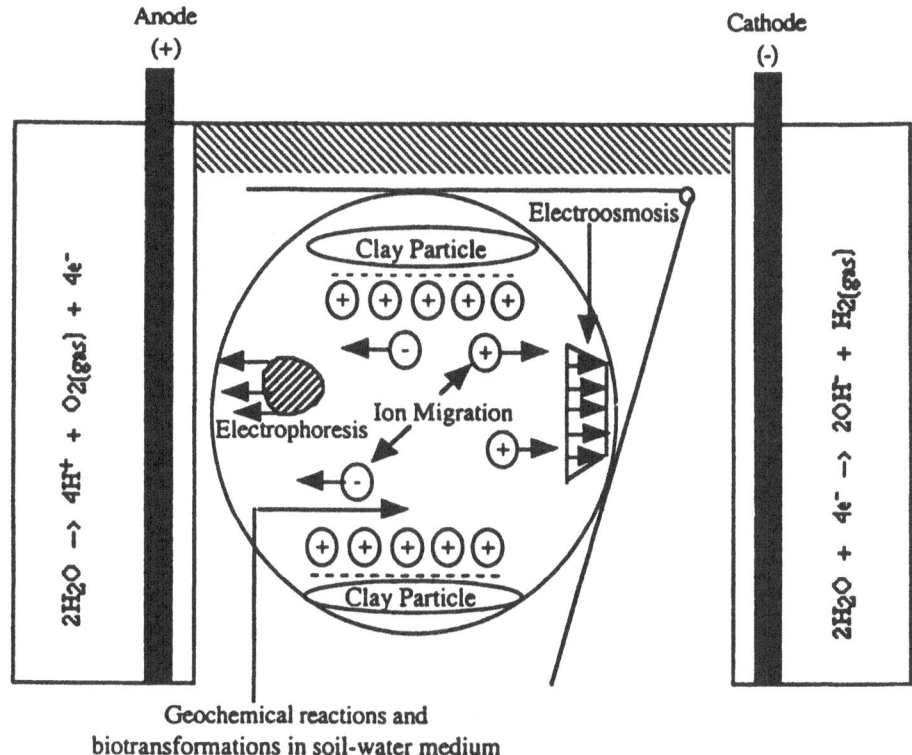

Figure 1. A schematic showing processes involved during the application of electric fields in soils.

the ground in an open flow arrangement. The groundwater in the pore media or an externally supplied fluid is used as the conductive medium for ions transport. The low-level direct current results in species transport by coupled and uncoupled condition phenomena, including electroosmosis and ion migration, in the porous media. Electroosmosis mobilizes the pore fluid in the soil, usually from the anode (+ve electrode) toward the cathode (−ve electrode), while ion migration effectively separates the anions (negative ions) and cations (positive ions) by their transport to the anode and cathode, respectively. Studies on transport under electric fields have generally focused on heavy metals. Bench-scale tests and limited field studies demonstrated the use of electric fields for extraction of heavy metal and radionuclides from soils (Lageman *et al.*, 1989; Hamed *et al.*, 1991; Pamukcu and Wittle, 1992; Acar and Alshawabkeh, 1993; Lageman, 1993; Probstein and Hicks, 1993; Runnels and Wahli, 1993; Eykholt and Daniel, 1994; Alshawabkeh and Acar, 1996; Pamukcu *et al.*, 1997). Electric fields have also been used for electroosmotic extraction of organic contaminants from soils (Shapiro *et al.*, 1989; Acar *et al.*, 1992; Bruell *et al.*, 1992). Recently, the Department of Energy, Environmental Protection Agency (EPA), Monsanto, General Electric, and Dupont applied electric fields for electroosmotic extraction of TCE from a site in Paducah, Kentucky using layered horizontal electrodes in what is called the "Lasagna" process (Ho *et al.*, 1997).

A further innovation is the use of direct electric currents to stimulate and enhance in situ bioremediation of organics in low permeability soils. Besides generating an electric field, applications of electric currents will results in several environmental changes in the soil pore fluid, including changes in pH, ionic strength, dissolved oxygen concentration, and temperature. Most of these environmental changes result from electrolysis reactions at the electrodes. It is well established that application of electric currents through electrodes immersed in water cause oxidation at the anode and reduction at the cathode as described in the following reactions,

$$2H_2O \quad\quad \Leftrightarrow \quad 4H^+ + O_2 + 4e^- \quad \text{(Anode)}$$

$$2H_2O + 2e^- \quad \Leftrightarrow \quad 2OH^- + H_2 \quad\quad \text{(Cathode)}$$

Acid is produced at the anode, dropping the pH to around 2, and a base is introduced at the cathode, increasing the pH to around 10–12. Unless these pH changes are neutralized, extreme environmental conditions are produced, limiting microbial growth and biotransformation. Further, oxygen gas is one of the products of electrolysis reaction at the anode, which will also affect the anaerobic microbial growth. Substrate utilization is usually an indication of microbial growth. The conventional theory of substrate utilization by microorganisms is based on the bioavailability of contaminants, presence of the appropriate microbial consortia, temperature, pH and adequate time of contact. Optimum pH for bacterial growth is near neutral, with pH range for growth between 5 and 9 (Gaudy and Gaudy, 1988). Acar et al. (1996, 1997) showed that it is possible to inject ammonium and sulfate ions (as electron acceptors) in fine sand and kaolinite at rates of about 5 to 20 cm/day under electric fields and also neutralize the electrolyte pH by addition of basic/acidic species at the electrodes. A combination of several of these techniques may offer some potential for in situ bioremediation under electric fields.

Most microorganisms grow rapidly at temperatures between 20 °C and 45 °C and are capable of growing over a range of 30 °C to 40 °C (Gaudy and Gaudy, 1988). An increase in temperature above 45 °C will significantly limit the growth of most microbes (some could survive high temperatures). Temperature increase due to current application will depend upon field strength and resistivity of the medium. Acar and Alshawabkeh (1996) reported 10 °C increase in temperature in an unenhanced (no additives were used to control electrolyte pH changes) large-scale test on extraction of lead from kaolinite.

Additionally, the dependency of reversible sorption of bacteria on charge/electrostatic interactions indicates that an applied electric field may play a significant role in bacterial adhesion and transport in subsurface environments. As microbes are generally negatively charged, they will migrate under electric fields by electrophoresis. The role of bacterial surface itself (e.g., lipid content and surface charge) could influence the electrophoretic mobility of bacteria in porous media. However, it should be noted that transport of microorganisms might not be strictly governed by electrophoresis alone, since microorganisms, as living entities, may be subject to other influences or "attractors." Studies that investigate bioremediation by electric fields are rare. DeFlaunn and Condee (1997) demonstrated the possibility of pure bacterial cultures transport in bench-scale soil samples under application of an electric current. While it is clear that bacteria could be transported by electrophoresis, it is necessary to assess what electric field intensity they can sustain and what effects the electric field may have on the activity of the microorganisms.

While complex environmental changes in the soil pores under electric fields will impact microbial behavior; this study was targeted at investigation of the impact of electric fields on the activity of unacclimated anaerobic consortia. The selection of anaerobic consortium of bacteria was based on the hypothesis that if an appropriate anaerobic bacterial consortium may be transported to the site of contamination (as well as the ionic nutrients), the need to transport an electron acceptor such as oxygen (which is needed for aerobic biotransformations) is eliminated. It is recognized here that other additives, such as nutrients and electron donors will also need to be transported to the site of anaerobic degradation of the contaminants in the subsurface. The primary goal of this study was to determine changes in activity as a function of time of exposure to electric fields. It was also a goal of this research to identify some of the limiting factors and potential control strategies to mitigate their action. Specifically, the changes in pH and generation/control of oxygen production due to the application of the electric fields on the bio-electrolyte medium were studied.

MATERIALS AND METHODS

All seed organisms came from stock cultures, which were maintained in 15L containers throughout the duration of the research. Seed organisms were considered unacclimatized, since they were maintained for a duration of less than two retention times in the stock reactor before use in the experiments. All stock cultures came from a mesophilic digester at the Red Hook Water Pollution Control Plant, Brooklyn, NY. The stock cultures were fed acetate on a batch-fed basis at an organic loading rate of $0.5 \frac{g\,COD}{L\text{-}day}$. A solids retention time of 15 days was used in all stock cultures and experiments. Alkalinity, pH, volatile suspended solids (VSS), dissolved oxygen (DO), temperature were periodically measured to ensure stock cultures were healthy. The nutrient feed solution used for all stock cultures and all experiments is provided in Table 1. Additional detail on media preparation and maintenance of anaerobic cultures is available elsewhere (Maillacheruvu et al., 1993; Maillacheruvu and Parkin, 1996). Ferrous ion and sulfide were added separately, because these may have been oxidized and/or precipitated if added to the other chemicals in the medium. Sodium bicarbonate was used to

Table 1. Composition of the nutrient solution

Nutrient	Concentration (mg/L)
NH_4Cl	400
$MgCl_2$	400
KCl	400
$CaCl_2 \cdot 2H_2O$	25
$(NH_4)_2HPO_4$	80
$FeCl_2 \cdot 4H_2O$	40
$(NaPO_3)_2$	10
$CoCl_2 \cdot 6H_2O$	2.5
KI	2.5
$MnCl_2 \cdot 4H_2O$	2.5
NH_4VO_3	0.5
$ZnCl_2$	0.5
$Na_2MoO_4 \cdot 2H_2O$	0.5
H_3BO_3	0.5
$NiCl_2 \cdot 6H_2O$	0.5
Cysteine	10

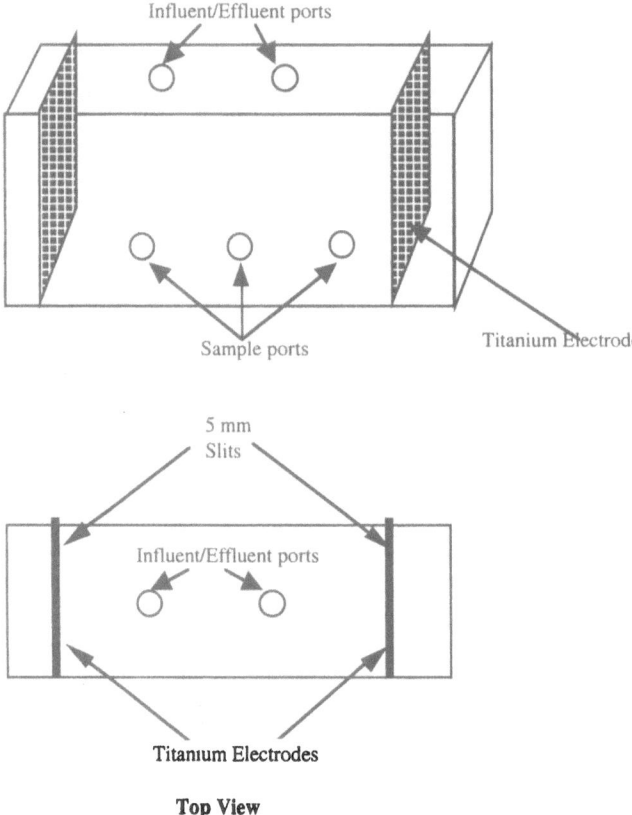

Figure 2. Schematic of experimental fed-batch reactor.

all cultures to maintain alkalinity ≥2,500 mg/L as CaCO$_3$. All chemicals were laboratory reagent grade.

Figure 2 shows the schematic of the batch experimental setup. Electrode plates were made of titanium, and were located on both ends of the reactors. The batch reactors were made of plexiglass with a total volume of 2.0 L and contained 1 L of anaerobic culture. Sample for sludge analysis, pH and/or alkalinity control, was withdrawn through the sample ports. Nutrients and acetate were fed using sample ports alocated at the top. All connections were sealed by silicone glue. All experiments were conducted using duplicates with a control that was not exposed to the electric field (EF).

The pH was determined using a Fisher Scientific Model 10 pH meter with a gel-filled electrode. A Shimadzu Model 5000A total organic carbon analyzer measured dissolved organic carbon (DOC). Alkalinity was determined by the potentiometric titration described in Standard Methods (1992). VSS was measured according to Section 2540 of Standard Methods (1992). DO was measured using a YSI Model 50B dissolved oxygen meter. DC current was applied to each experimental reactor using a Hewlett Packard Model E3610A power supply.

Various experimental conditions used in this study are described in Table 2. The anaerobic culture was .mixed thoroughly to achieve uniform solution by using Corning Model PC-131 stirrers. 2.5 M Na$_2$SO$_3$ was used to control DO production which is a

Table 2. Experimental Conditions

Experiment	Description		
	Electric Field Strength	Na_2SO_3	$NaHCO_3$
Control A	None	YES	YES
Control B	None	NO	YES
A.1	3.0 V/cm	YES	YES
A.2	6.0 V/cm	YES	YES
A.3	6.0 V/cm	NO	YES
B.1	1.5 V/cm	YES	YES
B.2	4.5 V/cm	YES	YES
B.3	4.5 V/cm	NO	YES
C.1	1.5 V/cm	NO	YES
C.2	1.5 V/CM	YES	NO

byproduct of the electrochemical reactions taking place in an electrolyte medium. After a brief startup period of five-six days (although not enough to reach steady state), batch reactor experiments were begun. The current, pH, and the electric potential differences were monitored on a continuous basis. After one day of current application, the anaerobic culture that was pumped out using a B-D 10-ml syringe for in situ pH, VSS, and DOC analysis was separated into two parts. One part of the sample was immediately analyzed (designated the "t_0 sample") for DOC concentration while the other part allowed to "recover" for a period of 24 hours in an evacuated glass vial to preserve anaerobic conditions. After 24 hours had elapsed the DOC was measured again. The second part of the sample was designated the "t_{24} sample". This process was carried out throughout the duration of the experiments, both in the presence and in the absence of electric current. Filtered samples were acidified using 2.0 N HCl to a pH of about 3, prior to DOC analysis. All DOC analyses were conducted using a Shimadzu TOC 5000A analyzer.

RESULTS AND DISCUSSION

The effect of electric current on environmental conditions such as pH and dissolved oxygen was evaluated. It was found that even in a completely mixed batch reactor system used in this study, the pH decreased rapidly at first and reached a value of between 4.5 and 4.8 after 48 hours of exposure. This is significant since the pH needs to be buffered around 7.0 for most biological systems, and particularly anaerobic bacteria other than acid-formers. A concomitant drop in the alkalinity was also observed during this period. These results are interesting since the hydroxyl ion production and hydrogen ion (H^+) production would be expected to cancel each other out and result in a near-zero change in pH. One explanation for this result may be that the hydroxyl ions are either complexed or eliminated from the solution before reacting with the hydrogen ion. In the experimental reactors, pH buffering was achieved through use of $NaHCO_3$ and keeping the alkalinity levels at around 2,500 to 3,500 mg/L as $CaCO_3$. In actual soils, depending on the buffering capacity available as carbonate salts, additional amounts of bicarbonate or other appropriate buffer needs to be used (injected). This is consistent with the results obtained on acid front movement in soils being treated with electrokinetic techniques.

Dissolved oxygen was produced in the reactor and was found to be proportional to the amount of electric current. This is consistent with the theory of electrokinetic

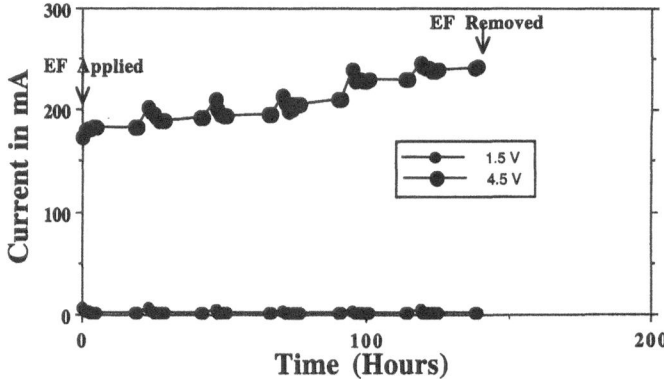

Figure 3. Variation in electric current in the reactor as a function of exposure time to electric field.

phenomena. The highest DO concentration observed in reactors where DO was not controlled was around 0.2 mg/L. The color also turned from a jet black color to a grey/light brown indicating that the culture was adversely impacted. At this level, it is likely that anaerobic activity was inhibited in the experimental reactors. However, in an actual soil, there may exist niches which are devoid of oxygen where anaerobic bacteria may survive. Further studies in soils are needed to evaluate this hypothesis. In many reactors, DO levels were under 0.1 mg/L. It is possible that at these levels, some anaerobic microorganisms (particularly sulfate-reducing bacteria or SRB) may indeed have survived.

Electric current in the reactor generally increased slightly with time for the different electric field strengths tested (1.5 V/cm, 3.0 V/cm, 4.5 V/cm and 6.0 V/cm). From Fig. 3, the increase ranged from 5% at 1.5 V/cm to about 15%–18% at 4.5 V/cm for an exposure period of about 140 hours. While this suggests that the ionic composition of the bio-electrolyte medium changed over time, the impacts on DOC removal efficiency and microbial activity do not appear to be very different, even with a three-fold difference in electric field strength.

The effect of electric currents on DOC removal is illustrated in Figs. 4 and 5. Figure 4 shows a comparison between DOC removal for buffered batch reactors at 1.5 V/cm and 4.5 V/cm in terms of ppm DOC and Fig. 5 shows the same data in terms of fraction removed. In both these experiments, the DO was controlled using sodium sulfite. The

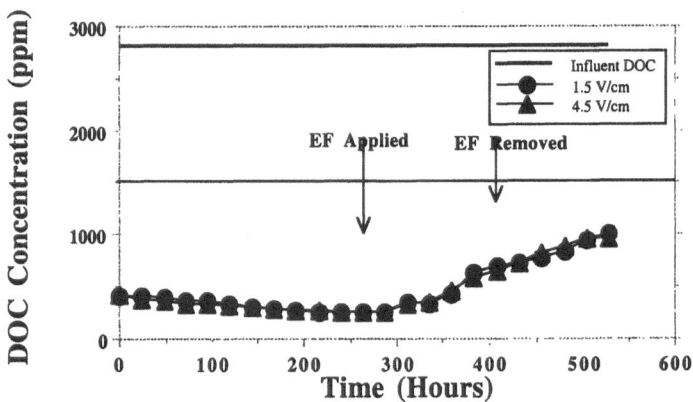

Figure 4. DOC removal in fed-batch reactor at different under 1.5 and 4.5 V/cm.

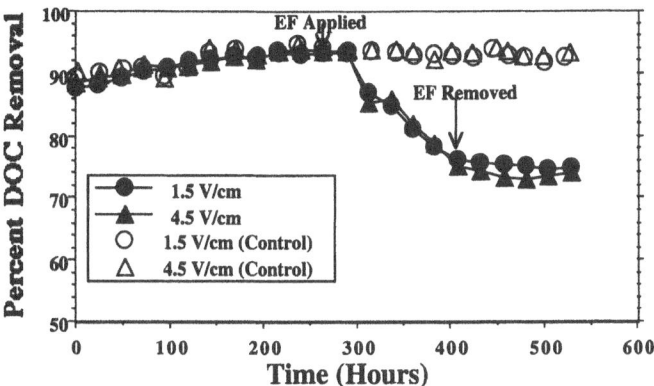

Figure 5. Percent DOC removal in fed-batch reactors under different electric field strengths.

DOC removal followed the same pattern for both these systems that contained unacclimated anaerobic culture. These data suggest that the magnitude of the electric current does not impact the DOC removal capacity of the anaerobic bacteria, for the range of electric fields used in this study. It was also found that sodium sulfite itself provided some buffering capacity. Results from 3.0 V/cm and 6.0 V/cm showed the same trends in terms of DOC removal and were in fact very close to the DOC removal at 1.5 V/cm and 4.5 V/cm experiments. These data suggest that DOC removal is neither accelerated nor impeded if electric field strength itself (and thereby electric current) are changed, for the range of the electric field strengths used in this study.

The impact of electric current on the microbial activity is of critical concern when evaluating the potential of a bacterial culture to be successful in an in situ bioremediation effort. Microbial activity was measured by the capacity of the culture to recover from exposure to electric currents, and retain its capacity to degrade a readily biodegradable substrate—acetate. The procedure is discussed in the section on materials and methods. Figure 6 shows the microbial activity data for experiments at electric field strengths of 1.5 V/cm and 4.5 V/cm and which were buffered using bicarbonate and to which sodium sulfite was added to scavenge any oxygen produced. These data indicate that in general, before the electric current was applied, the t_{24} sample showed more removal (about 20%) of DOC than the t_o sample. Indeed this is to be expected in the absence of electric fields since it is indicative of an active culture. In other words, although the cultures were not yet fully acclimated to the lab conditions, the anaerobic cultures in the experimental reactors were still active.

From Fig. 6, it is apparent that the initial "shock" exposure to the electric current results in a decrease in activity t_{24} sample as compared to the t_o sample. The implication of these data is that 24 hours is insufficient time for the culture to recover from exposure to electric current. This trend continued for several hours during which period the total DOC percent removal in the experimental reactor also dropped by about 10 to 20% in all the experiments tested. However, after a period of several hours of exposure, the t_{24} sample gradually showed an increase in microbial activity by exhibiting higher DOC removal than the t_o sample. This is an interesting phenomenon since it indicates a certain degree of acclimation of the culture to electric current, and a tendency for the culture to recover from the initial shock load in terms of changes in environmental conditions due to application of the electric current. Once the electric current application was stopped, the rate of DOC percent removal in the experimental reactors gradually improved. Eventually, after the removal of the electric current, the 24 hour sample showed about

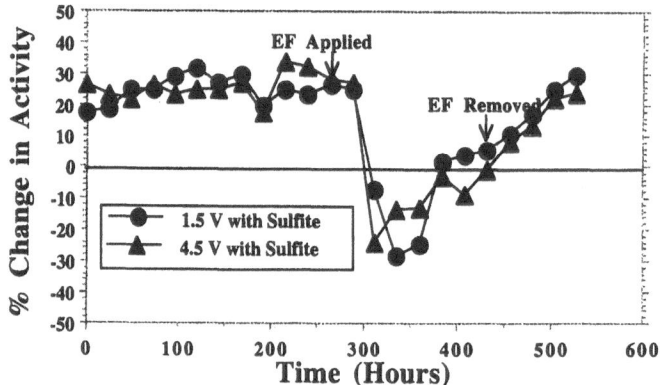

Figure 6. Percent change in activity over a 24 hour recovery period.

20% higher removal of DOC as compared to the t_o sample. These data also suggested that there was essentially no difference in recover of microbial activity 1.5 V/cm and 4.5 V/cm experiments. Results from 3.0 V/cm and 6.0 V/cm showed the same trends.

SUMMARY

The impact of electric current on the environmental conditions and the microbial activity in completely mixed fed-batch reactors was studied at various electric field strengths. The pH dropped continuously in non-buffered systems even though the conditions in the experimental reactors were completely mixed. Sodium bicarbonate was effective in buffering the system for the range of electric field strengths studied. Dissolved oxygen was observed in small quantities in the experimental reactors where no oxygen scavengers were used. Dissolved oxygen was, however, easily controlled by addition of sodium sulfite to the system. Finally, even for an unacclimated anaerobic culture, the microbial activity recovered after exposure to electric fields suggesting some degree of acclimation of the anaerobic culture to electric fields. Studies in progress to evaluate anaerobic enrichment cultures that have acclimated under laboratory conditions, both in completely-mixed systems as well as in soils.

ACKNOWLEDGMENTS

This work was initiated while both researchers were at that the Department of Civil and Environmental Engineering, Polytechnic University, Brooklyn, NY. The analysis of data was continued at Bradley University and Northeastern University where additional experiments are in progress. Partial support from Caterpillar Research Foundation, Peoria, IL is also acknowledged.

REFERENCES

Acar, Y.B. and Alshawabkeh, A. (1993) "Principles of electrokinetic remediation," Env. Sci. & Technol., Vol. 27, No. 13, pp. 2638–2647.

Acar, Y.B. and Alshawabkeh, A. (1996) "Electrokinetic remediation: I. Pilot-scale tests with lead spiked kaolinite," J. of Geotech. Eng.," ASCE, Vol. 122, No. 3, pp. 173–185.

Acar, Y.B. and Zappi, M. (1995) "Infrastructural needs in waste containment and environmental restoration," J. of Infrastructural Systems," ASCE, Vol. 1, No. 2, pp. 82–91.

Acar, Y.B., Li, H., and Gale, R.J. (1992) "Phenol removal from kaolinite by electrokinetics," J. of Geotech. Eng. Vol. 118, Nov. 1992, pp. 1837–1852.

Acar, Y.B., Ozsu, E.E., Alshawabkeh, A.N., Fazle R.M., and Gale, R. (1996) "Enhanced soil Bioremediation with electric fields," CHEMTECH, American Chemical Society, April 1996, Vol. 26, No. 4, pp. 40–44.

Acar, Y.B., Gale, R.J., Putnam, G., and Hamed, J.T. (1989) "Electrochemical Processing of Soils: Its Potential Use in Environmental Geotechnology and Significance of pH Gradients," 2nd International Symposium on Environmental Geotechnology, Shanghai, China, May 14–17, Envo Publishing, Bethlehem, PA, Vol. 1, pp. 25–38.

Alshawabkeh, A.N. and Acar, Y.B. (1992) "Removal of Contaminants from Soils by Electrokinetics: A Theoretical Treatise," J. of Env. Sci. & Health, A 27 (7), pp. 1835–1861.

Alshawabkeh, A. and Acar, Y.B. (1996) "Electrokinetic remediation: II. Theoretical model, J. of Geotech. Eng., ASCE, Vol. 122, No. 3, pp. 186–196.

Bruell, C.J., Segall, B.A., and Walsh, M.T. (1992) "Electroosmotic removal of gasoline hydrocarbons and TCE from clay," J. of Env. Eng. v. 118, Jan./Feb. 1992, pp. 68–83.

DeFlaunn, M. and Condee, C. (1997) "Electrokinetic transport of bacteria," J. of Hazardous Materials. v. 55, Aug. 1997, pp. 263–277.

DOE/EM-0232 (1995) "Estimating the Cold War Mortgage, Volume I" The 1995 Baseline Environmental Management Report, US Department of Energy, March 1995.

Eykholt, J. and Daniel, D.E. (1994) "Impact of System Chemistry on Electroosmosis in Contaminated Soil," ASCE, J. of Geotech. Eng., V. 120 (5), May, pp. 797–815.

Gaudy, A.F. and Gaudy, E.T. (1988) Elements of Bioenvironmental Engineering, Engineering Press, Inc., San Jose, California, 592p.

Hamed, J., Acar, Y.B., and Gale, R.J. (1991) "Pb(II) Removal from Kaolinite Using Electrokinetics," ASCE, Journal of Geotechnical Engineering, Vol. 112, February, pp. 241–271.

Ho, S.V., Athmer, P.W., Sheridan, P.W., and Shapiro, A. (1997) "Scale-up Aspects of the Lasagna Process for in-situ Soil Remediation," J. of Haz. Materials, Special Edition on Electrochemical Decontamination of Soil and Water, Edited by Yalcin B. Acar and Akram N. Alshawabkeh, pp. 39–60.

Lageman, (1993) "Electro-Reclamation," Environ. Sci. & Technol., V61. 27, No.13, pp. 2648–2650.

Maillacheruvu, K.Y. and Parkin, G.F. (1996) "Kinetics of growth, substrate utilization and sulfide toxicity for propionate, acetate and hydrogen utilizers in anaerobic systems," Water Env. Res., 68 (7), 1099–1106, November–December, 1996.

Maillacheruvu, K.Y., Parkin, G.F., Peng, C.Y., Kuo, W.C., Oonge, Z.I., and Lebduschka, V. (1993) "Sulfide toxicity in anaerobic systems fed sulfate and various organics," Water Environ. Res., 65 (2), 100–109, March–April, 1993.

Mitchell, J.K. (1993) Fundamentals of Soil Behavior, John Wiley and Sons, New York, 437p.

NRC (1993) In-situ bioremediation: when does it work? National Research Council, National Academy Press, Washington D.C., 1993, 207p.

Probstein, R.F. and Hicks, R.E. (1993) "Removal of Contaminants from Soils by Electric Fields," Science, 1993, 260, pp. 498–504.

Runnels, D.D. and Wahli, C. (1993) "In Situ Electromigration as a Method for Removing Sulfate, Metals, and Other Contaminants from Groundwater," Groundwater Monitoring Review, Winter 1993, pp. 121–129.

Shapiro, A.P., Renauld, P., and Probstein, R. (1989) "Preliminary studies on the removal of chemical species from saturated porous media by electro-osmosis," Physicochemical Hydrodynamics, 11, No. 5/6, pp. 785–802.

Standard Methods for the Examination of Water and Wastewater (1992) 18th ed., American Public Health Association, Washington D.C.

Zappi, M., Gunnison, D., Pennington, J., Cullinane, J., Teeter, C.L., Brannon, J.M., and Myers, T. (1993) "Technical approach for in-situ biological treatment research: bench-scale experiments," US Army Corps of Engineers, Waterways Experiment Station, Vicksburg, MS, August 1993, Technical Report No. IRP-93-3.

8

MECHANSTIC FACTORS AFFECTING FENTON OXIDATIONS IN NATURAL WATERS

Matthew A. Tarr and Michele E. Lindsey

Department of Chemistry
University of New Orleans
New Orleans, Louisiana 70148

1. ABSTRACT

The use of *Fenton's reagent ($Fe(II) + H_2O_2$ yields hydroxyl radical*) has been applied to remediation of *contaminated soil* sites and treatment of *industrial waste streams.* However, degradation of dissolved organic pollutants by Fenton's reagent is strongly affected by the presence of other dissolved species. *Natural organic matter (NOM)* exerts three main influences on hydroxyl radical mediated oxidation of pollutants: 1) reduction of hydroxyl radical concentration through *scavenging*, 2) reduction in *hydroxyl radical formation rate* and efficiency through iron binding or redox coupling, and 3) *sequestering* of pollutants away from hydroxyl radical through *pollutant-NOM binding*. Sequestering of pollutants away from hydroxyl radical appears to be a significant mechanism for reducing *degradation efficiency*. All three effects must be accounted for in developing models for *in situ degradation*. Furthermore, the effects observed in the presence of natural organic matter may be similar to effects of non-pollutant organic compounds present in industrial waste streams.

2. INTRODUCTION

For decades, the use of chemicals for industrial, agricultural, and personal applications has been widespread, resulting in many benefits to society. Unfortunately the potential for enormous harm to the environment was not recognized until recently, and consequently, severe contamination exists in thousands of sites worldwide. Only in recent years has awareness of such problems risen, and as a result, considerable effort has been put forth in the development of new technologies for the remediation of contaminated sites. Examples of such sites are superfund sites, agrochemical distribution and application sites, industrial sites, harbors, and military sites. At many of these locations, contamination of soil and water may consist of mixtures of environmentally refractory, toxic compounds at relatively high concentration. Remediation of such sites often requires specialized methods due to the diversity of contaminants present.

Emerging Technologies in Hazardous Waste Management 8, edited by Tedder and Pohland
Kluwer Academic/Plenum Publishers, New York, 2000.

Many techniques are currently in use for remediating contaminated sites. Among these are washing or extraction, incineration, vitrification, bioremediation, and chemical degradation.

Bioremediation has become one of the more promising techniques for treatment of non-volatile organic contaminants. However, several limitations have prevented its widespread application. Among the drawbacks to bioremediation are inhibition by non-pollutant metabolites, toxicity of contaminants at high concentration, and an inability to deal with contaminant mixtures.[1]

Such limitations of bioremediation have been illustrated in many studies. Haggblom,[2] et al. showed that the biodegradation of six chlorinated aromatics could be carried out individually, but no conditions could be found to achieve degradation of a mixture of all six compounds. Other researchers obtained similar results with 2,4,5-trichlorophenoxyacetic acid (2,4,5-T) and 2,4-dichlorophenoxyacetic acid (2,4-D),[3] and with chlorinated phenols.[4] Furthermore, some compounds are biodegraded at low concentrations, but become toxic at higher concentrations.[5,6]

Treatment of dense non-aqueous phase liquids (DNAPLs) has also been problematic. Methods for DNAPL removal have generally relied on intact removal of the DNAPL by pumping or extraction. However, pumping and extraction can be time consuming, requires additional treatment of the recovered material, and can leave residual material.[7] The residual material remaining in the soil may result in continuous release of dissolved DNAPL into aquifers.[8] Other remediation techniques involve the addition of alcohols, surfactants, or organic solvents to promote dissolution or displacement of the DNAPL.[9,10] In addition to possible limitations in remediation efficiency, these techniques have the added drawback of introducing potentially unwanted chemicals into the contaminated site.

Chemical oxidation of pollutants is another widely investigated technique for decontamination. These techniques rely on the addition or formation of reactive species which then degrade the pollutants present. Among these methods are ozone treatment, ultraviolet photocatalysis, supercritical water oxidation, and Fenton type catalytic degradation with hydrogen peroxide. These techniques rely on generation of reactive species such as the hydroxyl radical ($HO^{\cdot}$), the hydrated electron [$e^{-}(aq)$], singlet oxygen ($^{1}O_2$), and superoxide (O_2^{-}), and other species.

Hydroxyl radical is one of the strongest oxidants known, and reacts rapidly with a wide variety of compounds. It is the key oxidant in the degradation of pollutants by ultraviolet/ozone treatment,[11] ultraviolet/peroxide treatment,[11] titanium dioxide photocatalysis,[12] iron/hydrogen peroxide treatment,[13] and sonolysis.[14] The highly reactive, non-specific nature of the hydroxyl radical makes it useful in degrading a wide variety of species. Although photolytic and photocatalytic methods can by very effective, they are limited to optically transparent systems. Iron/H_2O_2 and sonolysis systems are not limited in this manner, and may be applicable to soil systems, as well as aqueous systems.

The Fenton reaction (Equation 1) is part of a catalytic cycle which produces hydroxyl radical from hydrogen peroxide.[13]

$$H_2O_2 + Fe^{2+} \rightarrow Fe^{3+} + HO^{-} + HO^{\cdot} \tag{1}$$

In addition, some strongly oxidizing iron complexes may also be formed. These radicals results in degradation of a wide array of compounds. Reaction of $HO^{\cdot}$ with alkanes proceeds with a rate constant of $10^7 - 10^9\, l/(mol \cdot s)$, and reactions with alkenes and aromatics have rate constants on the order of $10^9 - 10^{10}\, l/(mol \cdot s)$.[15]

Ferric species are also capable of producing hydroxyl radical in the Haber-Weiss process[13]:

$$H_2O_2 + Fe^{3+} \rightarrow Fe^{2+} + H^+ + HO_2^{\cdot} \tag{2}$$

$$H_2O_2 + Fe^{2+} \rightarrow Fe^{3+} + HO^- + HO^{\cdot} \tag{3}$$

$$H_2O_2 + HO^{\cdot} \rightarrow H_2O + HO_2^{\cdot} \tag{4}$$

$$HO_2^{\cdot} + Fe^{3+} \rightarrow O_2 + Fe^{2+} + H^+ \tag{5}$$

$$HO_2^{\cdot} + Fe^{2+} \rightarrow HO_2^- + Fe^{3+} \tag{6}$$

Iron/peroxide systems have been used to degrade a number of contaminants, mostly in the aqueous phase. Degradation of the herbicides 2,4-D and 2,4,5-T in aqueous solution by $Fe(II)/H_2O_2$ or $Fe(III)/H_2O_2$ has been reported.[16] Reactions were pH sensitive, and acid conditions were necessary for iron solubility. Vella and Munder[17] used $Fe(II)/H_2O_2$ for the degradation of phenolic compounds in water. Again, a low pH (4) was used. Although complete elimination of the parent compound could be achieved for chlorinated and other phenols, the presence of phosphate significantly hindered degradation.

In order to allow reaction at near neutral pH, researchers have utilized iron chelating agents.[18,19] In these studies, several iron chelates were found to be active in degrading pollutants, although the chelator was also degraded at a slower rate.[18] This may be advantageous, since added organic species will not persist in the treated material. Three chelators were found to be particularly useful in the degradation of several pesticides in aqueous solution,[19] including the natural product gallic acid. Although no reference to experimental results is given, the authors note that these iron complexes may be useful for remediation in soil systems.

Fenton reagents have also been used for the degradation of perchloroethylene (PCE) and polychlorinated biphenyls (PCB's) adsorbed on sand.[20] These species cannot be successfully biodegraded due to their toxicity to microorganisms. The treatment required adjustment of the pH to 3 and resulted in significant decomposition of PCE; however, PCB treatment resulted in the formation of chlorinated degradation products. When the pH was raised to 7, degradation was severely limited.

Work on the use of Fe/H_2O_2 systems for remediation continues,[21-23] with additional studies involving biphasic systems (soil/water or immiscible solvent/water).[24-27] One possible drawback to the use of iron/H_2O_2 systems for aqueous and soil remediation is the interaction between iron and non-contaminant species present in the system. As mentioned above, phosphate has been observed to inhibit Fenton type degradation, and some iron complexes are inactive towards H_2O_2 mediated degradation. Reports are available showing both inhibition[28] and acceleration[29] of $HO^{\cdot}$ production in the presence of additional chemical species. Another important drawback is the scavenging of hydroxyl radical by non-pollutant species.

Although success of Fe/H_2O_2 in aqueous and simple soil systems has been demonstrated, fundamental information is critically lacking, preventing the widespread application of these techniques. Investigation of the microchemical factors in Fe/H_2O_2 systems is necessary to allow engineering of this technique in real-world sites. Once such factors have been determined, this technique may provide a low cost, highly efficient alternative for *in situ* remediation of contaminated soils, especially where high contaminant concentrations or mixtures preclude the use of bioremediation. Furthermore, this method would provide a relatively generic procedure, and would be applicable to a wide variety of wastes in diverse matrices, possibly including soil, sediment, groundwater, and industrial waste streams.

3. EXPERIMENTAL

In this study, several types of water were used: high purity water (Barnstead Nano-pureUV), high purity water with Suwannee River fulvic acid or humic acid added, water collected from a West Pearl River tributary at Crawford Landing (CL), Louisiana, and water collected from a small water body connecting Lake Pontchartrain and Lake Maurepas (LM), Louisiana. Natural water samples were filtered through pre-combusted (450 °C) glass fiber filters with a nominal cut-off of $0.5\,\mu m$ (Rundfilter MN GF-2, Macherey-Nagel, Düren, Germany), and subsequently stored in the dark at 4 °C. Standard Suwannee River fulvic acid and humic acid were obtained from the International Humic Substances Society (*http://www.ihss.gatech.edu*).

Pentachlorophenol (99+%), *benzoic acid* (99.5%), p-hydroxybenzoic acid (99%), *o-cresol* (99+%), *phenanthrene* (99+%), and *pyrene* (99%, optical grade) were purchased from Aldrich. *Phenol* (99+%) was purchased from Fisher and Fe(II) perchlorate hexahydrate (99+%) was purchased from Alfa. H_2O_2 was purchased from EM Science. All reagents were used as received.

High performance liquid chromatography was performed using a Hewlett-Packard 1090 liquid chromatograph. All separations were carried out with a Spherisorb ODS-2 column, $25 \times 4.6\,cm$, $2\,\mu m$ particle size (Alltech), and detection was performed with an absorbance detector at 254 nm. Pentachlorophenol (PCP) was analyzed using a mobile phase of 80/20 acetonitrile/water (1% acetic acid) with a flow rate of $1\,mL\,min^{-1}$ and a $20\,\mu L$ injection loop. Benzoic acid (BA) and its hydroxylated products were analyzed with gradient elution and on-column pre-concentration.[30] BA samples were acidified to pH ~ 3 prior to injection using a 1.5 mL loop. The eluent consisted of 15/85 acetonitrile/water (pH 2 with HCl) for 3 minutes followed by a linear gradient to 80/20 acetonitrile/water (pH 2 with HCl) in 13 minutes.

Degradation studies were typically carried out in 10–15 mL solutions. To each solution a small aliquot of aqueous $Fe(ClO_4)_2$ was added, followed by addition of a small aliquot of H_2O_2 to initiate degradation. The added concentrations of iron and peroxide in each reaction were adjusted by varying the volume and concentration of the added aliquots. All reactions were carried out in the dark and were shaken on an orbital shaker. Reactions were quenched by adding 0.5 mL of n-propanol. Iron solutions were prepared fresh each day and H_2O_2 was standardized against thiosulfate on a regular basis.

4. RESULTS AND DISCUSSION

Studies in this laboratory have demonstrated a dramatic decrease in oxidation efficiency in the presence of natural organic matter (NOM). Figure 1 presents data for the oxidation of pentachlorophenol in pure water and in CL natural water (10.9 mg/L dissolved organic carbon). The initial degradation rate and the total amount degraded are both lower in the presence of DOM. Furthermore, similar results were obtained with samples of LM natural water, indicating that this phenomenon is general to natural waters. Three causes for the decreased degradation are likely: scavenging of hydroxyl radical by NOM, alteration of the HO· production rate by iron-NOM interactions,[18,19,28,29]

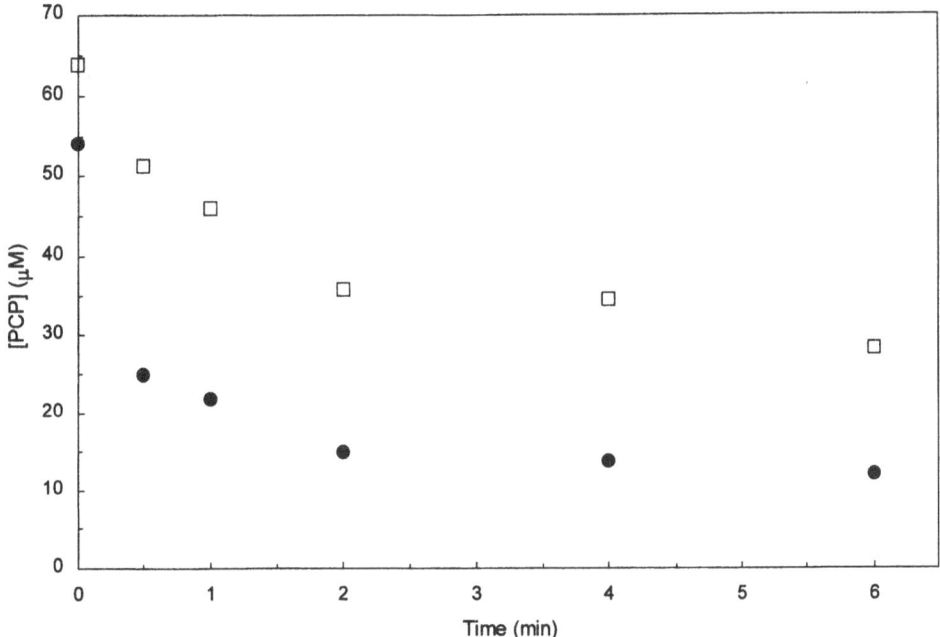

Figure 1. Loss of PCP in pure water (●) and CL natural water (□) after addition of 0.2 mM Fe(ClO$_4$)$_2$ and 0.4 mM H$_2$O$_2$.

and sequestering of pollutant to protective sites within macromolecular matrices of the NOM. Our data indicate that all three of these processes are occurring, and a discussion of each process follows below.

Hydroxyl radical production rates and steady state concentrations were measured using benzoic acid as a probe. Since hydroxyl radical concentration is not at steady state after a single addition of hydrogen peroxide to an Fe(II) solution, previous calculations[31] based on steady state concentrations are not applicable. Instead, the following approach was used to calculate either hydroxyl radical concentration or total hydroxyl radical formation.

The change in [HO·] with time is described by:[31]

$$\frac{d[HO\cdot]}{dt} = F - k_p[HO\cdot][P] - \sum k_{Si}[HO\cdot][S_i] - k_{HO}[HO\cdot]^2$$

where F is the hydroxyl radical formation rate, and the three negative terms are loss due to reaction with the probe molecule (P), reaction with scavengers (S), and self reaction, respectively. As [P] increases, the term—$k_p[HO\cdot][P]$ will dominate, and the total moles of P reacted will approximate the total moles of HO· formed (provided that reaction products do not compete effectively for hydroxyl radical).

If [P] is kept small so that $k_p[HO\cdot][P] \ll \sum k_{Si}[HO\cdot][S_i] + k_{HO}[HO\cdot]$, then d[HO·]/dt will be relatively unaffected by the probe. Furthermore, over short times, [HO·] does not change significantly. Therefore, [HO·] over that time period can be approximated by:

$$[HO\cdot]_{avg} = \frac{R_p}{k_p[P_{avg}]}$$

where R_p is the rate of probe reaction.

Since changes in [BA] are often small in these measurements, we used p-hydroxy-benzoic acid (p-HOBA) formation rate to quantitate BA loss rate.[32] We assumed that one p-HOBA is formed for every 5.9 BA reacted, based on the data of Zhou and Mopper.[30] Since changes in [BA] and [HO·] were small in the time windows used, the p-HOBA formation rate was pseudo zero-order, and the rates were determined by the change in [p-HOBA] divided by the time interval.

Figure 2 illustrates the HO· trapping efficiency as a function of benzoic acid concentration in pure water and diluted LM natural water. The natural water sample was diluted with pure water to minimize interferences of the NOM with the chromatographic analysis of BA and p-HOBA. At sufficiently high BA concentrations (>2mM), essentially all of the HO· is trapped. Under these conditions, the total moles of BA reacted is approximately equal to the total moles of HO· produced, and HO· formation rates can be estimated. Data for hydroxyl radical formation in pure and natural waters are presented in Fig. 3. Estimated time-dependent production rates are presented in Table 1. The pure water showed initial HO· production rates about 14% higher than the natural water. In both samples, rates of HO· production dropped off rapidly with time, with a residual production of around 0.3 nmol/s for at least 20 minutes. Furthermore, the natural water showed a lower yield of HO· per mole peroxide (Table 1). The lower production in the natural water is likely due to interaction of the natural water matrix with iron.

At low BA concentrations, only a portion of the total hydroxyl radical is trapped. For the natural water, 0.5 mM BA trapped only about 4% of the hydroxyl radical. For pure water, under the same conditions (0.5 mM BA), 41% of the hydroxyl radical is trapped, while at 0.2 mM BA, 25% of the radical is trapped. Because of the high degree

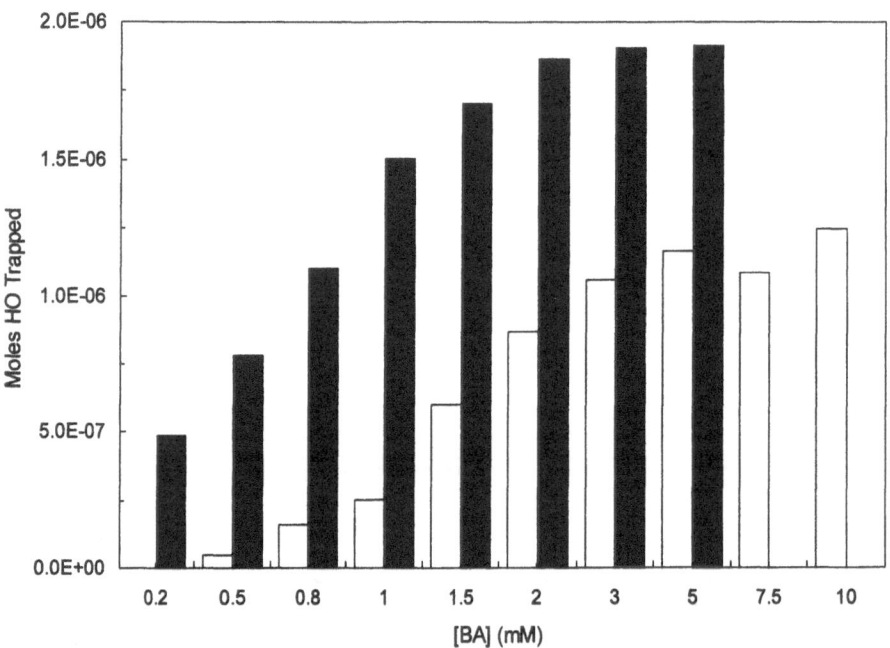

Figure 2. Hydroxyl radical trapping efficiency at varying benzoic acid concentrations in pure water (black bars) and in 33% LM natural water (white bars). Iron and peroxide were added at time zero at 0.53 mM each. Total HO· trapped was measured at t = 10 min. Natural water was diluted by addition of pure water.

Table 1. Hydroxyl radical formation rates (nmol s^{-1})

Time Period (s)	Pure Water	LM Natural Water
0–20	191	167
20–40	75	58
40–60	25	25
60–80	16	37
80–240	5.6	4.1
240–1,200	3.0	3.6
mol HO· formed after 10 min per mol H$_2$O$_2$ added	0.36	0.22

of hydroxyl radical trapping, in this study benzoic acid was not used to determine [HO·] in pure water.

As expected, higher natural water matrix content resulted in lower concentrations of hydroxyl radical, presumably due to scavenging by DOM. Figure 4 illustrates the time dependent [HO·] after a single addition of iron and peroxide to three dilutions of natural water. Initial hydroxyl radical levels are high; however, after two minutes, the [HO·] decays to a much smaller residual value. A general trend of decreasing HO· with increasing natural water content is apparent, indicating an inverse relationship between [NOM] and [HO·]. This finding is consistent with the assumption that NOM lowers HO· by interaction with iron and/or scavenging of hydroxyl radical. The rapid decay in hydroxyl radical is in good agreement with the time dependent degradation profile observed for PCP (Fig. 1). Residual PCP degradation was observed beyond 2 minutes, likely due to additional slow HO· production from remaining low levels of H$_2$O$_2$.

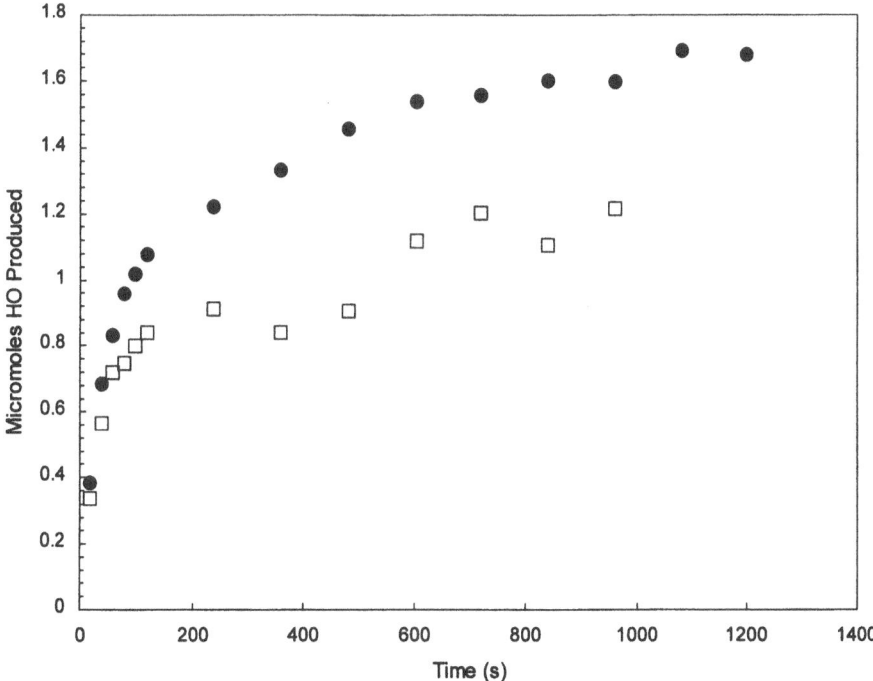

Figure 3. Time dependent HO· formation in pure water (•) and 33% LM natural water (□). [BA] = 9.0 mM. Natural water was diluted by addition of pure water.

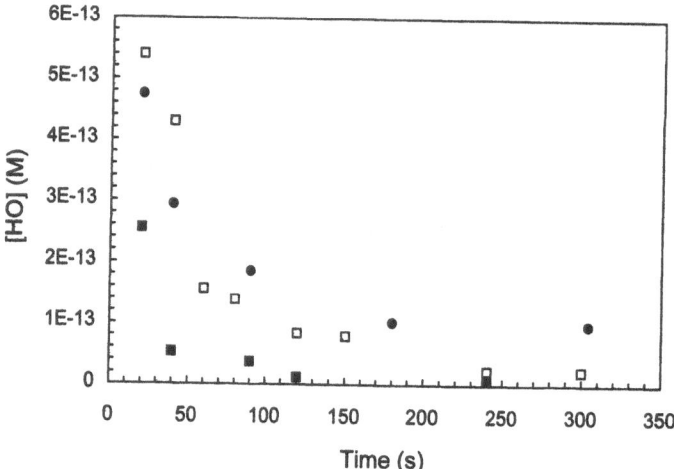

Figure 4. Time dependent hydroxyl radical concentration in diluted LM natural water. Iron and peroxide were added at time zero at 0.53 mM each. 20% LM natural water (●); 33% LM natural water (□); 75% LM natural water (■). All samples diluted with pure water.

Although PCP degradation showed complex kinetics after a single addition of iron and peroxide, the observed kinetics were dramatically simplified by making multiple peroxide additions. At time zero, both iron and peroxide were added to a PCP solution (in pure water). Every 2 minutes thereafter, an equivalent aliquot of peroxide was added. The PCP degradation followed pseudo first-order decay. These data are presented in Fig. 5, and indicate a time-averaged steady state HO· concentration. This type of kinetic data should be useful in designing more efficient methods for peroxide addition (e.g.—amount and time of addition). Furthermore, these data indicate that the iron remains catalytic over five consecutive peroxide additions. The integrity of the iron catalysis over longer time periods is of critical importance in devising effective *in situ* remediation technologies.

Given clear evidence from the literature showing that hydrophobic pollutants partition to natural organic matter (NOM),[33] it is likely that PCP partitioning to hydrophobic NOM sites is partly responsible for the observed reduction in oxidation rates. In this study, we illustrate this phenomenon qualitatively. We used pyrene as a model compound for hydrophobic pollutants. Pyrene is sensitive to microenvironmental polarity,[34] and provides a means of measuring its molecular-level surroundings. Our pyrene fluorescence studies have shown a change in microenvironment upon interaction with NOM. Data are shown in Fig. 6 as pyrene fluorescence vibronic band ratios. Changes in this ratio indicate changes in the local environment, generally associated with binding to NOM. The observed changes in microenvironment are small in this sample, but indicate that the pyrene is *partitioning* to *microenvironmental sites* within the NOM. Fluorescence lifetime data also indicate similar strong associations with NOM for two distinct natural water samples (Table 2). Both the steady state and lifetime fluorescence measurements indicate a strong interaction between hydrophobic compounds and NOM. By analogy with sorption to particulate matter,[35] binding to NOM likely plays exhibits a predominant influence on hydroxyl radical oxidation of hydrophobic compounds. In the case of sorption to particulate matter, sorbed molecules are essentially unreactive,[35] and oxidation only takes place upon desorption and re-partitioning to the bulk aqueous phase. For

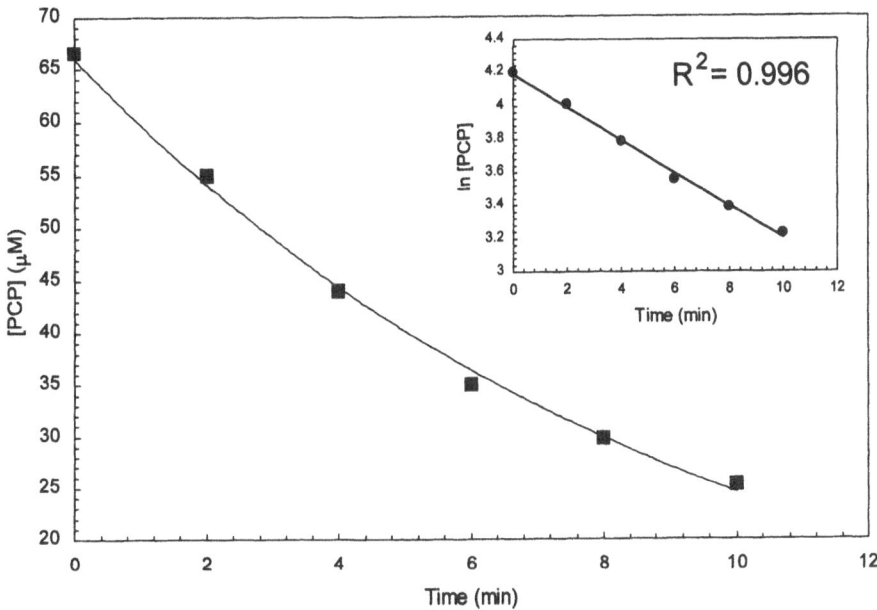

Figure 5. Pseudo first-order loss of PCP with multiple peroxide additions. Iron and peroxide were added at time zero at 0.4 mM each. Additional 0.4 mM aliquots of peroxide were added at 2, 4, 6, and 8 minutes.

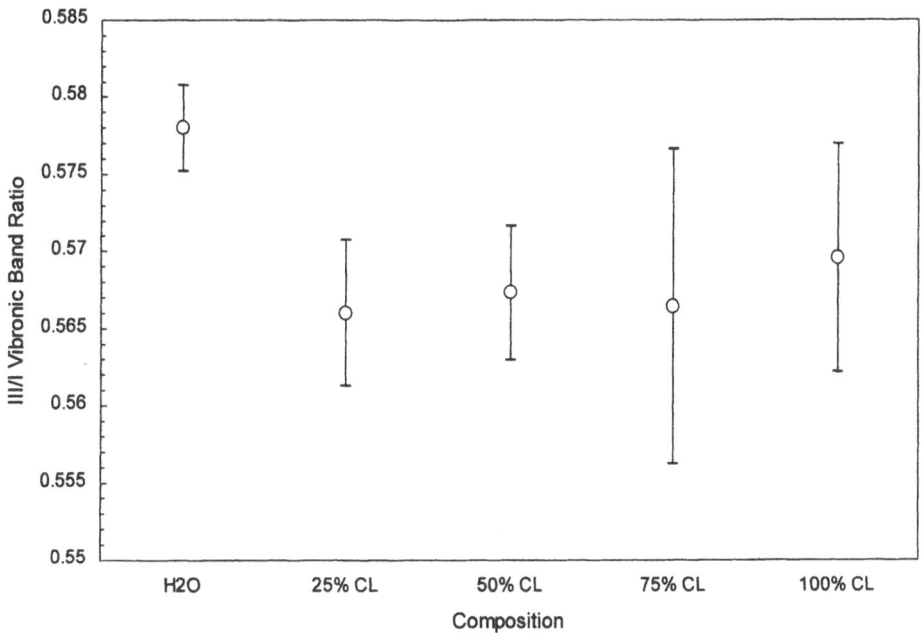

Figure 6. Pyrene fluorescence vibronic band ratios in pure water and mixtures of pure water with Crawford Landing natural water. Band I was measured at 373 nm and band III at 383 nm. Emission spectra were corrected for inner filter effects caused by NOM absorbance.

Table 2. Pyrene fluorescence lifetimes in water samples

Pure Water	CL Natural Water	LM Natural Water
139 ± 1 ns	82.6 ± 0.8 ns	82.01 ± 0.02 ns

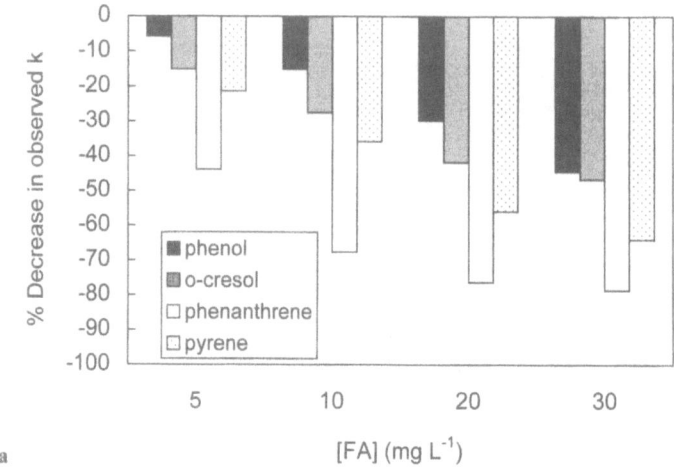

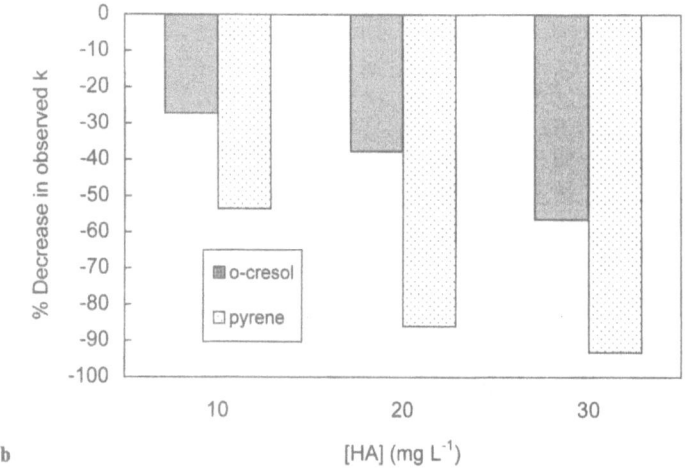

Figure 7. Observed decrease in rate constant for reaction with hydroxyl radical as a function of added Suwannee River fulvic acid (FA) or humic acid (HA) concentration. All rate constants were determined under steady state hydroxyl radical concentration (see ref. 36).

DOM solutions, the partitioning of iron to metal binding sites and the partitioning of pollutants to hydrophobic sites both play a role in protecting hydrophobic pollutants from degradation by Fenton's reagent.[36,37] Studies using a series of *aromatic hydrocarbons* indicate that these protective mechanisms are strongest for more *hydrophobic compounds*. Figure 7 illustrates the observed decrease in *rate constants* for several compounds upon addition of Suwannee River fulvic acid and humic acid. The extent of protection, however, is not completely explained by binding of the pollutant to DOM. It is therefore believed that binding of iron in separate hydrophilic sites also plays a role in decreasing pollutant degradation rates. These effects are due to spatial separation of the pollutant from the formation site of hydroxyl radical and will be important even when total radical formation and concentration are unaffected.

5. CONCLUSION

To date, our studies have shown that all three factors studied (altered HO· production rate, increased HO· scavenging, and binding to NOM) are occurring during Fenton oxidation in the presence of natural organic matter. These three factors are the key issues in determining the extent and efficiency of oxidation. Although hydroxyl radical formation and scavenging rates change upon addition of natural organic matter, the microenvironmental effects that result in spatial separation of pollutants and hydroxyl radical appear to be the most important factors for hydrophobic compounds. Clearly, all three factors most be accounted for in order to accurately predict degradation rates and to determine the peroxide loading required for complete degradation. The research to date illustrates that the effects of dissolved organic matter are complicated and require additional study in order to effectively engineer efficient *in situ* remediation or waste stream treatment based on Fenton's reagent.

ACKNOWLEDGMENT

This work was supported by the Louisiana Board of Regents under grant LEQSF(1998-01)-RD-A-34.

REFERENCES

1. "Biotechnology in Bioremediation of Pesticide Contaminated Sites," J.S. Karns in Pesticide Waste Management, J. Bourke, A.S. Felsot, T.J. Gilding, J.K. Jensen, and J.N. Seiber, Eds., ACS Symposium Series #510, 1992, pp. 148–156.
2. Haggblom, M.M., Rivera, M.D., and Young, L.Y. Appl. Environ. Microbiol. 1993, 59, 1162–67.
3. Haugland, R.A., Schlemm, D.J., Lyons, R.P. III, Sferra, P.R., and Chakrabarty, A.M. Appl. Environ. Microbiol. 1990, 56, 1357–62.
4. Smith, J.A. and Novak, J.T. Water Air Soil Pollution 1987, 33, 29–42.
5. Crawford, R.L. and Mohn, W.W. Enz. Microb Technol. 1985, 7, 617–20.
6. Racke, K.D., Coats, J.R., and Titus, K.R. J. Environ. Sci. Health 1988, B23, 527–39.
7. Michalski, A., Metlitz, M.N., and Whitman, I.L. Ground Wat. Mon. Remed. 1995, 15, 90–100.
8. Butcher, J.B. and Gauthier, T.D. Ground Wat. 1994, 32, 71–78.
9. Brandes, D. and Farley, K.J. Wat. Environ. Res. 1993, 65, 869–78.
10. Martel, R., Gelinas, P.J., Desonoyers, J.E., and Masson, A. Ground Wat. 1993, 31, 789–800.
11. "Comparative Aspects of Advanced Oxidation Processes," D.F. Ollis in Emerging Technologies in Hazardous Waste Management III, D.W. Tedder and F.G. Pohland, Eds., ACS Symposium Series #518, 1993, pp. 18–34.
12. Ollis, D.F., Pelizzetti, E., and Serpone, N. Environ. Sci. Technol. 1991, 25, 1523–29.
13. Walling, C. Acc. Chem. Res. 1975, 8, 125–131.
14. Riesz, P. and Kondo, T. Free Rad. Biol. Med. 1992, 13, 247–70.
15. Buxton, G.V., Greenstock, C.L., Helman, W.P., and Ross, A.B. J. Phys. Chem. Reference Data 1988, 17(2), 513–886.
16. Pignatello, J.J. Environ. Sci. Technol. 1992, 26, 944–51.
17. "Toxic Pollutant Destruction," P.A. Vella and J.A. Munder in Emerging Technologies in Hazardous Waste Management III, D.W. Tedder and F.G. Pohland, Eds., ACS Symposium Series #518, 1993, pp. 85–105.
18. Sun, Y. and Pignatello, J.J. J. Agric. Food Chem. 1992, 40, 322–27.
19. Sun, Y. and Pignatello, J.J. J. Agric. Food Chem. 1993, 41, 308–12.
20. "Decomposition of Perchloroethylene and Polychlorinated Biphenyls with Fenton's Reagent," C. Sato, S.W. Leung, H. Bell, W.A. Burkett, and R.J. Watts in Emerging Technologies in Hazardous Waste Management III, D.W. Tedder and F.G. Pohland, Eds., ACS Symposium Series #518, 1993, pp. 343–356.

21. Lipczynska-Kochany, E., Sprah, G., and Harms, S. Chemosphere 1995, 30(1), 9–20.
22. Gau, S.H. and Chang, F.S. Water Sci. Technol. 1996, 34(7–8), 455–462.
23. Kim, Y.K. and Huh, I.R. Environ. Eng. Sci. 1997, 14(1), 73–79.
24. Yeh, C.K. and Novak, J.T. Water Environ. Res. 1995, 67, 828–34.
25. Pignatello, J.J. and Chupa, G. Environ. Toxicology and Chem. 1994, 13, 423–27.
26. Ronen, Z., Horvathgordon, M., and Bollag, J.M. Environ. Toxicology and Chem. 1994, 13, 21–26.
27. Watts, R.J., Udell, M.D., and Monsen, R.M. Water Environ. Res. 1993, 65, 839–44.
28. Croft, S., Gilbert, B.C., Smith, J.R.L., Stell, J.K., and Sanderson, W.R. J. Chem. Soc. Perkin Trans. 1992, 2, 153.
29. Puppo, A. Phytochem. 1992, 31, 85–88.
30. Zhou, X. and Mopper, K. Mar. Chem. 1990, 30, 71–88
31. "Reaction of Oxygen Species in Natural Waters," N.V. Blough and R.G. Zepp, in Reactive Oxygen Species in Chemistry, C.S. Foote and J.S. Valentine, Eds., Chapman and Hall, 1994.
32. Lindsey, M.E. and Tarr, M.A. Chemosphere 1999, in press.
33. McCarthy, J.F. and Jimenez, B.D. Environ. Sci. Technol. 1985, 19, 1072–76.
34. Kalyanasundaram, K. and Thomas, J.K. J. Am. Chem. Soc. 1977, 99, 2039.
35. Sedlak D.L. and Andren, A.W. Wat. Res. 1994, 28, 1207–15
36. Lindsey, M.E. and Tarr, M.A. Environ. Sci. Technol. 1999, submitted.
37. Lindsey, M.E. and Tarr, M.A. Wat. Res. 1999, submitted.

9

INNOVATIVE SURFACTANT/COSOLVENT TECHNOLOGIES FOR REMOVAL OF NAPL AND SORBED CONTAMINANTS FROM AQUIFERS

Chad T. Jafvert and Timothy J. Strathmann

Purdue University
School of Civil Engineering
Purdue University
1284 Civil Engineering Building
West Lafayette, Indiana 47907

ABSTRACT

Conventional pump-and-treat removal of nonaqueous phase liquids and sorbed contaminants from groundwater has proved unsuccessful in many cases for a variety of reasons. To overcome some of the limitations of these methods, innovative technologies are currently under development which attempt to increase subsurface contaminant removal efficiencies through the addition of surfactants and/or cosolvents. The rationale behind the use of these substances for improving groundwater contaminant removal, as well as several case studies are reviewed.

INTRODUCTION AND DESCRIPTION

Conventional "pump-and-treat" remediation of dissolved, sorbed, and/or non-aqueous phase contaminants (NAPLs) from the subsurface has proved often times to be very inefficient and costly. Low recoveries often have resulted from: (*i*) slow dissolution of NAPLs into the ground-water, (*ii*) slow diffusion of contaminants from low conductivity zones to high conductivity zones, (*iii*) slow desorption of sorbed contaminants, and (*iv*) hydrodynamic isolation in dead-end zones. The first three factors result because many organic compounds in the subsurface have a greater affinities for organic phases (including natural organic matter and NAPL phases) which decreases effective-dissolution, -diffusion, and -desorption rates.

Upon reaching the water table of an unconfined aquifer, light nonaqueous phase liquids (LNAPLs) typically spread on the water-table surface. Dense nonaqueous

Emerging Technologies in Hazardous Waste Management 8, edited by Tedder and Pohland
Kluwer Academic/Plenum Publishers, New York, 2000.

phase liquids (DNAPLs), on the other hand, tend to sink into the ground-water creating a plug or volume of often times reduced hydraulic conductivity with respect to the surrounding water. When NAPL phases occur, three general zones of contamination may be described. These are the source or residual zone, the concentrated plume (where the center of mass in the ground-water plume occurs), and the dilute ground water plume.

Direct pump-and-treat of NAPL phases often results in "fingering" of the non-aqueous phase liquid as water moves through or around it to the well head. Conventional pump-and-treat, with reinjection of the local groundwater behind the plug, in an attempt to mobilize the NAPL often fails because water passes through or around the NAPL plug due to the high interfacial tension between the NAPL and water. The NAPL acts as a long-term continuous source of contaminant, through slow chemical dissolution into the ground water. High interfacial tension at the NAPL-water interface, and the resulting capillary forces, assure that oil droplet release to the mobile water phase does not occur. At some sites, the initial pump-and-treat recoveries (with local ground water) and the estimated NAPL or sorbed residual volumes in the subsurface suggest that remediation times of decades to centuries would be required for greater than 90% contaminant removal. To overcome these limitations, innovative technologies are currently under development which enhance flushing efficiency through the use of surfactants, cosolvents, or other chemical agents. Several of these technologies and initial field demonstrations of their application are discussed in this chapter.

SURFACTANTS

One class of chemical substances that has been proposed for enhancing pump-and-treat remediation of NAPL and sorbed contaminants from the subsurface are surfactants. The goal of surfactant flushing is to decrease the required flushing volume by *mobilizing* and/or *solubilizing* nonaqueous liquid phases and *solubilizing* sorbed contaminants. These phenomena occur because surfactants:

1) may lower the NAPL—water interfacial tension, thereby decreasing capillary forces within the porous media
2) may be used to create a Winsor type III middle phase microemulsion (see Fig. 1)
3) may form micelles or single phase microemulsions that can solubilize contaminants (with added cosurfactants, usually alcohols)

When removal of NAPL or sorbed compounds is the intent, anionic and/or nonionic surfactants are utilized to minimize loss of surfactant to the aquifer solids. Cationic, and to a lesser extent zwitterionic, surfactants strongly sorb to aquifer solids making them impractical for such uses. Several researchers, interested in retarding the migration of contaminant plumes, however, are examining the potential of cationic surfactants for increasing the carbon content of the aquifer solids, thereby reducing subsurface transport of hydrophobic contaminants. Others are interested in facilitating the bioremediation of contaminants through the addition of surfactants.

Surfactant molecules possess both polar (hydrophilic) and nonpolar (hydrophobic) regions. The polar region is commonly composed of a sulfate, sulfonate, carboxylate, or polyethoxylate group, and often contains a succinate or sorbitan group. The nonpolar region generally takes the form of a linear hydrocarbon chain; however, sometimes the hydrocarbon chain is branched and may contain phenolic or other aromatic groups. Because of their unique structure, surfactants are amphiphilic (both loving) in nature. The polar group has a high affinity for (high solubility in) polar solvents, such as water, whereas the nonpolar group has a high affinity for nonpolar or hydrophobic solvents or phases, which include most organic liquids. Upon addition to water, the large energy requirement for solubilization of the nonpolar portion of the surfactant molecule in the water is minimized by either transference of this region of the molecule to a nonpolar solvent through the liquid-liquid interface, or by self assemblage of these groups into aggregates known as micelles.

Overall partitioning of the surfactant molecules at the water-NAPL interface reduces the thermodynamic energy required to form this interface. Without surfactant present, the large amount of energy required to form a water-NAPL interface is compensated for physically by reducing the interfacial surface area. This results in a "rigid" or "taut" boundary between the two phases which is characterized by a high interfacial tension. By reducing the energy requirement to create the interface, the added surfactant increases the interfacial area and decreases "tautness", resulting in a lower interfacial tension. An increase in interfacial area may result in spontaneous formation of droplets, of very thin alternating sheets of water and NAPL separated by surfactant layers (lamella), or of other structures which *mobilize* the NAPL material. NAPL-in-water droplets which form spontaneously (i.e., no mixing energy required) are referred to as "single phase" microemulsions (SPME). "Single phase" refers to the fact that only one aqueous phase (i.e., the microemulsion phase) is present during their development, as opposed to "middle phase" microemulsions (MPME) (also referred to as Winsor type III microemulsions), which are formed in the presence of an additional aqueous phase.

Micelles

Self assembly of the nonpolar groups results in the formation of surfactant aggregates known as micelles. In micelle formation, individual surfactant molecules (monomers) orient themselves so that the hydrophobic groups fill the core of a sphere, while the exterior polar groups are in contact with the surrounding aqueous phase water molecules. Formation of these aggregates does not require a second phase or the addition of another organic compound; however, other organic molecules may be solubilized within the micelles. Typical surfactant micelles are composed of 50 to 100 monomers. Micelle formation is surfactant concentration dependent. At surfactant concentrations below a "critical micelle concentration" (cmc), virtually no micelles exist. Surfactant doses in excess of the cmc produce micelles at a concentration equal to the difference between the total surfactant concentration and the cmc. The term "critical micelle concentration" is really a misnomer, as this value is the "specific *monomer* concentration" that is in equilibrium with the micelles. The concentration of micelles has little effect on the cmc. This is analogous to precipitation reactions, where the amount of solid in equilibrium with a saturated aqueous phase is not dependent on the aqueous saturation concentration. Surfactant micelles are sometimes considered to be a pseudo-phase because they can not be separated from the aqueous phase in which they reside without destroying them. They are dynamic entities, forming and disaggregating spontaneously and nearly instantaneously in water upon a change in surfactant concentration. Due to counter-ion association on charged micelles, cmc values of anionic surfactants decrease with increasing ionic strength (*e.g.*, increasing salinity). Ionic strength has little effect on the cmc values of nonionic surfactants. The cmc values of common nonionic surfactants tend to be less than those of common anionic surfactants, with nonionic/anionic surfactant mixtures having intermediate cmc values. Micelles may *solubilize* aqueous phase organic compounds, resulting in further dissolution of a NAPL phase. No discrete delineation exists between micelles (containing a few non-surfactant organic molecules), swollen micelles (containing many non-surfactant organic molecules), and single-phase microemulsions (containing even more non-surfactant organic molecules). In addition to phase and interfacial properties, surfactants with or without additional additives may affect other water chemistry or aquifer properties; most notably ground-water viscosity, salinity, and solid phase carbon content. Polymers, for example, often have been added to decreasing preferential flow in porous media.

Based on the specific NAPL or sorbed contaminant and site characteristics, the specific surfactant that is utilized may be selected or precisely formulated as a mixture to create an optimum MPME, SPME, or micellar system in the subsurface. The selection/formulation process is best described by inspection of a Winsor type diagram,[1] as depicted in Fig. 1. In this figure, each line representing a surfactant molecule has two parts: The straighter portion represents the nonpolar group, and the curved portion represents the polar group. This figure shows the possible phase behavior within a 1:1 (v:v) mixture of water and an immiscible organic liquid. The specific phases that exist are dependent on several water chemistry and surfactant properties. For example, if a single surfactant is utilized, the hydrophobic-lipophilic balance (HLB) of that surfactant largely determines the phase behavior. If a water soluble (hydrophilic) surfactant is added, most of the surfactant will reside within the water phase, possibly creating a micellar pseudophase within the aqueous layer. Some, but not much, of the organic phase may be solubilized in the aqueous micelles, reducing the organic phase volume. If cosurfactants are added (usually linear alcohols) a SPME may result with organic liquid droplets of 0.01 to 0.1 mm in diameter.

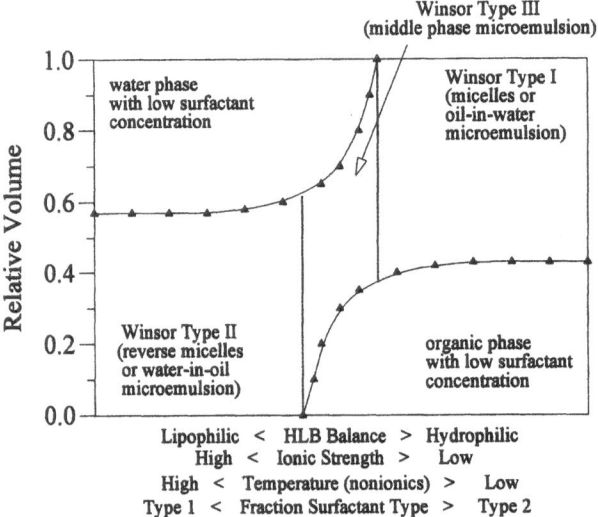

Figure 1. Typical Winsor diagram of equal volumes of organic phase with surfactant(s) and possibly cosurfactants.

Reverse Micelles

At the other extreme, if an organic soluble (hydrophobic) surfactant is added to the system, most of the surfactant will reside within the organic phase, creating a Winsor type II organic rich "reverse" micelle phase. A small amount of water may be solubilized within the reverse micelles. Classic type II systems should be quite impractical for aquifer restoration.

Middle Phase Microemulsions

If the hydrophilic and lipophilic functional groups of the surfactant are closely balanced, a Winsor type III "middle phase" microemulsion (MPME) will result. The term "middle phase" refers to the actual location of the phase in static layered solutions where clear top and bottom layers exist, separated by the "middle" phase. In the case of a DNAPL, the bottom layer is the excess DNAPL and the top layer is the excess water, as relationally depicted on Fig. 1. Also shown on Fig. 1 are some other controlling parameters, besides the HLB of a specific surfactant. The most notable is surfactant mixture ratio. Because of the wide variety of possible chemical compositions within NAPL spills, the probability of finding an existing inexpensive high volume surfactant that produces stable MPMEs for any given NAPL is quite low. However, the HLB value of an overall surfactant mixture can be varied quite easily by adjusting the ratio of surfactant species within a mixture. For example, it has been shown that MPME systems are possible for the halogenated solvent tetrachloroethylene, using mixtures of the branched nonionic ethoxylated surfactant Aerosol OT and sodium mono- and di-methyl naphthalene sulfonate.[2] Both MPME and SPME systems form spontaneously on water, surfactant (and possible cosurfactant), and NAPL contact, as opposed to larger diameter emulsion systems which require input of mixing energy.

Solubilization

Solubilization by micelles or microemulsions is a phase distribution process. A liquid phase, solid phase, or sorbed contaminant phase may exist in equilibrium with micellar or microemulsion solubilized compounds. The flushing rate of a NAPL from the subsurface, therefore, is dependent upon the solubility limit of the organic phase constituents within the micelle or microemulsion droplets and the partition coefficient for each species. Because a large fraction of a microemulsion can be the organic phase (*i.e.*, NAPL) itself, predicting the solubility behavior, *a priori*, essentially is impossible, and trial and error is used to formulate the correct surfactant mixture. SPME systems commonly contain a mixture that includes an anionic sulfonated surfactant, a nonionic ethoxylated surfactant, a linear alcohol, and often additional hydrocarbon compounds. Micellar systems are generally simple formulations of individual anionic or nonionic surfactant mixtures, or an anionic *and* nonionic mixture. A good estimate of the solubilization potential in simple micellar solutions can generally be made prior to any experimental work. Figure 2, for example, shows the values of micelle-water partition coefficients in relationship to the corresponding octanol-water partition coefficients for a series of aromatic compounds (PAHs).[3] In this case, the micelle-water partition coefficient, K_m, is described by the following relationship:

$$K_m = [PAH]_{mic}/[PAH]_{aq} \qquad (1)$$

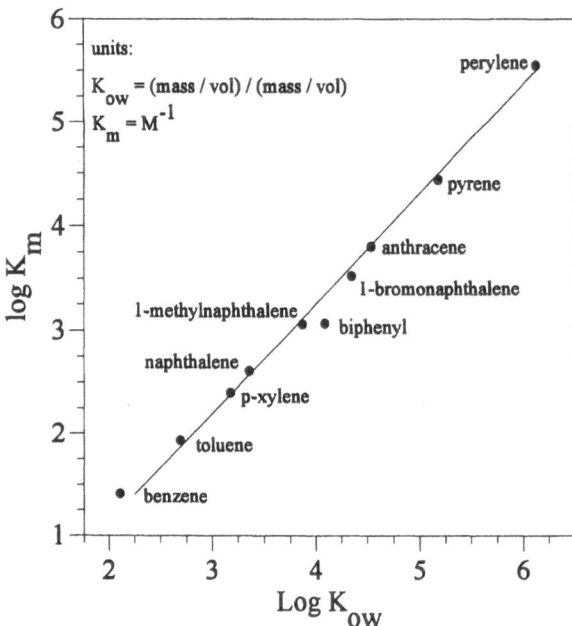

Figure 2. The relationship between the micelle-water partition coefficient (K_m) and the octanol-water partition coefficient (K_{ow}) for the anionic surfactant sodium—dodecylsulfate (from (3)).

where $[PAH]_{mic}$ has units of mass (mole) contaminant per mass (mole) of surfactant and $[PAH]_{aq}$ has units of mass (mole) contaminant per volume (L) of water. This relationship can be somewhat deceiving. It would appear that those compounds which are the least water soluble have the greatest solubility within the micelles. This is not true as these values represent partition constants only, and not absolute solubilities. Through this series of compounds, as the K_{ow} value increases, water solubility decreases (recall this is a log scale). For example, the solubility of perylene in water is only 0.0002 mg/L, whereas benzene's solubility is 1,800 mg/L, a difference of more than 6 log units. To produce a 4 log unit increase in the micelle-water partition coefficient, the solubility within the micelles must actually decrease approximately 2 log units.

Indeed, at saturation under standard pressure and temperature, approximately 35 benzene molecules can be solubilized by one dodecylsulfate micelle composed of 60 to 70 surfactant molecules, whereas, on average much less than one perylene molecule is solubilized within each micelle. Micellization of contaminants, therefore, generally will require significant surfactant doses. The molar solubilization ratio (MSR) is the ratio of contaminant molecules solubilized to surfactant molecules within micelles at saturation. Hence, for benzene the ratio is approximately 0.5 for micelles composed of dodecylsulfate. The MSR for most halogenated ethenes and ethanes (PCE, TCE, 1,2-DCA, etc.) is of this same order (0.2 to 0.5), meaning that for every ten surfactant molecules applied, approximately 2 to 5 molecules of chlorinated ethane or ethene may be recovered.

Solubilization within SPME systems is somewhat similar to simple micellar systems. In this case, however, because of the much larger volume of the microemulsion phase, more NAPL is solubilized. These systems are more complex in that the stability of the phase is partly determined by the partitioning to the subsurface materials of the many individual constituents of the microemulsion. The stability of SPME systems over long distances in the subsurface is not known.

COSOLVENTS AND OTHER TECHNOLOGIES

Unlike the solubility enhancements caused by partitioning into micelles or microemulsions, solubility enhancements caused by cosolvent addition generally occur because of changes in the bulk properties of the isotropic solution. An example of this type of enhancement is shown in Fig. 3 for toluene in mixtures of water with methanol.[4] As with mixed surfactants, advantages are gained by using "mixed" cosolvents. Short chained linear alcohols are excellent in solubilizing small chlorohydrocarbons, whereas, larger hydrocarbon cosolvents work best for larger and more hydrophobic contaminants. Addition of short linear alcohols also enhances the solubility of the larger cosolvents.

In practice, the benefits of flushing with cosolvents may result from the above mentioned increase in solubilization within the aqueous phase or through the more efficient process of mobilization. Mechanisms responsible for mobilizing NAPL contaminants include: (*i*) creation of a single phase condition, (*ii*) decrease in the water-NAPL interfacial tension, and (*iii*) swelling of the NAPL by solubilization of the cosolvents within this phase. Mobilizing by creating a single phase is essentially the same as "solubilizing" the NAPL plug. Decreasing the interfacial tension occurs due to the changes in the surface tension of both phases (the cosolvent dissolves in both phases making their chemical and physical interfacial properties more similar). Similarly, swelling of the NAPL is accompanied by beneficial changes in phase viscosity, density, and other properties. As with surfactant systems, mobilizing, rather than solubilizing, results in: (*i*) use of less total material (surfactant or cosolvent), (*ii*) use of smaller volumes of material (i.e., fewer pore volumes), and (*iii*) shorter treatment times. However, precise formulations are required,

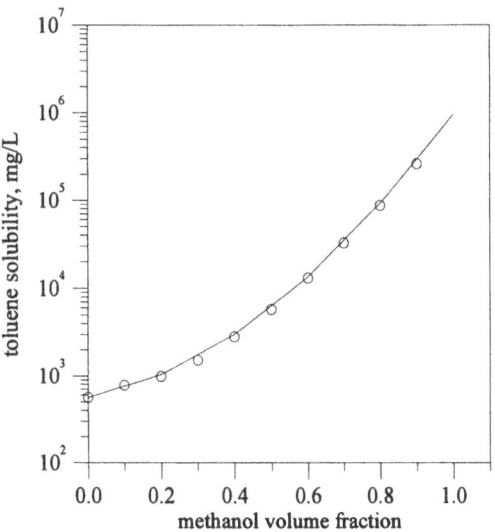

Figure 3. Solubility of toluene in water-methanol mixtures (from (4)).

resulting in the need for much more intensive laboratory bench and pilot-scale work before full scale remedial design can begin.

Rather than using surfactants or cosolvents, several investigators have proposed the use of other solubilizing agents that associate or "bind" contaminants. Mark Brusseau and coworkers at the University of Arizona have proposed the use of cyclodextrins as flushing agents. Cyclodextrin compounds are generally linear chains of glucose (sugar) molecules with the ends joined to form a cyclic structure. The contaminants partition into the center of the ring structure, resulting in the formation of "inclusion" complexes. Because the diameter of the ring can be controlled by the number of glucose molecules in the chain, specific cyclodextrins possibly could be used for specific contaminants. Suzanne Lesage and coworkers at Environment Canada have proposed using humic acids to enhance the removal of gasoline and diesel fuel constituents from subsurface environments. They have performed pilot-scale demonstrations. Although humic acids are quite recalcitrant to biodegradation, their use is advantageous because they are natural materials.

DESIGN CONSIDERATIONS

As with any remediation scheme, several factors need to be considered when designing or evaluating a system that utilizes surfactants or cosolvents. Although not an exhaustive list, the following general factors should be considered in the design or evaluation of such system.

1) The selection of a surfactant or cosolvent mixture is dependent on the anticipated removal mechanism: mobilizing versus solubilizing; with surfactant micelles, mixed micelles, micelles with cosurfactants, middle phase microemulsions, single phase microemulsions, or any of a semi-infinite variety of cosolvent mixtures. In the case of *mobilizing* systems, anionic/nonionic surfactants are added to injection wells at levels generally around 1 to 3% with or without additional chemicals (cosolvents, alcohols, inorganic salts, etc.).

2) The optimum removal mechanism will depend on interactions among the contaminants, the surfactant(s)/cosolvent(s), the subsurface porous media, and other subsurface constituents (e.g., dissolved Na^+, Ca^{2+}, Mg^{2+}, Fe^{2+}, and H^+). Hence, both batch (test tube) phase-behavior experiments and column flooding studies are required.

3) Hydrodynamic characterization of the site is both critical and necessary prior to modeling. The surfactant/cosolvent can be effective only if it reaches the contaminant zone and can be extracted from this zone. Hence, the configuration of injection/extraction wells is highly dependent on site characteristics. Initial demonstrations have employed various well designs, including single vertical circulation wells, trenches (for flooding) with extraction wells, and injection wells with separate extraction wells in all types of configurations (including horizontal wells). Prior to large field scale efforts, hydrodynamic modeling of the treatment flood is almost essential as potential problems may be discovered and averted. Low recovery efficiencies are often due to not capturing the material within the flush, rather than ineffective surfactant formulations. Preflushing with polymers or foams possibly may decrease preferential flow in heterogeneous formations.

4) The mass or concentration of material to be removed from the subsurface must be estimated prior to and after treatment. In the past, sample cores were often collected and analyzed. Due to aquifer heterogeneity, however, elaborate methods of data reduction were required to estimate the total mass or volume. More recently, inter-well partitioning tracer tests (IWPTT) have been used to estimate the mass or volume of material. Because these tests are generally performed using the same well arrays that are utilized in the surfactant/cosolvent flush, they provide an accurate measure of what is present and/or removed from the zone accessible by these wells. Even with these tests, however, dead-end zones may not be accessible, resulting in calculated recovery errors.

5) Treatment is not complete upon extraction. Large volumes of water, contaminants, and surfactants/cosolvents that are generated must be treated. Several research groups are or have investigated ultrafiltration as a means of separating and concentrating both contaminants and surfactants prior to further treatment. Solvent extraction, floatation, air stripping, vacuum steam stripping, foam fractionation, and photochemical treatment of the surfactant-contaminant waste-stream have also been investigated as possible separation processes. Following separation of surfactants and contaminants, the contaminants may be further treated.

TECHNOLOGY DEMONSTRATION

The first documented field tests of surfactant/cosolvent flushing were performed less than 10 years ago. Since that time, over 15 field demonstrations have been completed or are in progress. Many of these demonstrations are summarized here. In addition, because of the complexity and number of variations on the technology, many investigators are involved in either laboratory scale investigations or modeling efforts to address issues of surfactant or cosolvent mixture optimization, environmental fate, and toxicity. Table 1 summarizes studies in which surfactants or cosolvents have been injected in an attempt to facilitate subsurface contaminant removal. Some of these studies are summarized below. Summaries are based on information provided by the participating investigators, and content varies accordingly.

Bordon

The surfactant technology demonstration was conducted in a 3 m × 3 m × 3 m cell, composed of a clean surficial sand aquifer (<1% clay, <0.1% organic carbon).[5,6] The cell was constructed by driving sheet piling into the underlying clay aquitard. A second sheet piling wall was installed 1 m outside the inner wall for containment purposes. Five injection wells and five extraction wells were installed parallel to each other on opposite sides of the cell. Multi-level monitoring wells were installed as well. 271 L of a perchloroethylene (PCE) NAPL were introduced to the cell in a controlled release. The cell was separated into an upper and lower zone by a thin perched layer located 1 m below ground surface (BGS). 14.4 pore volumes (PVs) of a 2% surfactant solution were pumped through the cell. PCE in the extracted solution was removed by air stripping, allowing the investigators to recycle the surfactant solution.

Results indicate that the NAPL concentration was rapidly reduced in areas of high hydraulic conductivity within the cell. Concentrations in the top 1 m were reduced from an initial value of 10% NAPL saturation to a value of 1%. Below the perched layer, where the ground was initially 20 to 100% saturated with NAPL, concentration was reduced to 3%. Approximately 80% of the introduced material was recovered, with the remaining material lost to either volatilization or hydraulic dead zones within the formation.

Table 1. Surfactant or Cosolvent Studies or Demonstrations

Study name	Participants	Major contaminants	Process[a]	Refs.
Bordon	SUNY Buffalo Univ. Waterloo Centre. for Groundwater.	DNAPL PCE	surfactant flushing	5–6
Corpus Christi	SUNY Buffalo Dupont Corp. Remediation Group Ecosite, Inc.	DNAPL carbon tetrachloride	surfactant flushing	5–9
Estrie Region		NAPL hydrocarbon	surfactant flushing	10
General Motors	General Motors NAO R&D Center	oils and PCBs	surfactant flushing	11–16
Thouin Sand Quarry	Univ. Québec, Dept. Geological Eng. Canada Min. of Environment and Fauna	DNAPL, chlorinated solvents	surfactant mobilization	17–20
Traverse City	Univ. Oklahoma, Inst. App Surfactant Res. R.S. Kerr Environmental Research Lab U.S. Coast Guard, DOW Chemical	PCE and aviation fuel	micelle solubilization	21–24
DOE Portsmith	INTERA, Inc., SUNY Buffalo Univ. Texas at Austin	DNAPL TCE and PCBs	surfactant flushing	
Industrial NJ Site	New Jersey Inst. Of Technology GHEA Associates, Inc.	BTEX and chlorinated organics	surfactant flushing	25
Picatinny Arsenal	USGS & Univ. Virginia	sorbed TCE	micelle solubilization	26–28
Hill AFB		NAPL JP4 and chlorinated org.		
OU1, Test 1	Univ. Florida		cosolvent solubilization	29–33
Cell 3, OU1	Clemson Univ.		cosolvent mobilization	
Cell 5, OU1	Univ. Oklahoma		MPME	
Cell 6, OU1	Univ. Oklahoma		micelle solubilization	
Cell 8, OU1	Univ. Florida		SPME	34
OU2	INTERA, Inc., Radian, Inc. Univ. Texas at Austin		surfactant w/polymer	
OU2	Rice Univ.		surfactant foam flushing	

[a]If a specific process was not identified or engineered for (i.e., solubilization, SPME), the more general term "surfactant flushing" is used in the table.

Corpus Christi

Carbon tetrachloride (CTET) was present at the demonstration site in a shallow aquifer, underlain by a thick clay aquitard.[5-9] The target zone was a 12 ft thick fine grain sand lens (variable smectitic clay content 1 to 15%, 0.025 to 0.031% organic carbon, and 12,000 mg/L total dissolved solids) located between 12 to 24 ft BGS. Hydraulic conductivity was moderate ($\geq 10^{-3}$ cm/sec). An array of 6 injection wells and a single central extraction well were installed. Surfactant was recycled in subsequent flushing PVs during part of the test demonstration.

Prior to surfactant addition, concentrations of CTET in monitoring wells was between 200–300 mg/L. Concentrations jumped to 800–900 mg/L when surfactant concentrations reached 0.5% in monitoring wells. Approximately 73 gallons of CTET was removed using 12.5 PVs of surfactant solution.

Estrie Region

This commercial surfactant demonstration site was located beneath a machine shop in Estrie Region, in Québec, Canada.[10] The contamination zone extended from 2 to 4.3 m beneath a concrete slab floor. The volume of contaminated soil was estimated to be 1,800 m^3 with approximately 1 m of a free floating hydrocarbon phase. The soil was characterized as a fine sand (10–12% silt). The water table was located approximately 3 m BGS. Hydraulic conductivity in the zone of contamination was estimated at 10^{-4} cm/sec. 400 injection-extraction wells were installed in 3 distinct zones: A peripheral network to hydraulically isolate the contamination, a second zone used only to extract surfactant-contaminant solution, and a third zone which was used both to inject and extract material. The extract was processed in an on-site wastewater treatment system. Laboratory feasibility trials were performed to: Select the surfactant; evaluate biodegradability and toxicity of wash solutions; and plan hydraulic controls. A computerized system for data acquisition was installed to monitor ground-water movement. Nonionic biodegradable surfactants were selected for use.

Over a 50-day combined washing period (20 and 30 days in 1993 and 1994, respectively) 37.6 m^3 of free phase oil was recovered. 8 subsequent cycles of washing (60 day period of time) resulted in approximately 160,000 kg of oil recovered (mostly emulsified). Following extraction, high levels of hydrocarbon degrading microorganisms were detected in the soil.

General Motors

The test site for this demonstration was located on a five acre parcel surrounded by a previously installed containment wall which extends to a depth of 60 ft BGS.[11-16] Contamination was confined to the upper 15 ft of subsurface fill material. Contamination was estimated from soil cores to be approximately 6,000 ppm PCBs and 67,000 ppm miscellaneous oils. Surfactant solution was applied to a test plot within the test site, 10 ft in diameter by 5 ft deep. The extracted leachate contained both PCBs and other oils. Biotreatment of the leachate facilitated removal of both the surfactant and oil contaminants. The PCBs were then removed by adsorption on activated carbon. In other tests performed on the leachate, the surfactant solution was concentrated using a Romicon Model HF-Lab-5 ultrafiltration unit with either a PM500 or a XM50 membrane.

Approximately 10% of the contaminants were recovered using 5.7 PVs of surfactant solution. An additional 14% was recovered using 2.3 PVs during a similar test the following year. In the ultrafiltration experiments, 67% of the surfactant mass was recovered with the PM500 membrane, with 90 and 83% of the PCBs and oils being retained, respectively. The XM50 membrane did not capture the surfactant quite as efficiently (46%), but retained approximately the same fraction of PCB and oil (94% and 89%, respectively).

Thouin Sand Quarry

The site for this surfactant technology demonstration was located approximately 20 km northeast of Montréal, Québec, Canada.[17-20] The test site was characterized by a 2 m thick silty sand layer underlain by a 30 m thick deposit of silty clay. Waste oils and organic compounds dissolved in water were detected in nearby ditches and creeks. One central injection well, four recovery wells (spaced in a square), and 12 multilevel observation wells were installed in the 4.3 m × 4.3 m test plot. Recovery wells were located 3 meters from the injection well. Soil samples collected during well installation indicated that the initial NAPL concentration was approximately 55,000 mg/kg dry soil. During the test demonstration the plot was flooded in the following sequence: (1) 1.34 PV water, (2) 0.54 PV polymer, (3) 0.9 PV surfactant, (4) 1.6 PV polymer, (5) 1.4 PV water, (6) and finally with an injection of bacteria and nutrients to increase biodegradation of the remaining DNAPL.

In the zone swept by the washing solution, 86% of the residual NAPL was recovered using the rinse schedule described above. Insufficient flow in regions between the extraction wells resulted in low recoveries of NAPL in some areas. The use of a polymer solution before and after injection of the surfactant solution appeared to be beneficial in reducing preferential flow paths, despite soil heterogeneity.

Traverse City

This surfactant demonstration site consisted of a highly conductive sand formation with natural ground-water velocities between 3 and 5 ft/day.[21-24] The saturated zone was located less than 10 ft BGS. Soil contamination at the site consisted of PCE (1,000 μg/kg) and aviation fuel (1,000 mg/kg). Groundwater PCE levels were typically 10 μg/L. The objectives of the demonstration were to evaluate a vertical circulation well (VCW) system, and to maximize surfactant recovery. The VCW system was a single borehole well system with two 5 ft screen lengths separated by a 3 ft spacer. The screens were isolated from each other within the well with packers. Water/surfactant was injected through the top screen and water/contaminants/surfactant extracted through the bottom screen. 540 gallons of surfactant solution (Dowfax 8390) were added at a surfactant concentration 10 times the cmc value (3.8 wt%).

Due to the high hydraulic conductivity at the site, in order to capture both the surfactant and solubilized contaminants with the VCW, the extraction rate needed to exceed the injection rate by a factor of 10 to 15. After accounting for dilution water added, estimated recovery enhancements due to surfactant addition were 40 to 90 times of background water flushing alone.

Picatinny Arsenal

The site of this surfactant flushing demonstration project was located on a golf course.[26-28] TCE contamination was detected within a sand and gravel aquifer at soil concentrations of approximately 1 to 5 mg/kg and aqueous phase concentrations of 1 to 5 mg/L. The water table was approximately 10 ft BGS, with a lower confining layer (10 to 15 ft thick) located at 50 ft BGS. The test site was up gradient from an existing pump-and-treat system, installed as an interim remedy, as the site is a listed Superfund site. Three injection wells, 10 ft apart, were installed perpendicular to the natural gradient and tangential to the pump-and-treat extraction well, approximately 100 yards down-gradient from the zone of contamination. One monitoring well was located 30 ft up-gradient and three monitoring wells located down-gradient at 10 ft intervals from the injection wells. Clean water was pumped to all three injection wells for approximately 30 days at 3 gal/min. Water in the center injection well was replaced with 400 mg/L of surfactant (Triton X-100) at the same flow rate for 30 days while clean water continued in the outer wells. Clean water was pumped to all three wells for an additional 4 months. The purpose of this study was to investigate whether the surfactant would increase the rate of desorption of TCE from the solid phase.

Hill Air Force Base Sites

Nine test cells for various remediation demonstration projects were constructed at Operational Unit 1 (OU1) on Hill Air Force Base (HAFB), near Salt Lake City, Utah.[29-34] Studies are completed, underway, or in design stages at Operational Unit 2 (OU2). HAFB is a CERCLA site listed on the National Priorities List. In general approximately 6 to 9% residual NAPL existed within the saturated zone pore space of each cell. The NAPL was composed largely of JP4 jet fuel (approx. 90 to 95%), with lesser amounts of chlorinated organic solvents (chloroalkenes and chlorobenzenes) and even some PCBs. A nearby landfill contributed to the contamination. Technologies under investigation, each with a dedicated field cell, included: steam flushing for NAPL mobilization (Tyndall AFB and Praxis Environmental Technologies), air sparging (Michigan Tech.), flushing with cyclodextrins (U. of Arizona), flushing with co-solvents (Clemson), flushing with a surfactant/alcohol single phase microemulsion (SPME) (U. of Florida), flushing with a surfactant that "solubilizes" (U. of Oklahoma), flushing with a surfactant that "mobilizes" (U. of Oklahoma), and in-well sparging (same as in-well aeration) (U. of Arizona). In addition, at Operational Unit 2, two demonstrations using surfactants were performed.

All the test cells at OU1 were similar in design. Each cell (3 m × 5 m in surface area) was constructed by driving sheet piling into a clay layer approximately 30 BGS. A sand and gravel aquifer extends from approximately 15 ft BGS down to the clay aquitard. Four injection and three extraction wells were installed on opposite sides of each cell. 12 evenly spaced sampling wells, each with nested ports at 5 vertical depths, were installed in the interior of each cell. Prior to and following treatment of each cell, a partitioning tracer test was performed. The mix of tracers were designed according to the expected volume of NAPL within the cell before and after treatment. Some of the surfactant or cosolvent demonstrations are summarized below.

Test 1, OU1

The first demonstration carried out at HAFB consisted of treatment with a cosolvent mixture. An inter-well partitioning tracer test was conducted prior to introduction of cosolvent mixtures. Prior to all testing, the water table in the cell was raised to 5 m BGS to saturate the NAPL smear zone. In the cosolvent flushing test, approximately 40,000 L (10 PVs) of the cosolvent mixture (70% ethanol, 12% n-pentanol, and 28% water) was pumped through the cell over a 15 day period. The intent was to solubilize (not mobilize) the NAPL. An inter-well partitioning tracer test was carried out following the cosolvent flush. During tests, samples were collected from the 12 dedicated monitoring wells and the 3 extraction wells. Soil cores were also collected. The first tracer test indicated a NAPL content of 7% of the porosity. Results indicated that >90% of several target contaminants and >75% of the total NAPL mass was removed during treatment.

Cells 5, 6, and 8 at OU1

The intent of the demonstration project in cell 5 was to mobilize the NAPL within a middle phase microemulsion (MPME). The MPME was formulated using an aerosol OT/Tween series surfactant mixture with added $CaCl_2$. The intent in cell 6 was to solubilize the NAPL within surfactant micelles. An overall 3.6 wt% solution of Dowfax 8390 was employed to accomplish solubilization. The specific Dowfax surfactant was chosen largely because of its low sorption potential in the subsurface, resulting from the presence of dual negative charges on each monomer. Ten PVs of surfactant were used, followed by 5 PVs of water. This system is less efficient than the mobilizing system used in Cell 5. The intent with the demonstration in cell 8 was to solubilize NAPL constituents within a SPME system (Brij 97 with n-pentanol as a cosurfactant). Initial laboratory studies were performed in batch, column, and 2-D tanks using the NAPL and sedimentary material from the site and other materials. Pre- and post-treatment partitioning tracer tests were performed. Samples were collected from the 12 multiport sampling wells and the three extraction wells.

OU2 Test 1

The surfactant demonstration test at OU2 on HAFB took place in a shallow sand and gravel aquifer. 3 injection and 3 extraction wells were installed with a monitoring well midway between the injectors and extractors, and a fourth injection well located 10 ft behind the line of 3 injection wells. The project was scheduled in 2 phases. In the first phase a pilot test was performed: (1) To demonstrate that hydraulic control of injected fluids was possible, (2) to quantify the volume of NAPL by conducting a partitioning tracer test between the injector and extractor wells, (3) to show that surfactant could be injected and extracted at the designed rates, (4) to test a steam stripper on the effluent stream, and (5) to identify potential problems and gain some experience with the sampling and chemical analysis methods. The goal of the second phase was to attempt to remove as much of the DNAPL with a micellar flood as possible. A partitioning tracer test was carried out subsequent to the surfactant flood.

The following results were obtained during the first phase of the project. The partitioning tracer test indicated that approximately 800 gallons of NAPL is present in the test section of the aquifer. During the solubilization test, with an 8% surfactant injection over 0.6 days (0.6 pore volumes), the TCE concentration in the central monitoring well rose from approximately 600 mg/L prior to the injection to 40,000 mg/L after. The maximum DNAPL solubilization capacity was calculated to be 61,000 mg/L for this surfactant injection concentration. Problems with surfactant foaming in the steam stripper were encountered.

REFERENCES

1. Winsor, P.A., "Hydrotropy, solubilization, and related emulsification processes, Part I," *Trans. Faraday Soc.*, 54:376–399, 1948.
2. Shiau, B.J., Sabatini, D.A., and Harwell, J.H., *Ground Water* 32:561–569, 1994.
3. Jafvert, C.T., "Sediment and saturated-soil-associated reactions involving an anionic surfactant (dodecylsulfate). 2. Partitioning of PAH compounds among phases," Environ. Sci. Technol. 25:1039–1045, 1991.
4. Yalkowski, S.H., "Solubility of Organic Solutes in Mixed Aqueous Solvent," Final Report to the R.S. Kerr Research Lab., U.S. EPA, contract CR811852-01-0, 1985.
5. Fountain, J.C., Waddell-Sheets, C., Lagowski, A., Taylor, C., Frazier, D., and Byrne, M., "Chapter 13: Enhanced removal of dense nonaqueous-phase liquids using surfactants," in *Surfactant-Enhanced Subsurface Remediation: Emerging Technologies*, ACS Symposium Series 594, Dave A. Sabatini, Robert C. Knox, and Jeffrey H. Harwell, eds., 1995.
6. Fountain, J.C. and Hodge, D., "Project summary: Extraction of organic pollutants using enhanced surfactant flushing: Initial field test (part 1). NY State Center for Hazardous Waste Management, February, 1992.
7. Fountain, J.C., "Project summary: Extraction of organic pollutants using enhanced sur-factant flushing, part II," NY State Center for Hazardous Waste Management, November 1993.
8. Fountain, J.C. and Waddell-Sheets, C., "A pilot field test of surfactant enhanced aquifer remediation: Corpus Christi, Texas," Extended Abstract from ACS symposium in Atlanta Georgia, September 27–29, 1993.
9. Fountain, J.C., "A pilot scale test of surfactant enhanced pump and treat," in Proceedings of Air and Waste Management Association's 86th Annual Meeting, Denver, June 13–18, 1993.
10. Ross, A., Boulanger, C., and Tremblay, C., "In situ remediation of hydrocarbon contamination using an injection-extraction process," Remediation *Management*, March/April, pp 42–45, 1996.
11. Abdul, A.S. and Gibson, T.L., "Laboratory studies of surfactant-enhanced washing of polychlorinated biphenyls from sandy materials," *Environ. Sci. and Technol*, 25:565–670, 1991.
12. Ang, C.C. and Abdul, A.S., "Aqueous surfactant washing of residual oil contamination from sandy soil," *Ground Water Monitoring Review*, 11:121–127, 1991.
13. Ang, C.C. and Abdul, A.S., "A laboratory study of the biodegradation of an alcohol ethoxylate surfactant by native soil microbes," *J. of Hydrology*, 138:191–209, 1991.
14. Abdul, A.S., Gibson, T.L., Ang, C.C., Smith, J.C., and Sobczynski, R.E., "*In situ* surfactant washing of polychlorinated biphenyls and oils from a contaminated site," *Ground Water*, 30:219–231, 1992.
15. Abdul, A.S. and Ang, C.C., "*In situ* surfactant washing of polychlorinated biphenyls and oils from a contaminated field site: Phase II pilot study," *Ground Water*, 32: 727–734, 1994.
16. Ang, C.C. and Abdul, A.S., "Evaluation of an ultrafiltration method for surfactant recovery and reuse during *in situ* washing of contaminated sites: Laboratory and field studies," *Ground Water Monitoring and Remediation*, 1994.
17. Martel, R., Gélinas, P., Desnoyers, Jacques E., and Masson Anne, "Phase diagrams to optimize surfactant solutions for oil and DNAPL recovery in aquifers," *Ground Water*, 31:789–800, 1993.
18. Martel, R., Gélinas, P., and Laurent, S., "*In situ* recovery of DNAPL in sand aquifers: clean-up test using surfactants at Thouin Sand Quarry," presented at the 5th Annual Symposium on Ground-water and Soil Remediation, Toronto, Ontario Canada, Oct. 2–6, 1995.
19. Martel, R. and Gélinas, P., "Surfactant solutions developed for NAPL recovery in contaminated aquifers," *Ground Water*, 34:143–154, 1996.
20. Martel, R. and Gélinas, P., "Residual diesel measurement in sand columns after surfactant/alcohol washing," *Ground Water*, 34:162–167, 1996.
21. Knox, R.C., Sabatini, D.A., Harwell, J.H., West, C.C., Blaha, F., Griffin, C., Wallick, D., and Quencer, L., "Traverse City field test," presented at Workshop on In Situ Surfactant Use, Kansas City, MO, sponsored by the R.S. Kerr Environmental Research Laboratory, Ada OK, held September 20, 1995.
22. Sabatini, D.A., Knox, R.C., Harwell, J.H., Soerens, T.S., Chen, L., Brown, R.E., and West C.C., "Design of a surfactant remediation field demonstration based on laboratory and modeling studies," in review, *Ground Water*, submitted September 20, 1996.
23. Knox, R.C., Sabatini, D.A., Harwell, J.H., Brown, R.E., West, C.C., Blaha, F., and Griffin, S., "Surfactant remediation field demonstration using a vertical circulation well," in review, *Ground Water*, submitted September 20, 1996.

24. Sabatini, D.A., Knox, R.C., and Harwell, J.H. eds., *Surfactant Enhanced Subsurface Remediation: Emerging Technologies*, ACS Symposium Series, number 594, American Chemical Society, Washington, D.C., 312 pages, 1995.

25. Gotlieb, I., Bozzelli, J.W., and Gotlieb, E., "Soil and Water Decontamination by Extraction with Surfactants," *Seper. Sci. and Technol.*, 28:793–804, 1993.

26. Di Cesare, D. and Smith, J.A., "Effects of Surfactants on the Desorption Rate of Nonionic Organic Compounds from Soil to Water," *Reviews of Environmental Contamination and Toxicology*, 134:1–29, 1994.

27. Deitsch, J.J. and Smith, J.A., "Surfactant Enhanced Remediation of Ground Water at Picatinny Arsenal, New Jersey," in Morganwalp, D.W. and Aronson, D.A., eds., U.S. *Geological Survey Toxics Substances Hydrology Program-Proceedings of the Technical Meeting. Colorado Springs, Colorado-* September 20–24, 1993, U.S. Geological Survey Water Resources Investigations Report 94-4015, 1994.

28. Deitsch, J.J. and Smith, J.A., "Effect of Triton X-100 on the Rate of Trichloroethene Desorption from Soil to Water," *Environmental Science and Technology*, 29:1069–1080, 1995.

29. Annable, M.D., Rao, P.S.C., Hatfield, K., Graham, W.D., and Wood, A.L., "Use of Partit-ioning Tracers for Measuring Residual Napl Distribution in a Contaminated Aquifer: Preliminary Results from a Field-Scale Test," proceedings, 2nd Tracer Workshop, U. of Texas, Austin, TX.

30. Pope, G.A., Jin, M., Dwarakanath, V., Rouse, B., and Sepehrnoori, K., "Partitioning Tracer Tests to Characterize Organic Contaminants," proceedings, 2nd Tracer Workshop, U. of Texas, Austin, TX.

31. Augestijin, D.C.M. and Rao, P.S.C., "Enhanced Removal of Organic Contaminants by Solvent Flushing," ACS Symposium Series, submitted, 1995.

32. Rao, P.S.C., Lee, L.S., and Wood, A.L., "Solubility, sorption, and transport of hydrophobic organic chemicals in Complex Mixtures," U.S. Environmental Protection Agency, EPA/600-M-91-009, March, 1991.

33. Wood, A., Lynn, Bouchard, D., Brusseau, M., and Rao, P.S.C., "Cosolvent effects on sorption and mobility of Organic Contaminants in Soils," *Chemosphere*, 21:575–587, 1990.

34. Monthly Progress Reports available from the Advanced Applied Technology Demonstration Facility for Environmental Technology, at DOD/AATDF, Rice University, Energy & Environmental Systems Institute, M.S. 316, P.O. Box 1892 (or 6100 S. Main St.), Houston, Texas 77251 (or 77005).

10

PILOT-SCALE TREATMENT OF TNT-SPIKED GROUNDWATER BY HYBRID POPLAR TREES

Phillip L. Thompson[1]*, Liz A. Ramer[2], Pu Yong[2], and Jerald L. Schnoor[2]

[1]Department of Civil and Environmental Engineering
Room 524 ENGR
Seattle University
Seattle, Washington 98122
[2]Department of Civil and Environmental Engineering
116 ERF
University of Iowa
Iowa City, Iowa 52242

ABSTRACT

Groundwater contaminated with *TNT (2,4,6-trinitrotoluene)* is common at ammunition waste sites around the world. *Phytoremediation* or the use of plants to remediate environmental contamination is a promising means of addressing this problem. This paper describes a pilot-scale, greenhouse experiment that examined the irrigation of hybrid poplar trees (*Populus deltoides × nigra*, DN34) with TNT-spiked groundwater. *TNT* removal and metabolite production were monitored and showed that up to 90% of the *TNT* could be removed with only minor production of metabolites. Biomass production, water use and *TNT toxicity* were also evaluated under these more natural (non-laboratory) conditions. The results indicated that a poplar tree remediation system may be a feasible solution for low-levels of groundwater contamination.

1. INTRODUCTION

Large volumes of liquid wastes containing residual *TNT (2,4,6-trinitrotoluene)* are generated at munitions production facilities which often creates a red or pink waste stream.[1,2] These waste streams were once discharged into lagoons,[3] and although this disposal method is no longer practiced, thousands of acres across the United States are still highly toxic.[4]

---

* To whom correspondence may be addressed

Emerging Technologies in Hazardous Waste Management 8, edited by Tedder and Pohland
Kluwer Academic/Plenum Publishers, New York, 2000.

TNT has an estimated log K_{ow} (octanol-water partition coefficient) of 1.9,[5] and Pennington *et al.*[6] tested soils from 13 army ammunition plants finding an average soil partition coefficient (K_d) of 4.0 L/kg. Hence, *TNT* can be moderately mobile in the environment and can pose a significant threat to groundwater. Since one-half of the U.S. population drinks groundwater,[7] there is considerable interest and regulatory pressure to reduce health risks associated with such contamination. *TNT* is toxic to a variety of higher-order organisms[8] and can cause cytopenia (a disorder of the blood-forming tissues) resulting in reduced production of leukocytes, erythrocytes, and reticulocytes in mammals.[9]

Phytoremediation or the use of plants to remediate wastes may be relatively inexpensive when compared to more traditional clean-up methods such as excavation and incineration. A number of studies have evaluated the uptake and fate of *TNT* within plants,[10-22] clearly showing plants are capable of absorption and transformation. Our early studies[21] of *TNT* toxicity to 8-inch cuttings in a temperature-controlled laboratory indicated that concentrations greater than 5 mg/L were detrimental to hybrid poplar trees (*Populus deltoides × nigra*, DN34). The objective of this research was to evaluate the performance of large poplar trees exposed to *TNT* by monitoring the rates of transpiration and removal of *TNT* as well as develop an understanding of water use by these larger trees. Another important aspect of this study was the real-time measurement of environmental conditions (light intensity, temperature, humidity), because it reinforces that the conditions were not typical of the constant temperature and generally low light intensity found in most laboratory studies. This project was approached as if it were an actual groundwater treatment system. For example, large cuttings (greater than 1 meter in length) were allowed to grow for three months and then were exposed to 5 mg/L *TNT* for 4 weeks and 25 mg/L for the final week of the experiment. In addition to monitoring TNT concentrations, this study quantified 4-amino-2, 6-dinitrotoluene (4ADNT) and 2-amino-4, 6-dinitrotoluene (2ADNT) which have been identified as the primary transformation products of both microorganisms[8] and poplar trees.[22]

2. MATERIALS AND METHODS

2.1. Cutting Selection

A total of fifteen, 1.2 meter, poplar branches (*Populus deltoides × nigra*, DN34) were cut from dormant trees in Amana, Iowa. These branches were planted 23 cm deep in sand-filled pots, and 100 percent showed bud-break after a two-week period. Eleven of the trees were successfully maintained for the duration of the experiment, and six were chosen for data analysis, because they were of similar size and were transpiring at similar rates at the time dosing with *TNT* began.

2.2. Instrumentation

The temperature and humidity within the greenhouse was measured with a Perception II weather station (Davis Instruments Corp.). Temperature calibration was verified with an analog thermometer, and a psychrometer was used to calibrate the humidity sensor. Solar radiation in the photosynthetically active region (PAR) was measured at the canopy crown with an LI-190SA quantum sensor (LI-COR). Incident radiation was recorded on an Envirolink ENV-50 data logger (Sensormetrics). Gravimetric data was acquired with a Pennsylvania Instruments analytical balance with a range of 0.0 to 11.5 ± 0.001 kg.

2.3. Experimental Set-up

The pots (Duraco Products) used in this experiment were 27-L in volume. Approximately 7.7 kg of 2.5-cm gravel lined the bottom of each pot to permit adequate drainage. The remainder of each pot was filled with approximately 136 kg of washed, concrete (well-sorted) sand giving a field capacity of about 4-L for each pot.

A total of 14 pots were used for this experiment, and each vessel was supported on one-foot high risers. Drained water was collected in 26-L storage containers (Sterilite). Evaporation from each drainage collector was accounted for by measuring the water in evaporation controls. Three pots without trees were used to determine the evaporation from the surface of the pots. Transpiration was determined with the equation: In—Out—E1 + E2 where E1 is the evaporation from the surface of the pot and E2 is the evaporation from the drainage collector. The pots were organized in a randomized, 2 × 7 grid that is illustrated in Fig. 1.

2.4. Watering and Sampling

Trees were watered with groundwater from the 460-m deep Jordan aquifer. The water was pumped and chlorinated by the University of Iowa water plant. Chlorine was reduced with a granular activated carbon filter. The groundwater was determined to have a consistent quality and chlorine removal was verified by using N,N-diethyl-*p*-phenylene-diamine (DPD) with titration by ferrous ammonium sulfate (FAS).[23] The groundwater was supplemented with reagent grade (Fisher Scientific) macronutrients (nitrogen, phosphorus, magnesium, potassium) and micronutrients according to Epstein.[24] Both nutrient-enhanced groundwater and nutrient-enhanced groundwater-with-*TNT* were prepared 50 liters at a time in plastic carboys (Nalgene). The 5 mg/L *TNT* solution was prepared by adding 2 liters of 125 mg/L *TNT* (Chem Service) to 48 liters of nutrient water.

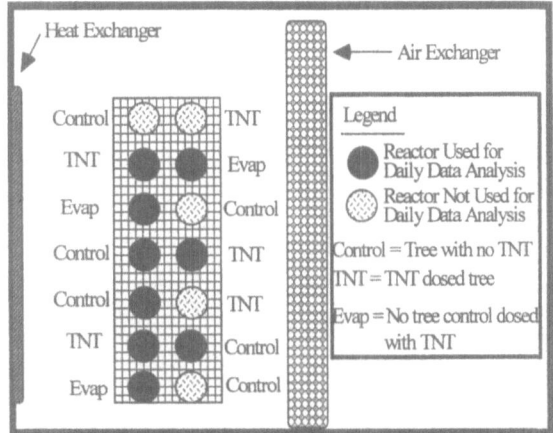

Figure 1. This schematic (plan view) illustrates the organization of the trees used in the greenhouse experiment. The trees that were used for daily data comparison had comparable transpiration rates before dosing with TNT had begun.

The 25 mg/L dose of *TNT* required the use of a supersaturated, 625 mg/L *TNT* solution. The supersaturated solution was prepared by adding 1.25 g of *TNT* to two liters of deionized water and heating at 80 °C overnight. Prior to *TNT* addition, all trees were watered daily with four liters of nutrient enhanced groundwater. After three months of growth, watering with 5 mg/L *TNT* began (twice daily, four liters each time). After five weeks, this concentration was increased to 25 mg/L.

Samples of the nutrient solution, raw tap water and *TNT* solution were taken for the morning and afternoon watering periods. Samples of pot drainage were taken immediately after watering thus minimizing photodecomposition.

2.5. Quantification

The measurement of *TNT* and its metabolites was performed with a Gilson 308 high performance liquid chromatograph (HPLC) equipped with a Spectra 100 UV detector set at 230 nm. A guard column protected a 25 cm Hypersil CPS LC-CN (Supelco) separation column, and the mobile phase consisted of a water-methanol-tetrahydrofuran (60.5:25:14.5, v/v/v) mixture fed at a flowrate of 1.0 ml-min^{-1}. The injection loop volume was 100 µl. *TNT* metabolite standards (1,3,5-trinitrobenzene (TNB), 2,4-dinitrotoluene, 4-amino-2,6-dinitrotoluene (4-ADNT), 2-amino-4,6-dinitrotoluene (2-ADNT) and 2,4-diamino-6-nitrotoluene (2,4-DANT) were supplied by Accustandard, and the method detection limit for TNT, 2-ADNT and 4ADNT was 5 ppb. Statistical significance was determined by using the two-tailed Student's t-test assuming unequal variances.

3. RESULTS AND DISCUSSION

3.1. TNT Removal

The results of the feed solution and drainage *TNT* concentrations for the pilot-scale project are presented in Table 1. The feed solution target concentration of 5 mg/L was determined to be 4.84 ± 0.7 mg/L while the 25 mg/L concentration was 24.5 ± 3.8 mg/L. The controls (pots without trees) indicated a slight but insignificant ($p < 0.05$) removal of *TNT* whereas the trees were able to achieve up to 90 percent *TNT*

Table 1. The removal of TNT by hybrid poplar trees exceeded that of controls by a factor of four. Two metabolites of TNT (2-ADNT and 4-ADNT) were detected in both systems. (ND = not detected, values are averages ± standard deviation, n = number of sample days)

Controls (no trees)	5 mg/L Dose (n = 28)	25 mg/L Dose (n = 7)
[TNT] Irrigated, mg/L	4.7 ± 1.3	24.4 ± 3.8
[TNT] drainage, mg/L	3.9 ± 1.0	18.9 ± 2.0
[4ADNT] Drainage, mg/L	0.2 ± 0.9	0.1 ± 0.2
[2ADNT] Drainage, mg/L	ND	ND
% TNT Removal	17.0	22.5
Trees		
[TNT] Irrigated, mg/L	4.7 ± 1.3	24.4 ± 3.8
[TNT] drainage, mg/L	0.4 ± 0.3	4.3 ± 1.3
[4ADNT] Drainage, mg/L	0.2 ± 0.4	2.2 ± 1.0
[2ADNT] Drainage, mg/L	ND	1.1 ± 1.0
% TNT Removal	91.5	82.4

Table 2. Environmental conditions for the experimental period. Photon flux is an average of the normally distributed photoperiod. (Note: ( ) = 1 standard deviation)

Mean Temperature °C	Mean % Humidity	Mean Photon Flux μmoles/m^2-s
25.6 (5.9)	40.2 (19.9)	465 (143)

removal. A sorption isotherm for *TNT* to the sand in the pots was also performed, and the results (not shown) indicated that sorption was not a significant removal mechanism.

At the time of this research, our methods could not detect non-radiolabeled *TNT* or its transformation products within plant tissues. Based on our experience with *[U-^{14}C]TNT* uptake by this particular poplar hybrid,[22] it is likely that most (up to 75%) of the applied *TNT* remained in root tissues with smaller fractions being translocated to the leaves. This greenhouse study had two to three times the light intensity of laboratory conditions, so it is possible that the behavior of *TNT* translocation and transformation could be different. Thus, it may prove beneficial to undertake the costly proposition of conducting an experiment of similar scale using *[U-^{14}C]TNT*, so that a more thorough understanding of *TNT* fate could be established.

The results clearly indicated that poplars and their associated microflora could remove TNT from solution relative to unplanted controls, hence it is conceivable that a groundwater treatment system using poplars planted in sand could be implemented. Recirculation of the drained water would probably result in the total removal of *TNT* and its metabolites. Harvested plant biomass would likely be incinerated or landfilled.

3.2. Biomass Production and Water Use

Table 2 summarizes the average environmental conditions for the five-week period of TNT dosing. Many of the trees had reached a height of 4 to 5 meters feet or more and were transpiring up to 9 liters of water per day at the time of harvest. Average biomass production and water use data are presented in Table 3. Although this data is specific to these experimental conditions, the above-ground biomass data provides an estimate for the magnitude of material that might be harvested for fuel or pulp production if such a treatment system were established.

The biomass production of controls and dosed trees was very similar over the entire experimental period as was average daily water use. However, these averages do not reflect any toxic effects that may have been experienced towards the end of the experiment as a closer look at the daily transpiration of the trees indicates.

Table 3. Average biomass production and water use for trees watered with TNT and controls. (Note: ( ) = 1 standard deviation)

	Total Dry Mass, kg	Above Ground Dry Mass, kg	Daily Water Use per Tree, kg
Controls	0.75 (0.22)	0.68 (0.19)	4.15 (2.04)
Dosed Trees	0.84 (0.32)	0.63 (0.22)	3.31 (1.29)

3.3. Decreased Transpiration

Transpiration can vary dramatically in a group of trees,[25] thus transpiration was monitored for one week prior to dosing with 5 mg/L *TNT* so that trees transpiring at similar rates could be established and this variation could be minimized. After monitoring the set of 11 trees, six were chosen as controls and five were selected as the trees that would be irrigated with *TNT*.

It was further determined that three of the controls and three of the dosed trees were transpiring at equal rates and could be compared on a daily basis. Figures 2 and 3 show that these two groups of trees continued to have comparable transpiration rates during the first two weeks of dosing, but towards the end of the third week, the transpiration of dosed plants began to decrease. Figure 4 illustrates that the difference between the controls and the dosed trees became more marked during the fourth week, and when the dose was increased to 25 mg/L the difference in transpiration rates approached two fold indicating a pronounced decrease in transpiration. Leaf chlorosis and leaf-drop further evidenced the onset of toxicity in the dosed plants. The decrease in transpiration for the controls on days 42 and 43 was probably related to a decrease in light intensity for those days as indicated by light measurement data (Fig. 5). Figure 6 summarizes the weekly transpiration data and indicates that 5 mg/L *TNT* concentrations may be toxic when the exposure period is greater than 3 weeks.

Biomass data (leaf, stem and root masses) was also collected and the transpiration values on the final day of the experiment were normalized to above-ground and total biomass respectively. Figure 7 illustrates the results of normalizing the transpiration data to biomass data and indicates that the differences between controls and dosed trees are significant ($p < 0.1$).

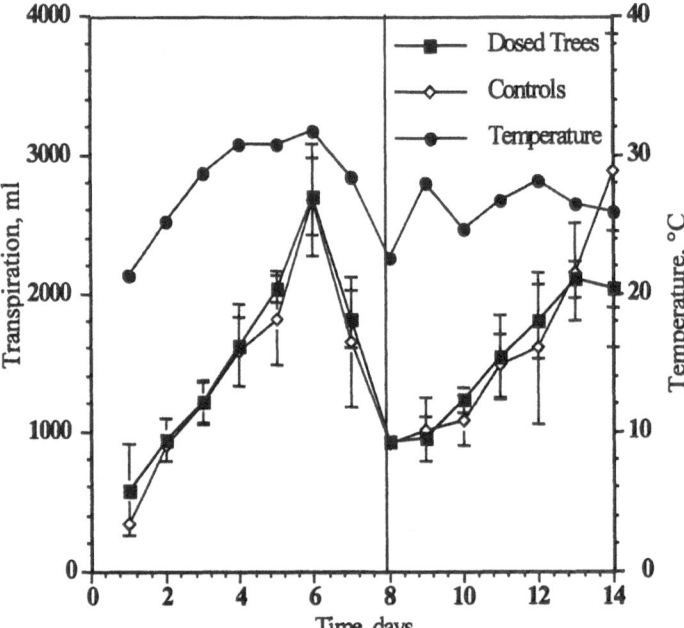

Figure 2. The daily transpiration for trees under greenhouse conditions for trees dosed with 5 mg/L TNT compared to controls (no TNT). The vertical line on day 8 indicates the start of dosing. (n = 3 for each group and error bars represent ±1 standard deviation.)

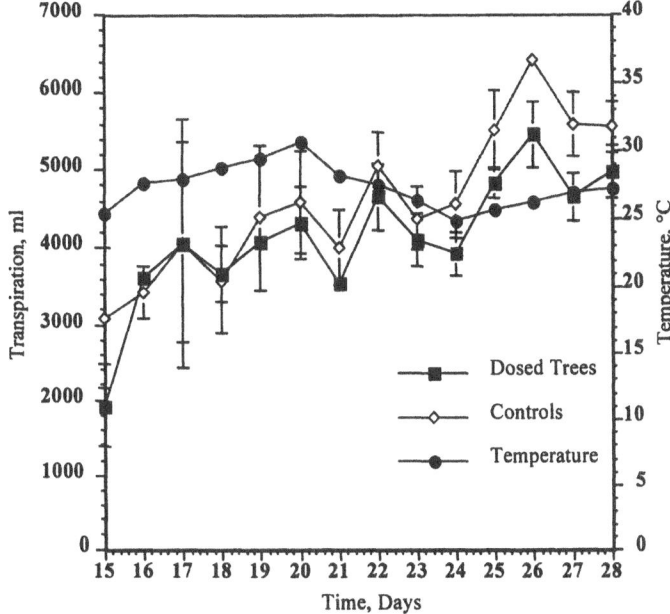

Figure 3. Second and third weeks of dosing with 5 mg/L TNT. Indications of toxicity begin to appear towards the end of the third week of exposure to 5 mg/L TNT. (n = 3 for each group and error bars represent ±1 standard deviation.)

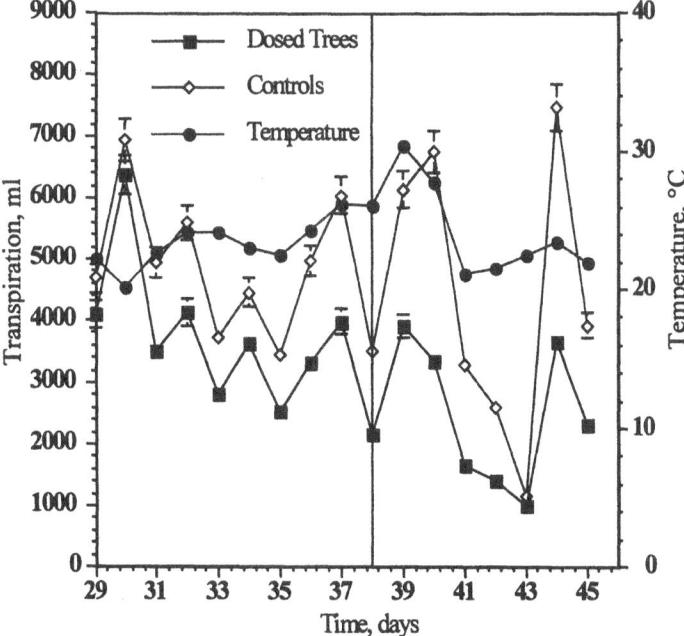

Figure 4. Transpiration of trees dosed with 5 mg/L TNT continues to drop during fourth week of exposure. The vertical line indicates the beginning of the 25 mg/L TNT dosing which significantly decreased transpiration relative to the controls. (n = 3 for each group and error bars represent ±1 standard deviation.)

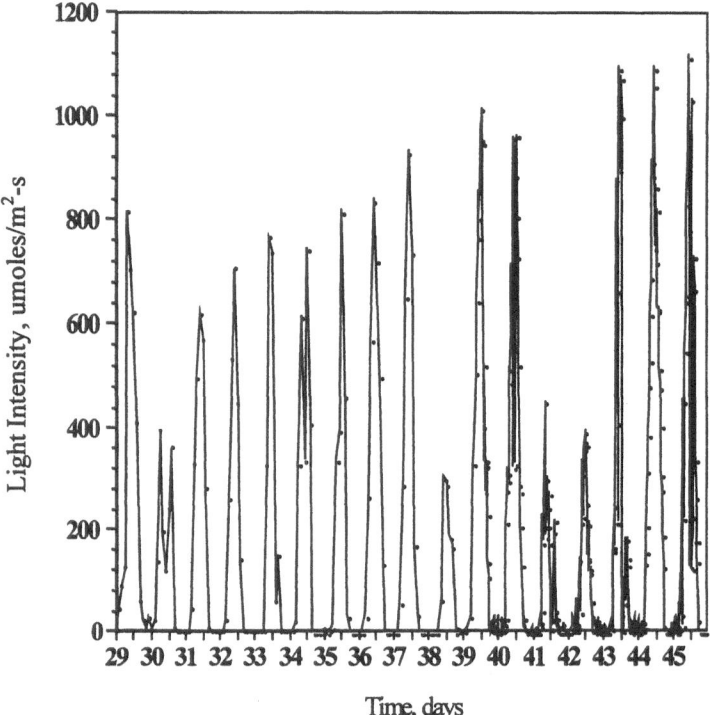

Figure 5. Daily light intensity in the photosynthetically active region (PAR) and the subsequent quantification of low levels of light explained the low transpiration rates of the controls on days 42 and 43 of the greenhouse experiment. Sampling rates were initially once every two hours but were adjusted to a final rate of twice an hour.

4. CONCLUSIONS

The results from a pilot-scale greenhouse project indicate that the irrigation of hybrid poplar trees with *TNT* contaminated groundwater may be a feasible remediation system. The trees achieve up to 90% removal of *TNT* which is concomitant with the uptake of volumes of water in the range of 4 to 9 liters per day. Although water use and biomass production between controls and trees irrigated with *TNT* were on the whole insignificant, and differences in transpiration appeared to increase with time indicating that 5 mg/L *TNT* concentrations may be chronically toxic. However, the small toxic effects at 5 mg/L *TNT* indicated that such a treatment scenario may be applicable to sites where groundwater *TNT* concentrations are in the ppb range. Concentrations of 25 mg/L TNT were found to clearly inhibit transpiration of the poplar hybrid relative to controls. Future research needs to address the fate of *TNT* in plants growing under natural conditions (e.g. full sunlight, fluctuating temperature, etc.), ideally with mass balances of *[U-¹⁴C]TNT*.

ACKNOWLEDGMENTS

The authors thank the United States Corps of Engineers for funding this project. We also thank Dr. Ronald Spanggord of SRI International for his assistance with the analytical chemistry.

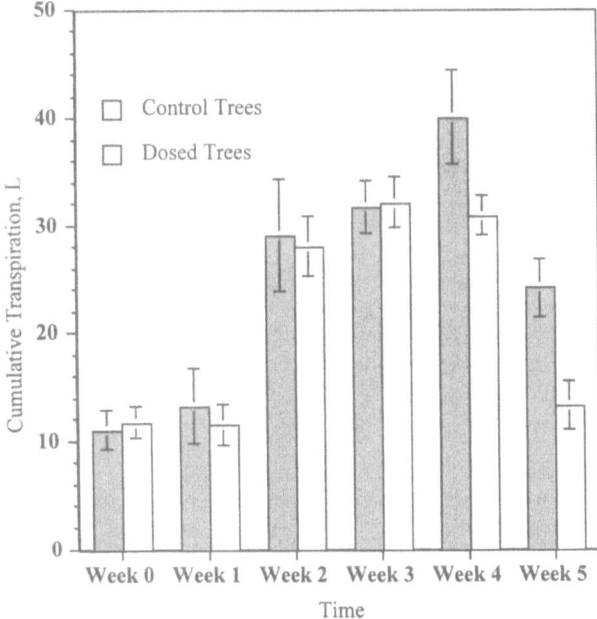

Figure 6. Large poplar cuttings planted in sand were grown under greenhouse conditions. After the first week of monitoring the transpiration of the trees (Week 0), dosing with 5 mg/L TNT began and was increased to 25 mg/L at the beginning of Week 5. The cumulative transpiration of the dosed trees did not become significantly ($p < 0.05$) less than that of the controls until Week 5. ($n = 3$ for each group and error bars represent ± 1 standard deviation.)

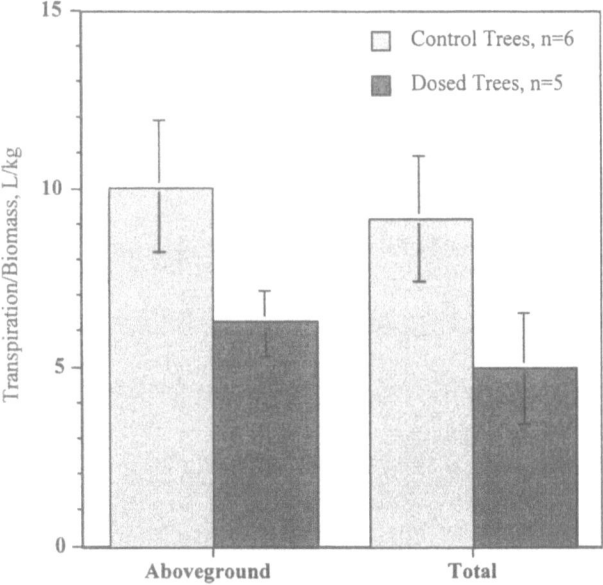

Figure 7. Large poplar cuttings planted in sand were grown under greenhouse conditions. The normalization of transpiration to aboveground and total biomass reflected a significant difference between controls and trees dosed with 25 mg/L TNT. ($n = 3$ for each group and error bars represent ± 1 standard deviation.)

REFERENCES

1. Urbanski, T., Chemistry and Technolgy of Explosives. MacMillan. New York, 1964.
2. Won, W.D., Heckly, R.J., Glover, D.J., and Hoffsommer, J.C., Appl. Microbio. 1974, 5, 513.
3. Higson, F.H., Adv. Appl. Microbio. 1992, 27, 1.
4. Funk, S.B., Roberts, D.J., Crawford, D.L., and Crawford, R.L., Appl. Environ. Microbio. 1993, 7, 2171.
5. Haderlein, S.B., Weissmahr, K.W., and Schwarzenbach, R.P., Environ. Sci. Technol. 1996, 30, 612.
6. Pennington, J.C. and Patrick, R.P., Journ. Environ. Qual. 1990, 19, 559.
7. Browner, C.M., Statement Before the U.S. Senate Committee on Environment and Public Works. March 5th, 1998.
8. Kaplan, D.L., Biotransformation Pathways of Hazardous Energetic Organo-Nitro Compounds. Portfolio Publishing Co, Houston, 1990.
9. Harris, J.W. and Kellermeyer, R.W., The red cell: production, metabolism, destruction, normal and abnormal, Harvard University Press: Cambridge, 1970.
10. Schott, C.D. and Worthley, E.G., "The Toxicity of TNT and Related Wastes to an Aquatic Flowering Plant, 'Lemna Perpusilla' Torr," Final Report. Edgewood Arsenal, Aberdeen Proving Ground, MD, report EB-TR-74016, 1974.
11. Leggett, D.C. and Palazzo, A.J., Journ. Environ. Qual. 1986, 15, 49.
12. Harvey, S.D., et al. Journ. Chrom. 1990, 518, 361.
13. Hughes, J.B., et al. Environ. Sci. Technol. 1997, 31, 266.
14. Bhadra, R.D., et al. Environ. Sci. Technol. 1999, 33, 446.
15. Goerge, E., et al. Environ. Sci. Pol. Res. 1994, 1, 229.
16. Medina, V.F. and McCutcheon, S.C., Bioremed. 1996, 7, 31.
17. Peterson, M.M., et al. Environ. Pollut. 1998, 99, 53.
18. Peterson, M.M., Environ. Pollut. 1996, 93, 57.
19. Scheidemann, P.A., et al. J. Plant Physiol. 1998, 152, 242.
20. Vanderford, M., Shanks, J.V., and Hughes, J.B., Biotechnol. Letters, 1997, 19: 277.
21. Thompson, P.L., Ramer L.A., Guffey, A.P., and Schnoor, J.L., Environ. Tox. Chem. 1997, 17, 902.
22. Thompson, P.L., Ramer, L.A., and Schnoor, J.L., Environ. Sci. Technol, 1998, 32, 975.
23. Greenberg, A.E., Clesceri, L.S., and Eaton, A.D., Standard Methods for the Examination of Water and Wastewater. American Public Health Association, Washington D.C., 1992.
24. Epstein, E., Mineral nutrition of plants: Principles and perspectives; John Wiley and Sons, Inc., New York, 1972.
25. Dye, P.J. and Olbrich, B.W., Plant, Cell and Environment, 1993, 16, 45.

11

IN SITU GROUNDWATER REMEDIATION USING TREATMENT WALLS

Radisav D. Vidic and Frederick G. Pohland

Department of Civil and Environmental Engineering
University of Pittsburgh
Pittsburgh, Pennsylvania 15261

ABSTRACT

Development of treatment wall technology for the clean up of contaminated ground-water resources has expanded in the past few years. The main perceived advantage of this technology over *ex situ* and other *in situ* ground-water remediation approaches is reduced operation and maintenance costs. Since the first commercial application of zero-valent iron using a funnel-and-gate system for the removal of chlorinated hydrocarbons in February, 1995, several field- and pilot-scale studies are evaluating the feasibility of this technology for treatment of both organic and inorganic contaminants.

Although, considerable design details have already been developed through field- and pilot-scale applications of this technology, some critical issues (e.g., establishing tested and proven design procedures, improving construction technologies, documenting long-term performance, and evaluating synergy with other ground-water remediation technologies) still remain to be resolved. Currently planned field-scale tests and many ongoing laboratory studies are designed to address these issues and facilitate wider implementation of this technology.

1. TECHNOLOGY DESCRIPTION

Treatment walls involve construction of permanent, semi-permanent, or replaceable units across the flow path of a contaminant plume. As the contaminated groundwater moves passively through the treatment wall, the contaminants are removed by physical, chemical and/or biological processes, including precipitation, sorption, oxidation/reduction, fixation, or degradation. These mechanically simple barriers may contain metal-based catalysts, chelating agents, nutrients and oxygen, or other agents that are placed either in the path of the plumes to prevent further migration or immediately downgradient of the contaminant source to prevent plume formation (Fig. 1). The reactions that take place in such systems depend on a number of parameters such as pH, oxidation/reduction potential, concentrations, and kinetics. Therefore, successful application

Emerging Technologies in Hazardous Waste Management 8, edited by Tedder and Pohland
Kluwer Academic/Plenum Publishers, New York, 2000.

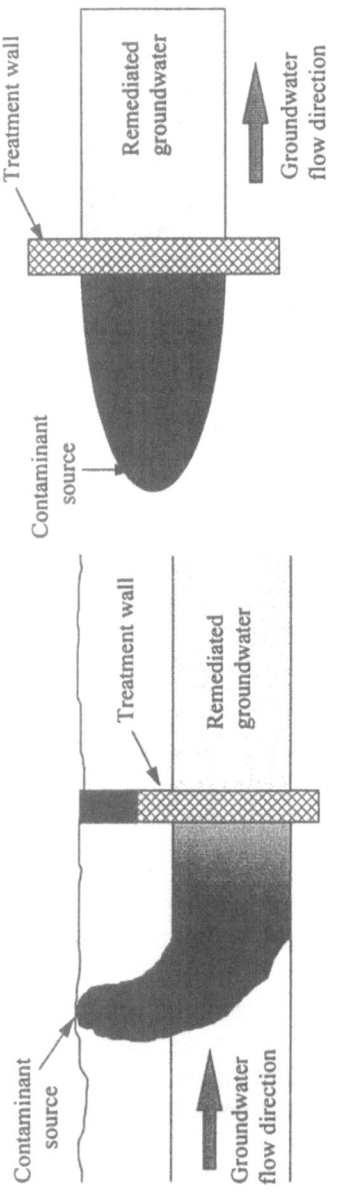

Figure 1. Schematic of a Simple Treatment Wall System.

of this technology requires a sufficient characterization of contaminants, ground-water flow regime and subsurface geology.

Permeable reactive walls potentially have several advantages over conventional pump-and-treat methods for ground-water remediation. Reactive walls can degrade or immobilize contaminants *in situ* without any need to bring them up to the surface. They also do not require continuous input of energy, because a natural gradient of ground-water flow is used to carry contaminants through the reaction zone. Only periodic replacement or rejuvenation of the reaction medium might be required after its capacity is exhausted or it is clogged by precipitants and/or microorganisms. Furthermore, technical and regulatory problems related to ultimate discharge requirements of effluents from pump-and-treat systems are avoided with this technology.

The key issues associated with the application of treatment walls are discussed below.

1.1. Site Characterization

Site characterization is the first step in assessing the potential applicability of treatment wall technology for ground-water remediation, and involves hydrological, geological, and geochemical description of the site and contaminant properties and distribution.

Hydrogeologic modeling and monitoring of the site define the basic dimensions of the contaminant plume, direction of the plume movement, and most appropriate location for the treatment wall. Site geologic characterization includes lithology, stratigraphy, grain size distribution and structural relationships, and should be documented in a set of geologic cross sections for the wall location. Hydrologic characterization of the site should include aquifer/aquitard boundaries, hydraulic conductivity, and ground-water gradient and flow direction.

Spatial distribution of the contaminant as well as its properties (solubility, vapor pressure, specific density, partitioning, etc.) and chemical relationship to site geology should be determined using literature information and analytical testing of soil and ground-water samples collected during site investigations.

Many of the above-mentioned parameters are difficult to determine with certainty which results in considerable variation in the level of contaminant mass flux. Therefore, treatment wall design must account for this inherent variability by incorporating features or safety factors capable of compensating for uncertainty.

1.2. Treatment Wall Design

Major issues associate with the design of a treatment wall include the selection of the reactive media (chemical makeup, particle size distribution, proportion and composition of admixtures, etc.), residence time in the reaction zone, and the reaction zone size for appropriate life span, as well as addressing issues like the effect of the reaction zone medium on ground-water quality and the ultimate fate or disposition of a treatment wall.

Selection of the reactive media is based on the type (i.e., organic vs. inorganic) and concentration of ground-water contaminants to be treated, ground-water flow velocity and water quality parameters, and the available reaction mechanisms for the removal of contaminants (i.e., sorption, precipitation, and degradation). Tables 1 and 2 in Section 3 provide useful information for the initial selection and effectiveness of various reactive media for different contaminants.

Typically, treatment wall system design is based on the results of treatability studies that can incorporate both batch reaction tests and laboratory- or field-scale column experiments. Batch tests are intended to obtain initial measures of media reactivity (i.e., degradation half life, sorption kinetics and capacity, etc.) that form the basis of the reactor design. Alternatively, literature information can be utilized to asses the initial information about media reactivity (e.g., Johnson, *et al.*, 1996 provide a comprehensive review of the kinetic data obtained for zero-valent iron degradation of halogenated hydrocarbons). Column tests are typically conducted by packing a column with the reactive medium and passing the contaminated groundwater through the column until steady-state performance of the reactor is obtained. Flow velocities are adjusted to simulate ground-water velocity and reactor residence time. In addition, information about geochemical reactions between the contaminated groundwater and the reactive medium as well as the impact of the treatment wall on ground-water quality can be assessed from these studies. Information about several batch and column studies can be found later in the text.

Life span of sorption and precipitation barriers is limited by the ultimate capacity of the medium to facilitate appropriate removal reactions. Once the ultimate capacity of the medium is exhausted, contaminant breakthrough will occur. In addition, contaminant release or resolubilization may occur after the plume or reactive medium is expanded. In the case of sorption and precipitation barriers for treatment of radioactive contaminants (i.e., Sr, U), an important issue is the possibility of exceeding the limits for Class A low-level nuclear waste as a result of excessive accumulation of these materials on the surface of the reactive medium. This might mean that the wall would have to be replaced at that time, regardless of the fact that the ultimate capacity of the medium might not be exhausted, because low-level nuclear waste above Class A must be solidified/stabilized. Alternatively, it might be possible to rejuvenate the media by *in situ* leaching methods. These issues are generally not of concern for the treatment walls designed for contaminant degradation.

1.3. Installation and Construction

Several methods have been conceived for the installation of permeable treatment walls (MSE, 1996). Most experience with installation of these walls pertains to relatively shallow emplacements (less than 10 m) using standard geotechnical design and construction approaches, although a few technologies for deeper installations have been identified.

In the simplest case, a trench of the appropriate width can be excavated to intercept the contaminated strata and backfilled with reactive material (Fig. 2). This method would normally be limited to shallow depths in stable geologic materials. More often, steps like shoring of the trench and use of an appropriate slurry or steel sheet piling are required for excavation to greater depths. Unlike conventional construction approaches for ground-water cutoff walls that utilize a soil-bentonite slurry (or cement or cement-bentonite slurry), installation of permeable treatment walls requires use of biodegradable polymers instead of bentonite or cement to avoid the problems of plugging the wall with residual slurry material.

Frequent criticism that the life expectancy of the reactive media in a treatment wall may degrade with time has been addressed by developing a construction approach

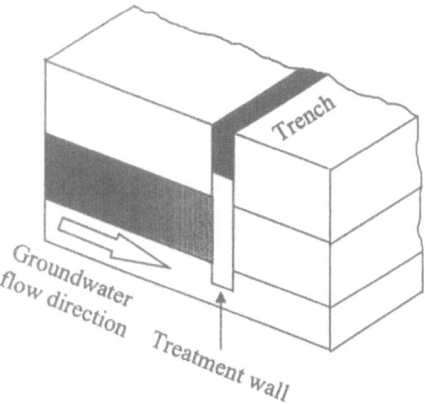

Figure 2. Treatment Wall in a Trench.

whereby the reactive media is placed in the subsurface in removable cassettes (MSE, 1996). A temporary sheet pile box or a large diameter caisson is installed into the subsurface and the screen panels are placed on the up- and downgradient side, while impermeable panels are placed on the lateral sides. Steel rail guides for the cassettes are installed within this interior compartment and the temporary sheet piles or caisson are removed. The cassette is a steel frame box (2.5-m long, 1.5-m wide and 0.5-m thick) with two opposing screened sides and two impermeable sides which is filled with the reactive media and lowered into the cavity. By allowing replacement of cassettes with depleted reactive media, the full-scale remediation system operation life can be extended nearly indefinitely.

Specialized trenching methods require the use of trenching machines that have been developed for installing underground utilities and constructing french drains and interceptor trenches. The most widely available utility trenching machines have depth capability of less than 7 m, while some specialized machines used for interceptor well construction can excavate up to 8–10 m. These machines incorporate a mechanism to temporarily shore the trench behind the cutter in a more or less continuous operation until the drain pipe and backfill are placed. Excavation rates on the order of 0.3 m of trench per minute to a depth of 7 m achieved with these specialized machines may lower the cost of treatment wall installation.

Soil mixing processes that are commercially used in solidification and stabilization of soils and sludges rely on soil augers to drill into the soil and inject and mix reagents. Commercially available equipment can penetrate soils up to 12 m with 2.5 to 3.5-m diameter augers, or up to 45 m in soils with a 1-m diameter auger, and has been used to form soil-cement ground-water cutoff walls by augering in an overlapping, offset pattern. The particular advantage of this method compared to traditional excavation approaches is that there is no need for handling of the excavated material as possible hazardous waste. Other specialized technologies like jet grouting, mandrel-based emplacement, and vibrating beam technology also have the potential to be adopted for continuous treatment wall installation.

Creation of treatment zones in place of treatment walls, that are confined within strict boundaries, can be accomplished with injection wells (Fig. 3) or by hydraulic fracturing. Well systems typically involve injection of fluids or fluid/particulate mixtures

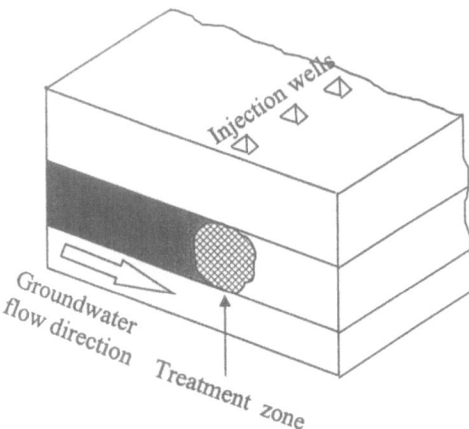

Figure 3. Injected Treatment Zone.

for distribution into a treatment zone within the target area of the aquifer. Potential advantages of this approach are that there is no need to construct a trench and possible aquifer access at greater depths. However, there is a question of reliability of injection for creating homogeneous treatment zones. Horizontal hydraulic fracturing is capable of creating propped fractures generally less than 2.5-cm thick and 7 to 12 m in diameter that can be filled with reactive material. However, this technology is typically used at shallower depths (3 to 12 m), and there is no current record of a field application to treatment zones. On the other hand, depending on the soil type, vertical hydrofracturing (MSE, 1996) can create fractures up to 20-cm wide, which may be suitable for treatment wall applications.

The funnel-and-gate system for *in situ* treatment of contaminated plumes consists of low hydraulic conductivity (e.g., 1×10^{-6} cm/s) cutoff walls with gaps that contain *in situ* reaction zones (Fig. 4). Cutoff walls (the funnel) modify flow patterns so that ground-water primarily flows through high conductivity gaps (the gates). The type of cutoff walls most likely to be used in the current practice are slurry walls, sheet piles, or soil admixtures applied by soil mixing or jet grouting. Starr and Cherry (1994) provide a comprehensive modeling study of various alternative funnel-and-gate systems and guidance for optimizing the design of such systems.

1.4. Monitoring Requirements

Although it is desirable to preserve the utility of the property at which a ground-water remediation project is being conducted, which is one of the main potential advantages of permeable treatment walls installed below ground level, careful performance monitoring is required during the operation of both pilot- and full-scale systems. Parameters requiring monitoring to assess performance include:

- contaminant concentration and distribution;
- presence of possible by-products and reaction intermediates;
- ground-water velocity and pressure levels;
- permeability assessment of the reactive barrier;

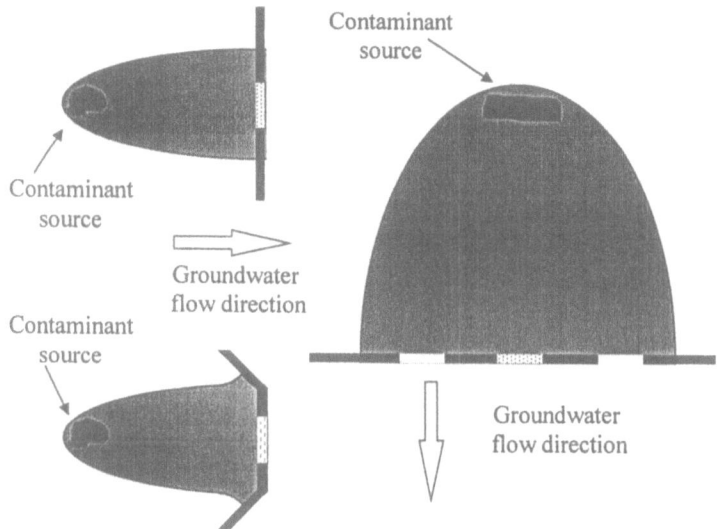

Figure 4. Funnel-and-Gate System.

• ground-water quality parameters (e.g., pH, redox potential, alkalinity); and
• dissolved gas (e.g., oxygen, hydrogen, carbon dioxide) concentrations.

Monitoring wells would have to be installed on both sides (upgradient and downgradient) of the treatment zone in order to obtain information about the long-term performance of the technology. In addition, several monitoring methods (i.e., tracer, nuclear, and electromagnetic) are being developed to evaluate the existence, size, and location of breaches in a subsurface barrier as well as to monitor the barrier longevity (MSE, 1995).

2. TECHNOLOGY PERFORMANCE

Results of numerous studies on the use of reactive permeable barriers to treat both organic and inorganic contaminants in ground-water are presented below. Additional information can be found in EPA, 1998.

2.1. Full-Scale Zero-Valent Iron Walls

The main contaminants resulting from the semiconductor manufacturing process at a site in Sunnyvale, CA (Yamane *et al.*, 1995; Shoemaker *et al.*, 1996) are trichloroethylene (TCE) (0.05–0.2 mg/L), cis-1,2-dichloroethylene (*cis*-1,2-DCE) (0.45–1.0 mg/L), vinyl chloride (VC) (0.1–0.5 mg/L), and CFC-113 (0.02–0.06 mg/L). The treatment wall system consists of a 75-m slurry wall on either side of 12-m long, 1.2-m wide, and 6-m deep permeable wall that is charged with 100% granular iron (i.e., no sand mixture) to a total depth of about 3.5 m. Total construction and reactive media costs were $720,000. Ground-water velocity through the wall is approximately 30 cm/day, which provides a residence time of 4 days. Ground-water samples collected from performance monitoring wells within the wall showed no volatile organic compounds (VOCs) above the detection limit of 0.5 µg/L.

Treatment wall installed at Moffet Federal Airfield, CA consists of two wing walls, 6.5 m long, constructed of interlocking sheet piles which channel the groundwater into a 3.2-m wide, 3.2-m thick, and 8.2-m deep reaction cell. There is 0.6 m of concrete beneath the wall to prevent ground-water infiltration from below. Groundwater flowing into the wall encounters 0.6 m of pea gravel, 2 m of 100% granular iron and another 0.6 m of pea gravel. Pea gravel was used to ensure adequate distribution of groundwater, because there are sand channels in the area. Preliminary estimates place the cost of the wall at approximately $300,000. Ground-water quality sampling was performed in June, 1996 and indicated that the influent TCE concentration of 850–1,180 µg/L was degraded to 3–320 µg/L within 0.3 m into the iron wall, and to 11–36 µg/L within 1.3 m into the wall. Samples taken from the downgradient pea gravel section of the wall indicated TCE concentrations of 16–45 µg/L, which were assumed to be residual TCE contamination within the pea gravel remaining from the wall installation. Although there was no VC in the influent groundwater, 3 µg/L was detected 0.3 m into the iron wall, while no VC was detected within 0.6 m of the iron wall.

A treatment wall consisting of an iron-sand mixture was installed at Canadian Forces Base, Borden, Ontario (EPA, 1995; Shoemaker et al., 1996) approximately 5 m downgradient from the source of contaminant plume. The plume was about 2 m wide and 1 m thick, with maximum concentrations along the axis of about 250 mg/L of TCE and 43 mg/L of PCE. The wall was constructed using sealable joint sheet piling to a depth of about 10 m, with a total width of 1.5 m and length of 5.5 m (transverse to the flow). The reactive media backfill consisted of 22% by weight iron grindings collected from a local machine shop, and 78% concrete sand to ensure sufficient porosity of the wall. Ground-water flow velocity was about 10 cm/day, which provided a residence time in the wall of approximately 15 days. During the four-year monitoring period, the wall consistently removed 90% of TCE and 86% of PCE in the incoming water. The principal product detected at a monitoring point located downgradient of the wall was 1,2-DCE, with a maximum concentration of 0.2 mg/L. No vinyl chloride was detected in the treated groundwater. No visible precipitate formed on the surface of iron grindings, although losses of 185 mg/L and 82 mg/L of calcium and bicarbonate, respectively, were measured across the wall. Subsequent laboratory studies aimed at simulating the performance of the treatment wall revealed that a higher percentage of iron grindings in the wall might have resulted in complete removal of TCE and PCE across the wall.

Zero-valent iron has been evaluated for the removal of chromate through a treatment wall constructed at the U.S. Coast Guard Air Station, Elizabeth City, NC as a series of large-diameter augered holes in a staggered 3-row array (Powell et al. 1995; Puls et al., 1995). Twenty one 20-cm columns were installed to a depth of 6.7 m throughout a 5.5 m² area. The mixed waste contaminant plume was between 4.2 and 6 m below the surface, and the water table ranged from 1.5 to 2 m below the surface. Columns were charged with a mixture of 50% iron filings (two types), 25% clean coarse sand and 25% aquifer material, to a depth between 3 and 8 m below the surface. Untreated groundwater contained 1–3 mg/L of Cr(VI) and a total concentration of TCE and DCE of 6.5 mg/L that were reduced to below 0.01 mg/L and 1.5 mg/L in the treated water, respectively. Most of the organic reduction was due to a decrease in TCE concentration of about 75%, while 1,2-DCE showed almost no change. Dissolved iron in the groundwater increased from

0.05 mg/L to 1–20 mg/L, while dissolved oxygen decreased from 0.6 mg/L to below 0.1 mg/L with a slight increase in alkalinity.

A funnel-and-gate system installed at the Lowry Air Force Base, CO consists of a 3.5-m wide and 1.6-m thick gate with 5-m long cutoff walls, oriented at a 45° upgradient angle, that were installed on each side of the reaction zone (Duster *et al.*, 1996). Ground-water level is approximately 2.6 m below the surface, and the top of the weathered claystone bedrock confining layer is approximately 5.7 m below the surface. The predominant contaminant in the incoming groundwater was TCE (source located 100 m upgradient from the wall location) at a concentration of 850 µg/L, while the average total concentration of all other chlorinated hydrocarbons (VC, 1,1-DCE, *trans*- and *cis*-1,2-DCE, 1,1,1-trichloroethane (TCA), 1,1-dichloroethane (DCA), 1,2-DCA, and PCE) was about 300 µg/L. Results during a six-month monitoring period indicate complete degradation of the chlorinated hydrocarbons within the first 0.3-m of the wall (9 hours residence time). Only *cis*-1,2-DCE was present at significant concentrations (10 µg/L) after 9 hours of residence time inside the wall, having the highest calculated half-life among all contaminants of 2.2 hours. By approximately 18 hours of residence time (0.6 m into the wall) all analytes had degraded to their respective analytical quantitation limits. Redox potential in groundwater dropped to about −500 mV while pH increased from 6.5 to 10.0 across the wall. Total alkalinity decreased rapidly with cross-sectional distance into the wall from about 550 mg/L to below 50 mg/L, while Fe^{+2} and Fe^{+3} were not detected at concentrations above 1 mg/L due to precipitation of iron salts. Cost analysis estimated a decrease in treatment costs from \$440,000 per kilogram of contaminants removed from groundwater during the first year of operation down to \$50,000 per kilogram of contaminants removed for a 10 year treatment period.

2.2. Pilot-Scale Studies with Zero-Valent Iron

A pilot-scale study conducted at the SGL Printed Circuit Site, Wayne, NJ utilized commercial granular iron for the removal of 4–12 mg/L of PCE, 1 mg/L of TCE, and 0.15 mg/L of 1,2-DCE from contaminated groundwater (Vogan *et al.*, 1995; Shoemaker *et al.*, 1996). Design of the pilot-scale reactor was based on half-lives, determined using 100% granular iron, of 0.4–0.6 hr for PCE, 0.5–0.7 hr for TCE, 1.5–3.7 hr for DCE, and 1.2–0.9 for vinyl chloride. In order to achieve a New Jersey standard of 0.01 mg/L for 1,2-DCE, residence time in the fixed-bed reactor was set at 24 hr. The reactor was built in a 2.4-m diameter fiberglass tank filled with granular iron to a depth of 1.7 m. Concentration profiles for PCE, TCE, and *cis*-1,2-DCE after 30 and 60 days of operation showed nondetectable levels in the effluent. In fact, disappearance of all three constituents occurred about midway through the reactor. Based on laboratory testing, calcium carbonate, siderite, and possibly iron hydroxide could have precipitated in the reactor.

Three different reactive media: (1) fine grade iron filings (40-mesh) from Master-Builder, Inc., (2) stock iron filings (−8 + 50-mesh) from Peerless Metal Powders & Abrasive, Inc., and (3) palladized iron filings obtained by chemically plating palladium (at 0.05% of iron) on a 40-mesh iron filings from Fisher Scientific, were tested in parallel treatment trains for the removal of TCE from a contaminated groundwater at Portsmouth Gaseous Diffusion Plant in Piketon, OH (Liang *et al.*, 1996). Each treatment train consisted of three 55-gal drums packed with a total of 488 L of iron filings except in the case of palladized iron where only one drum packed with 45 L was used due to much faster TCE degradation rates on palladized iron observed in the laboratory studies.

This pilot plant was constructed on February 29, 1996 and operations began on March 5, 1996. During the initial start-up phase of this study, problems with maintaining gravity flow through the system (1.1–1.5 L/min), encountered due to the build-up of the gas (believed to be hydrogen produced by reductive dissociation of water by iron filings) in the drums and manifolds, were resolved by installing pressure release valves on the drums. During first week of operation, influent TCE concentrations on the order of 170 μg/L were reduced in all treatment trains to below detection limit (2 μg/L). After 51 days of operation, effluent TCE concentrations were still below detection limit in the treatment train (1), while they were 3 and 12 μg/L in the treatment trains (2) and (3), respectively. TCE half-lives increased from 1.4, 14, and 19 minutes after seven days of operation to half-lives of 4.1, 16, and 35 minutes after 51 days of operation for the treatment rains (3), (1), and (2), respectively. The more rapid deterioration of palladized iron filings may be attributed to the substantially higher number of pore volumes that have passed through this treatment train (about 3345 pore volumes compared to 331 pore volumes for the treatment trains (1) and (2)), as well as possible reactions of sulfides with palladium.

Contaminated groundwater at Hill AFB, UT was treated above ground in a 1.4-m long and 0.3-m diameter fiberglass canister filled with 100% Master Builder iron, "Blend B, GX-27" (Strongsville, OH) (Shoemaker et al., 1996). The reactor was operated in an upflow mode with a flow rate of 0.38 to 3.8 L/min. Influent pH was stable at about 7.5, while dissolved oxygen varied from 4 to 6 mg/L. Effluent pH rose above 9 while dissolved oxygen (DO) decreased to 1–2 mg/L. The majority of the influent TCE, at 2 mg/L, was degraded prior to reaching the first port, located about 22 cm into the canister. Concentrations of cis 1,2-DCE and VC were observed to increase within the canister, but were below 0.001 mg/L in the effluent, while ethane and ethene in the effluent accounted for 60% of the initial TCE mass in the influent. The experiment was terminated when the pressure drop across the canister increased from an initial value of 3.45 kPa to 6.89 kPa. X-ray diffraction tests indicated that precipitation of iron and calcium carbonate compounds caused clogging of the bed.

Pilot-scale tests using a contaminated groundwater from an electronics manufacturing facility located near Belfast, Northern Ireland, were performed in a 100-cm long acrylic column charged with 100% granular iron and equipped with several sampling ports along the column length (Thomas et al., 1995). The results of TCE analyses of the groundwater indicate that TCE is present at a maximum identified concentrations of up to 390 mg/L. Column tests at flow velocities of 109 and 54 cm/day were conducted until steady state contaminant profiles were established in each test. Using the flow velocity, the distance along the column was converted to time and the degradation rate constants were calculated for the organic compounds using a first-order kinetic model. Rapid declines in TCE concentration were observed for both tests and the detection limits were reached between 40 and 60 cm along the column. Calculated half-lives for TCE were 1.2 hours (flow velocity of 109 cm/day) and 3.7 hours (flow velocity of 54 cm/day). Concentrations of cis-1,2-DCE increased to between 10 and 22 mg/L due to dechlorination of DCE. Interpretation of cis-1,2-DCE profiles was difficult because of the presence of methyl ethyl ketone (MEK) in groundwater which eluted at the same time on the photo ionization detector as the target compound. Since MEK should be neither produced nor degraded in the presence of iron, the declines in the combined cis-1,2-DCE/MEK concentration was used to calculate cis-1,2-DCE half-lives of 12.5 and 23.9 hours for the flow rates of 109 and 54 cm/day, respectively. VC was produced due to TCE and cis-1,2-DCE degradation with a maximum concentration in the downstream portion of the column reaching 100–300 µg/L. No substantial increase in the pH over the influent levels was observed in these tests, most likely due to the high concentration of dissolved organics in the groundwater. Measured Eh declined from about 250 mV to about –200 mV and marked increases in iron were measured in the column effluent due to the corrosion of iron metal by water.

Four different iron millings or filings (70–99.9% iron in spherical particles, cylindrical pieces and polygons with densities ranging from 7.81–8.06 g/mL) from various iron fabrication processes were investigated for the removal of ground-water contamination at Moffet Federal Airfield, CA (PRC, 1996) using batch and column studies. The groundwater used in the batch studies contained approximately 22 mg/L of TCE, while the sample collected for column studies contained 1.2 mg/L TCE and 0.12 mg/L PCE. TCE removal efficiency in batch tests ranged from 64% to 100% for different iron types, while the column tests revealed a TCE half-life of 0.63 hr and a PCE half-life of 0.29 hr. In addition, half-lives of the reaction by-products 1,2-DCE and VC were estimated at 3.1 hr and 4.7 hr, respectively.

Four different treatment zones were installed below surface in the path of the uranium mill tailing drainage at UMTRA Site in Durango, CO. Two were in baffle style boxes about $1.8 \times 1 \times 1.2$ m and two were horizontal beds approximately 4.5×9 m with 0.3 m of steel wool. Bi-metallic, zero-valent iron steel wool, and zero-valent iron foam from Cercona are the media being tested at the site. Influent concentrations have varied from May through July in the range of 2.9–5.9 mg/l for U, 0.9 for Mo and 27–32 mg/l for NO_3^-. The first test started in May 1996 with Cercona iron foam in the baffle design as the first media to be tested. The earliest results showed a decrease in U concentration to 0.4 mg/L and NO_3^- to 20 mg/L. The results after 3 months of operation showed a decrease in U and nitrate to below detection limits while Mo decreased to 0.02 mg/L. Although not confirmed, it is suspected that biological reduction enhanced by the high hydrogen environment produced by the contact between iron and water has resulted in the removal of nitrate from the system.

2.3. Oxygen Releasing Compounds

The study in Borden, Ontario, Canada evaluated the ability of a proprietary solid peroxide formulation (referred to as an oxygen-releasing compound or ORC) to provide sufficient dissolved oxygen and enhance biodegradation of benzene and toluene (Bianchi-Mosquera *et al.*, 1994). Benzene (3,947 ± 284 µg/L) and toluene (3,819 ± 264 µg/L) were injected through 16 1.5-m sections of 25-cm O.D. PVC well screen jetted to approximately 1.5 m below ground surface. Background samples were collected from a location approximately 4 m upgradient from the source, while the downgradient concentrations were monitored along the lines of four monitoring points (0.6-cm stainless steel sampling points with a 2.5-cm screen) located every 0.5 m from the source. ORC in briquette form raised the concentration of DO in the groundwater to as much as 15 mg/L, while benzene and toluene concentrations decreased below detection limits. After the injection of benzene and toluene ceased, the DO levels rose to 45.6 mg/L and the oxygen production continued for at least 10 weeks. ORC in the pencil form also released oxygen into the groundwater, but the levels of DO were not as high as with ORC in the briquette form, and benzene and toluene concentrations remained above those measured when briquettes were used.

A total of 20 PVC wells was installed and loaded with a total of 342 oxygen releasing compound (ORC) socks to remediate a groundwater at a site in Belen, NM, where an unknown quantity of gasoline spill occurred for an unknown length of time (Koenigsberg *et al.*, 1995; Johnson and Methvin, 1996). The aquifer is shallow, unconfined and comprised mainly of well sorted sands, with a groundwater level at 1.5 m bellow ground surface and the average groundwater gradient of 0.0015. The range of interstitial velocities at the site was estimated at 3.0–3.3 cm/day. The average background concentration of DO and BTEX were approximately 1 and 2 mg/L, respectively. Less than two weeks after installation, dissolved oxygen mass increased an order of magnitude (maximum levels in excess of 18 mg/L) and remained constant for at least another month. There was a 78% decrease in the total BTEX mass in the immediate viscinity of the barrier and 58% decrease in the broad study area of 36 × 30 m. After three months of operation, approximately half of the oxygen placed in the system was exhausted and a concomitant decrease in the BTEX mass was observed. After 279 days of operation, 47% of the socks were replaced with fresh ones, since an increase in BTEX levels was observed. BTEX levels were once again noted to decrease from about 10 mg/L to below detection limits in proportion to the available oxygen.

2.4. *In Situ* Microbial Filters

The study conducted at Chico Municipal Airport, CA demonstrated the effectiveness of resting-state (no nutrients) *in situ* microbial filters approach for remediating groundwater contaminated with TCE (Duba *et al.*, 1996). The first field test of this technology, conducted at the Wilson Corners site at the Kennedy Space Center, FL, was terminated prematurely because only 1–2 ppm of the total contaminant load was biodegraded. This low biodegradation rate was attributed to insufficient oxygen in the groundwater. A second field trial was conducted at Chico Municipal Airport in Chico, CA where the groundwater was contaminated with $425 \pm 50 \,\mu g/L$ of TCE and the dissolved oxygen was 7.0 mg/L. About 5.4 kg (dry weight) of a pure strain methanotrophic bacteria, *Methylosinus trichosporium* OB3b, was suspended in groundwater and injected into the aquifer through a single well at a depth of 28 m and at a rate of 3.8 L/min. Approximately 50% of the injected bacteria attached to the soil, forming an *in situ*, fixed-bed, quasi-spherical bioreactor with an average radius of about 1.2 m. Contaminated groundwater was subsequently withdrawn through the biofilter by extracting groundwater through the injection well at 3.8 L/min for 30 hr and then at 2 L/min for the remaining 39 days of the field experiment. During the first 50 hr of ground-water withdrawal, 98% of TCE was biodegraded. TCE concentration in the extraction well then gradually increased as biofilter degradation capacity and/or longevity were exceeded in various parts of the biofilter.

In situ nitrate attenuation by heterotrophic denitrification using an alternative septic system design that utilizes a treatment wall charged with an organic carbon source (sawdust) is currently being evaluated at three sites in Ontario (Long Point, Killarney, and Borden), Canada (Robertson and Cherry, 1995). Two barrier configurations (horizontal layer positioned in the vadose zone below a conventional septic-system infiltration bed and a vertical wall intercepting a horizontally-flowing downgradient plume) were evaluated in four field trials. During one year of operation, both barrier configuration have been successful in substantial attenuation (60 to 100%) of input NO_3^- levels of up to 125 mg/L as N. Mass balance calculations and preliminary results suggest that conveniently sized barriers have the potential to last for decades without replenishment of the reactive material.

3. TECHNOLOGY APPLICABILITY

3.1. Treatment Walls for Organic Contaminants

The current status of treatment wall technology applications for inorganic contaminants is indicated in Table 1. Included is the nature of the reactive medium, contaminants, level of investigation, selected references, location, and some distinguishing features.

Control of organic contaminants in groundwater with treatment walls has had considerably more field and full-scale applications than the control of inorganic contaminants. Zero-valent iron has been applied in a commercial system for the control of chlorinated hydrocarbons (Shoemaker *et al.*, 1996), while a number field and pilot-scale studies with this reactive media showed very encouraging results for this technology. Among other reactive media investigated for the control of organic contaminants in groundwater, resting-state microorganisms (Duba *et al.*, 1996) have been successfully applied in the field, while palladized iron (Muftikian *et al.*, 1995; Grittini *et al.*, 1995) and organobentonites (Smith and Galan, 1995) showed a good potential in pilot- and laboratory-scale studies.

3.2. Treatment Walls for Inorganic Contaminants

The current status of treatment wall technology applications for inorganic contaminants is indicated in Table 2. Included is the nature of the reactive medium, contaminants, level of investigation, selected references, location, and some distinguishing features.

Reactive media that have been implemented in pilot and field-scale studies for the treatment of inorganic contaminants include ferric oxyhydroxide for the control of U and Mo (Morrison *et al.*, 1995), dithionite for the removal of Cr, V, Tc, and U (Williams *et al.*, 1994), zero-valent iron for the control of U (Dwyer *et al.*, 1996) and Cr (EPA, 1995), and sawdust as a carbon source for biological removal of nitrate (Robertson and Cherry, 1995). Among other reactive media that have been investigated in laboratory studies, zeolites and hydroxyapatite showed a good potential for field applications.

4. COST OF TREATMENT WALLS

4.1. Capital Costs

The cost of the impermeable sections of the treatment wall system can be obtained from experiences with slurry walls or sheet pile installations. If the reactive media is zero-valent iron, the cost of the media can be estimated based on the density of about $2.83 \, kg/m^3$ and a cost of approximately \$440–500/tonne. A recent review by DuPont (Shoemaker *et al.*, 1996) suggested that installation costs between \$2,500 and \$8,000 per L/min of treatment capacity can be used as a rule-of-thumb for estimating the capital cost of these systems. Since zero-valent iron treatment walls is patented technology, a site licensing fee, which has been typically 15% of the capital costs (materials and construction costs), may also be required.

Table 1. Status of Treatment Wall Technology for Organic Contaminants

Reactive Media	Contaminants	Study Type	Reference	Location/Comments
zero-valent iron	Halogenated hydrocarbons	commercial	Shoemaker et al., 1996	Sunnyvale, CA Moffet Federal Air Filed, CA Coffeyville, KS Elizabeth City, NC
		field	EPA, 1995 Puls et al., 1995 Duster et al., 1996	Borden, Ontario Elizabeth City, NC Lowry AFB, CO
		pilot	Shoemaker et al., 1996 Shoemaker et al., 1996	SGL Site, NJ Hill AFB, UT Portsmouth, OH
	Halogenated hydrocarbons, Nitro aromatic compounds, Atrazine, DBCP, PCP, PCBs	laboratory	Agrawal and Tratnyek, 1996 Hardy and Gillham, 1996 Orth and Gillham, 1996 Roberts et al., 1996 Schlimm and Heitz, 1996 Weber, 1996	Chuang and Larson, 1995 Focht and Gillham, 1995 Pulgarin et al., 1995 Ravary and Kochany, 1995 Siantar et al., 1995 Gillham and O'Hannesin, 1994
palladized iron	Halogenated hydrocarbons, PCBs	laboratory	Muftikian et al., 1995 Grittini et al., 1995	Reaction completed in minutes; no accumulation of by-products
iron(II) porphyrins	Polychlorinated ethylenes, benzenes	laboratory	Gantzer and Wackett, 1992	B_{12} and coenzyme F_{430} catalyze complete reduction; hematin only to VC
dithionite	Halogenated hydrocarbons	laboratory	Amonette et al., 1994 Williams et al., 1994	90% of CCl_4 reduced in one week and less than 10% converted to THM
resting-state microorganisms	TCE TCE	field laboratory	Duba et al., 1996 Taylor et al., 1993	Chico Municipal Airport, CA Kennedy Space Center, FL
oxygen-releasing compound	Benzene, toluene	field	Bianchi-Mosquera et al., 1994	Borden, Ontario, Canada
zeolite	BTEX, TCA, PCA	laboratory	Bowman et al., 1994a,b	surfactant-modified zeolites sorbed organics through partition mechanism
surfactant modified silicates	PCE, naphthalene	laboratory	Burris and Antworth, 1992	no reduction in permeability; 2 orders of magnitude increase in capacity
organobentonites	Benzene, CCl_4, TCE, naphthalene, 1,2-dichlorobenzene	laboratory	Smith and Jaffe, 1994a,b Smith and Galan, 1995	selection of quaternary ammonium group influences sorption mechanism; 4% addition into the liner increases retention of benzene

Table 2. Status of Treatment Wall Technology for Inorganic Contaminants

Reactive Media	Contaminants	Study Type	Reference	Location/Comments
zero-valent iron	Cr	commercial		Elizabeth City, NC
	Cr	field	EPA (1995)	Elizabeth City, NC
	U	field	Dwyer *et al.* (1996)	UMTRA site, Durango, CO
sawdust	nitrate	field	Robertson and Cherry (1995)	Long Point, Killarney, and Borden, Canada
peat	Ni, Cr, Cd	laboratory	Ho *et al.* (1995)	low pH enhances capacity;
	U, Mo	laboratory	Morrison and Spangler (1992)	equilibrium reached in less than 2 hr
bentonite	Cs	laboratory	Oscarson *et al.* (1994)	clay compaction decreases sorption capacity
ferric oxyhydroxide	U, Mo	laboratory	Morrison *et al.* (1993, 1995)	require neutral pH; effluent
	U, Mo	pilot		below 0.05 mg U/L Monticello, UT
zeolite	Pb, Cd	laboratory	Ouki (1993)	pretreatment of chabazite
	Sr	laboratory	Fuhrmann *et al.* (1995)	and clinoptilolite with NaCl
modified zeolites	As, Cr, Cd, Pb	laboratory	Haggerty and Bowman (1994) Bowman *et al.* (1994a,b)	or surfactants enhances capacity
chitosan beads	V, Cd, Hg	laboratory	Rorrer *et al.* (1993), Kawamura *et al.* (1993), Charrier (1996)	slow rate of uptake; capacity = 500 mg V/g
hydroxyapatite	Pb	laboratory	Ma *et al.* (1993, 1994)	complete removal in 30 min; pH = 5–6; other metals inhibit removal of Pb
dithionite	Cr, V, Tc, U	field	Amonette *et al.* (1994), Williams *et al.* (1994)	18–30 m treatment zone to react for 5–30 days at Hanford 100H Area, WA
lime or limestone	acid mine drainage	commercial	Kleinmann *et al.* (1983)	Various sites

Table 3 summarizes capital costs for treatment walls (mostly zero-valent iron as a reactive media) that are already built or for which costs have been estimated (augmented table of Vogan and Kwicinski, 1996).

4.2. Operation and Maintenance Costs

A principal advantage of the permeable treatment walls technology over other ground-water remediation approaches is the reduced operation and maintenance (O&M) costs. Other than ground-water monitoring, the major factor affecting operating and maintenance costs is the need for periodic removal of precipitates from the reactive media or periodic replacement or rejuvenation of the affected sections of the permeable wall. It is currently difficult to predict the magnitude of inorganic precipitate formation prior to site-specific trials. A recent review by DuPont (Shoemaker *et al.*, 1996) suggested O&M costs between $1.3 and $5.2 per 1,000 L of treated water can be used as a rule-of-thumb for estimating the O&M costs of these systems.

Table 3. Capital Cost Summary for Treatment Walls

Location	Dimensions	Contaminants	Cost ($)		
			Construct.	Media	Total
Sunnyvale, CA (full-scale) (built)	75-m slurry wall on either side of 12 m-long treatment section; 6 m deep; 3.5-m vertical and 1.2-m flowthrough thickness of iron	1–2 mg/L VC, cis-1,2-DCE, and TCE	550,000	170,000	720,000
Moffet Federal Air Field, CA (full-scale) (built)	6.5-m long interlocking sheet piles on either side of a 3.2-m wide, 3.2-m thick, and 8.2-m deep reaction cell.	1 mg/L TCE			300,000
Coffeyville, KS (pilot-scale) (built)	150-m slurry wall on either side of 6-m long gate; 9 m deep; 3.8-m vertical and 1-m flowthrough thickness of iron	100's μg/L TCE	350,000	50,000	400,000
Elizabeth City, NC (full-scale) (built)	45-m long, 5.5-m deep, and 0.6-m wide zero-valent iron wall	TCE, chromium	220,000	200,000	420,000
Lowry AFB, CO (pilot-scale) (built)	3.5-m wide, 1.6-m thick, and 2.9-m deep gate with 5-m long cutoff walls on each side of the reaction zone.	850 μg/L TCE, 220 μg/L cis-1,2-DCE	105,000	32,500	137,500
New Hampshire (full-scale) (estimate)	400-m long wall with several gates; 9 m deep	100's μg/L TCE, VC, cis-1,2-DCE	1,200,000	900,000	2,100,000
Michigan (full-scale) (estimate)	90-m long with 3 gates; 6 m deep	10–100 mg/L TCE	300,000	135,000	435,000
Canada (full-scale) (estimate)	45-m long with 2 gates; 4.5 m deep	50–100 μg/L TCE	130,000	52,500	182,500

At the Intersil site in Sunnyvale, CA, it was estimated that the total O&M costs associated with ground-water monitoring and replacement of the entire reactive media in a 12-m long, 3.6-m deep and 1.2-m wide treatment wall every 10 years could be about $2 million over a 30-year period (Fairweather, 1996).

5. REGULATORY/POLICY REQUIREMENTS AND ISSUES

5.1. Regulatory Considerations

At the time of publication, the U.S. EPA was in the process of developing regulatory guidelines for the use of passive treatment wall technology. No State's regulations specifically addressing this technology are known to exist. Implementation of a treatment wall at a hazardous waste site, as with any remedial measure, requires the approval of appropriate State and/or federal regulatory agencies. Potential considerations to be addressed as part of this approval process involve site investigation, design and monitoring issues, including those listed below:

- Sufficient characterization of site geology, hydrology, contaminant distribution, and vectors impacting human health and the environment to permit adequate design of the treatment wall;
- Ability of the proposed design to account for uncertainties inherent in subsurface investigations/treatments;
- Ability of the proposed design to capture and adequately remediate the vertical and horizontal extent of the ground-water plume;
- Monitoring to measure concentrations of by-products in groundwater potentially produced through treatment wall reactions;
- Monitoring to measure potential releases of gaseous by-products; and
- Monitoring to characterize precipitate formation and wall clogging that may limit the effectiveness of the treatment method.

Implementation of passive treatment wall technology does not involve removal of groundwater or air from the subsurface. Therefore, unlike other remedial technologies such as pump-and-treat and soil vapor extraction, it does not require permits for discharges of groundwater or air to the environment.

5.2. Health and Safety Issues

Health and safety issues involved in the use of treatment wall technology are mainly associated with installation of the wall and will vary according to the method of installation used (trenching, drilling, injection, etc.). Health and safety concerns associated with these installation methods will generally be the same as for any other application of the particular installation technique. Exposure of workers to hazardous substances during installation and operation of the wall may be lower than with other conventional treatment technologies due to lack of direct contact with contaminated materials.

As stated above, monitoring is required to ensure that any groundwater and air releases that may result from use of treatment wall technology do not impact offsite receptors.

Impacts from treatment wall installation and maintenance may be less than with other technologies due to the placement of all treatment materials underground, with minimal disturbance to surface activities.

6. FUTURE NEEDS

6.1. Design and Construction

At present, there is a general need to establish tested and proven design procedures and protocols for treatment wall technology. Protocols for site characterization have been fairly well developed so far, but the criteria for selecting the location of a treatment wall is still quite empirical. Therefore, there is a need to develop better predictive models that can assist in determining optimal location and sizing of the wall. These models should not only include ground-water hydrology and hydraulics and contaminant transport and fate in the subsurface, but also chemical reactions occurring inside the wall. It is particularly important to account for by-product generation, precipitate formation and clogging of the wall, and loss of media reactivity.

A major potential limitation of treatment wall technology is the potential for constructing the wall at depths greater than 10 m. Wall installation at depths of 10 to 30 m presents additional difficulty and escalates construction costs to the point that they may become limiting for the implementation of this technology. Installations at depths greater than 30 m are only theoretically possible with current technologies. Moreover, the existence of surface obstructions (e.g., buildings, roads) and underground utilities represents additional challenges to the placement of treatment walls in the subsurface.

6.2. Implementation Issues

A particular concern with treatment wall technology is the question of long-term performance under variable conditions that are commonly associated with contaminated groundwaters (e.g., seasonal variations in ground-water flow velocity and patterns, variations in the contaminant speciation and concentration). Loss of permeability over time as a result of particle invasion, chemical precipitation, or microbial activity, and possible gradual loss of media reactivity as the reactant is either depleted or coated by reaction by-products, need to be resolved before this technology can be applied with wider confidence. Currently planned field-scale tests and many ongoing laboratory studies are designed to address some of these issues.

Investigations are needed to provide more fundamental understanding of the reactions pathways and possible by-products that might be generated by the reactions of contaminants with the reactive media. This could enable engineering manipulations to force certain, more desirable reaction pathways. In addition, development of novel reactive media and optimization of the existing ones could possibly improve competitiveness of treatment walls with other *in situ* and *ex situ* ground-water remediation technologies.

The reduction of chlorinated solvents by zero-valent iron requires contact between the organic compound and iron surface, which limits treatment with Fe(0) to water-soluble chemicals. Treatment of many soil contaminants which have functional groups that are reducible, but are strongly sorbed to sediments and soil samples (e.g., PCBs, dioxin, DDT, toxaphene, mirex, lindane, hexachlorobenzene), may not be feasible with this technology. This also applies for many other reactive media that are being investigated for use in treatment walls. Subsequently, there is a need to investigate the possibility of utilizing treatment sequences, e.g., combining *in situ* soil washing technologies with treatment walls.

REFERENCES

Agrawal, A. and Tratnyek, P.G. (1996) "Reduction of Nitro Aromatic Compounds by Zero-Valent Iron Metal." *Environ. Sci. Technol.*, 30:1, 153–160.

Amonette, J.E., Szecsody, J.E., Schaef, H.T., Templeton, J.C., Gorby, Y.A., and Fruchter, J.S. (1994) "Abiotic Reduction of Aquifer Materials by Dithionite: A Promising In-Situ Remediation Technology." 33rd Hanford Symposium on Health and the Environment, November 7–11, Pasco, WA, Battelle Press, Columbus, OH.

Bianchi-Mosquera, G.C., Allen-King, R.M., and Mackay, D.M. (1994) "Enhanced Degradation of Dissolved Benzene and Toluene Using a Solid Oxygen-Releasing Compound." *Groundwater Monitoring and Remediation*, 14:1, 120–128.

Bowman, R.S., Haggerty, G.M., Huddleston, R.G., Neel, D., and Flynn, M.M. (1994) "Sorption of Nonpolar Organic Compounds, Inorganic Cations, and Inorganic Oxyanions by Surfactant-Modified Zeolites." Proceeding of the 207th ACS National Meeting, San Diego, CA, March 13–17, Chapter 5, 54–64.

Bowman, R.S., Flynn, M.M., Haggerty, G.M., Huddleston, R.G., and Neel, D. (1994) "Organo-Zeolites for Sorption of Nonpolar Organics, Inorganic Cations, and Inorganic Anions." Proceedings of 1993 CSCE-ASCE National Conference on Environmental Engineering, July 12–14, Montreal, Quebec, Canada, 1103–1109.

Burris, D.R. and Antworth, C.P. (1992) "In Situ Modification of an Aquifer Material by a Cationic Surfactant to Enhance Retardation of Organic Contaminants." *J. Contaminant Hydrology*, 10, 325–337.

Burris, D.R., Campbell, T.J., and Manoranjan, V.S. (1995) "Sorption of Trichloroethylene and Tetrachloroethylene in a Batch Reactive Metallic Iron-Water System." *Environ. Sci. Technol.*, 29:11, 2850–2855.

Charrier, M.J., Guibal, E., Roussy, J., Delanghe, B., and Le Cloirec, P. (1996) "Vanadium(IV) Sorption by Chitosan: Kinetics and Equilibrium." *Water Research*, 30:2, 465–475.

Chuang, F.W. and Larson, R.A. (1995) "Zero-Valent Iron-Promoted Dechlorination of Polychlorinated Biphenyls (PCBs)." Proceedings of 209th ACS National Meeting, Anaheim, CA, April 2–7, 771–774.

Duba, A.G., Jackson, K.J., Jovanovich, M.C., Knapp, R.B., and Taylor, R.T. (1996) "TCE Remediation Using In Situ Resting-State Bioaugmentation." *Environ. Sci. Technol.*, 30:6, 1982–1989.

Duster, D., Edwards, R., Faile, M., Gallant, W., Gibeau, E., Myller, B., Nevling, K., and O'Grady, B. (1996) "Preliminary Performance Results from a Zero Valence Metal Reactive Wall for the Passive Treatment of Chlorinated Organic Compounds in Groundwater." Presented at Tri-Service Environmental Technology Workshop, May 20–22, Hershey, PA.

Dwyer, B.P., Marozas, D.C., Cantrell, K., and Stewart, W. (1996) "Laboratory and Field Scale Demonstration of Reactive Barrier Systems." Proceedings of the 1996 Spectrum Conference in Seattle, WA, August 18–23.

Environmental Protection Agency (1995) In Situ Remediation Technology Status Report: Treatment Walls, United States Environmental Protection Agency, EPA542-K-94-004, Washington, DC.

Environmental Protection Agency (1998) Permeable Reactive Barrier Technologies for Contaminant Remediation, United States Environmental Protection Agency, EPA/600/R-98/125, Washington, DC.

Fairweather, V. (1996) "When Toxics Meet Metal." *Civil Engineering*, 66:5, 44–48.

Focht, R.M. and Gillham, R.W. (1995) "Dechlorination of 1,2,3-Trichloropropane by Zero-Valent Iron." Proceedings of 209th ACS National Meeting, Anaheim, CA, April 2–7, 741–744.

Fuhrmann, M., Aloysius, D., and Zhou, H. (1995) "Permeable, Subsurface Sorbent Barrier for ^{90}Sr: Laboratory Studies of Natural and Synthetic Materials." Proceedings of Waste Management '95, February, 26–March 2, 1995, Tuscon, AZ.

Gantzer, C.J. and Wackett (1992) "Reductive Dechlorination Catalyzed by Bacterial Transition-Metal Coenzymes." *Environ. Sci. Technol.*, 25:4, 715–722.

Gillham, R.W. and O'Hannesin, S.F. (1994) "Enhanced Degradation of Halogenated Aliphatics by Zero-Valent Iron." *Ground Water*, 32:6, 958–967.

Grittini, C., Malcomson, M., Fernando, Q., and Korte, N. (1995) "Rapid Dechlorination of Polychlorinated Biphenyls on the Surface of Pd/Fe Bimetallic System." *Environ. Sci. Technol.*, 29:1, 2898–2900.

Haggerty, G.M. and Bowman, R.S. (1994) "Sorption of Chromate and Other Inorganic Anions by Organo-Zeolite." *Environ. Sci. Technol.*, 28:3, 452–458.

Hardy, L.I. and Gillham, R.W. (1996) "Formation of Hydrocarbons from the Reduction of Aqueous CO_2 by Zero-Valent Iron." *Environ. Sci. Technol.*, 30:1, 57–65.

Ho, Y.S., Wase, D.A., and Forster, C.G. (1995) "Batch Nickel Removal from Aqueous Solution by Sphagnum Moss Peat." *Water Research*, 29:5, 1327–1332.

Johnson, T.L., Scherer, M.M., and Tratnyek, P.G. (1996) "Kinetics of Halogenated Organic Compound Degradation by Iron Metal." *Environ. Sci. Technol.*, 30:8, 2634–2641.

Kawamura, Y., Mitsuhashi, M., Tanibe, H., and Yoshida, H. (1993) "Adsorption of Metal Ions on Polyaminated Highly Porous Chitosan Chelating Resin." *Ind. Eng. Chem.*, 32, 386–391.

Kleinmann, R.L.P., Tiernan, T.O., Solch, J.G., and Harris, R.L. (1983) "A Low-Cost, Low-Maintenance Treatment System for Acid Mine Drainage Using Sphagnum Moss and Limestone." *National Sympo-*

sium on Surface Mining, Hydrology, Sedimentology and Reclamation, University of Kentucky, Lexington, KY.

Koenigsberg, S., Johnson, J., Odenkrantz, J., and Norris, R. (1995) "Enhanced Intrinsic Bioremediation of Hydrocarbons with Oxygen Release Compound (ORC)" Sixth West Coast Conference on Contaminated Soils and Groundwater, March 11–14, Newport Beach, CA.

Liang, L., West, O.R., Korte, N.E., Goodlaxson, J.D., Anderson, F.D., Welch, C.A., and Pelfry, M. (1996) "A Field-Scale Test of Trichloroethylene Dechlorination using Iron Filings." Interim Report on the X-749/X-120 Groundwater Treatment Facility submitted to Department of Energy, Piketon OH, May, 1996.

Ma, Q.Y., Traina, S.J., Logan, T.J., and Ryan, J.A. (1993) "In Situ Lead Immobilization by Apatite." *Environ. Sci. Technol.*, 27:9, 1803–1810.

Ma, Q.Y., Traina, S.J., Logan, T.J., and Ryan, J.A. (1994) "Effects of Aqueous Al, Cd, Cu, Fe(II), Ni, and Zn on Pb Immobilization by Apatite." *Environ. Sci. Technol.*, 28:7, 1219–1228.

Matheson, L.J. and Tratnyek, P.G. (1994) "Reductive Dehalogenation of Chlorinated Methanes by Iron Metal." *Environ. Sci. Technol.*, 28:12, 2045–2053.

Morrison, S.J. and Spangler, R.R. (1992) "Extraction of Uranium and Molybdenum from Aqueous Solution: A Survey of Industrial Materials for Use in Chemical Barriers for Uranium Mill Tailings Remediation." *Environ. Sci. Technol.*, 26:10, 1922–1931.

Morrison, S.J. and Spangler, R.R. (1993) "Chemical Barriers for Controlling Groundwater Contamination." *Environ. Progress*, 12:3, 175–181.

Morrison, S.J., Spangler, R.R., and Tripathi, V.S. (1995) "Adsorption of Uranium(VI) on Amorphous Ferric Oxyhydroxide at High Concentrations of Dissolved Carbon(IV) and Sulfur(VI)." *J. Contaminant Hydrology*, 17, 333–346.

Morrison, S.J., Tripathi, V.S., and Spangler, R.R. (1995) "Coupled Reaction/Transport Modeling of a Chemical Barrier for Controlling Uranium(VI) Contamination in Groundwater." *J. Contaminant Hydrology*, 17, 347–363.

MSE Technology Applications (1995) "Subsurface Barriers Monitoring and Verification Technologies" Report for U.S. Department of Energy (TTP #PE1-5-10-06).

MSE Technology Applications (1996) "Analysis of Technologies for the Emplacement and Performance Assessment of Subsurface Reactive Barriers for DNAPL Containment" Report for U.S. Department of Energy (TTP #PE1-6-PL-31) under Contract No. DE-AC22-88ID12735.

Muftikian, R., Fernando, Q., and Korte, N. (1995) "A Method for the Rapid Dechlorination of Low Molecular Weight Chlorinated Hydrocarbons in Water." *Water Research*, 29:10, 2434–2439.

Orth, W.S. and Gillham, R.W. (1996) "Dechlorination of Trichloroethene in Aqueous Solution Using Fe^0." *Environ. Sci. Technol.*, 30:1, 66–71.

Oscarson, D.W., Hume, H.B., and King, F. (1994) "Sorption of Cesium on Compacted Bentonite." *Clays and Clay Minerals*, 42:6, 731–736.

Ouki, S.K., Cheesman, C., and Perry, R. (1993) "Effects of Conditioning and Treatment of Chabazite and Clinoptilolite Prior to Lead and Cadmium Removal." *Environ. Sci. Technol.*, 27:6, 1108–1116.

PRC Environmental Management, Inc. (1996) Final Iron Curtain Bench-Scale Study Report, Department of the Navy Contract No. N62474-88-D-5086.

Pulgarin, C., Schwitzguebel, J.P., Peringer, P., Pajonk, G.M., Bandara, J., and Kiwi, J. (1995) "Abiotic Degradation of Atrazine on Zero-valent Iron Activated by Visible Light." Proceedings of 209th ACS National Meeting, Anaheim, CA, April 2–7, 767–770.

Puls, R.W., Powell, R.M., and Paul, C.J. (1995) "In Situ Remediation of Ground Water Contaminated with Chromate and Chlorinated Solvents Using Zero-Valent Iron: A Field Study." Proceedings of the 209th ACS National Meeting, Anaheim, CA, April 2–7, 788–791.

Ravary, C. and Lipczynska-Kochany, E. (1995) "Abiotic Aspects of Zero-Valent Iron Induced Degradation of Aqueous Pentachlorophenol." Proceedings of 209th ACS National Meeting, Anaheim, CA, April 2–7, 738–740.

Roberts, A.L., Totten, L.A., Arnold, W.A., Burris, D.R., and Campbell, T.J. (1996) "Reductive Elimination of Chlorinated Ethylenes by Zero-Valent Metals." *Environ. Sci. Technol.*, 30:8, 2654–2659.

Robertson, W.D. and Cherry, J.A. (1995) "In Situ Denitrification of Septic-System Nitrate Using Reactive Porous Media Barriers: Field Trials." *Ground Water*, 33:1, 99–111.

Rorrer, G.L., Hsien, T.Y., and Way, J.D. (1993) "Synthesis of Porous-Magnetic Chitosan Beads for Removal of Cadmium Ions from Waste Water." *Ind. Eng. Chem.*, 32, 2170–2178.

Schlimm, C. and Heitz, E. (1996) "Development of Wastewater Treatment Process: Reductive Dehalogenation of Chlorinated Hydrocarbons by Metals." *Environmental Progress*, 15:1, 38–47.

Siantar, D.P., Schrier, C.G., and Reinhard, M. (1995) "Transformation of the Pesticide 1,2-Dibromo-3-Chloropropane (DBCP) and Nitrate by Iron Powder and by $H_2/Pd/Al_2O_3$." Proceedings of 209th ACS National Meeting, Anaheim, CA, April 2–7, 745–748.

Shoemaker, S.H., Greiner, J.F., and Gillham, R.W. (1996), in *Assessment of Barrier Containment Technologies: A Comprehensive Treatment for Environmental Remediation Applications*, R.R. Rumer and J.K. Mitchell, Eds., Chapter 11: Permeable Reactive Barriers, report prepared for US DOE, US EPA, and DuPont Company.

Smith, J.A. and Jaffe, P.R. (1994) "Benzene Transport thorough Landfill Liners Containing Organophilic Bentonite." *J. Env. Engineering*, ASCE, 120:6, 1559–1577.

Smith, J.A. and Jaffe, P.R. (1994) "Adsorptive Selectivity of Organic-Cation-Modified Bentonite for Nonionic Organic Contaminants." *Water, Air and Soil Pollution*, 72, 205–211.

Smith, J.A. and Galan, A. (1995) "Sorption of Nonionic Organic Contaminants to Single and Dual Organic Cation Bentonites from Water." *Environ. Sci. Technol.*, 29:3, 685–692.

Starr, R.C. and Cherry, J.A. (1994) "In Situ Remediation of Contaminated Ground Water: The Funnel-and-Gate System." *Ground Water*, 32:3, 465–476.

Taylor, R.T., Hanna, M.L., Shah, N.N., Shonnard, D.R., Duba, A.G., Durham, W.B., Jackson, K.J., Knapp, R.B., Wijesinghe, A.M., Knezovich, J.P., and Jovanovich, M.C. (1993) "In situ Bioremediation of Trichloroethylene-Contaminated Water by a Resting-Cell Methanotrophic Microbial Filter." *J. Hydrological Sciences*, 38:4, 323–342.

Thomas, A.O., Drury, D.M., Norris, G., O'Hannesin, S.F., and Vogan, J.L. (1995) "The *In-Situ* Treatment of Trichloroethene-Contaminated Groundwater Using a Reactive Barrier—Results of Laboratory Feasibility Studies and Preliminary Design Considerations." *Contaminated Soil '95*, W.J. van den Brink, R. Bosman, and F. Arendt, Eds., Kluwer Academic Publishers, Netherlands, 1083–1091.

Vogan, J.L., Gillham, R.W., O'Hannesin, S.F., Matulewicz, W.H., and Rhodes, J.E. (1995) "Site Specific Degradation of VOCs in Groundwater Using Zero Valent Iron." Proceedings of 209th ACS National Meeting, Anaheim, CA, April 2–7, 800–804.

Vogan, J. and Kwicinski, L. (1996) "Iron Walls Treat Groundwater." *Pennsylvania's Environment: Business, Technology & The Environment*, 1:12, 10–11.

Weber, E.J. (1996) "Iron-Mediated Reductive Transformations: Investigation of Reaction Mechanism." *Environ. Sci. Technol.*, 30:2, 716–719.

Williams, M.D., Yabusaki, S.B., Cole, C.R., and Vermeul, V.R. (1994) "In-Situ Redox Manipulation Field Experiment: Design Analysis." Thirty-Third Hanford Symposium on Health and the Environment, November 7–11, Pasco, WA, Battelle Press, Columbus, OH.

Yamane, C.L., Warner, S.D., Gallinati, J.D., Szerdy, F.S., Delfino, T.A., Hankins, D.A., and Vogan, J.L. (1995) "Installation of a Subsurface Groundwater Treatment Wall Composed of Granular Zero-Valent Iron." Proceedings of 209th ACS National Meeting, Anaheim, CA, April 2–7, 792–795.

12

REMOVAL OF *NITROAROMATIC* COMPOUNDS FROM WATER THROUGH COMBINED *ZERO-VALENT METAL* REDUCTION AND ENZYME-BASED *OXIDATIVE COUPLING* REACTIONS

H. R. Monsef[1], D. A. Michels[1], J. K. Bewtra[2], N. Biswas[2], and K. E. Taylor[1]*

[1]Chemistry & Biochemistry and
[2]Civil & Environmental Engineering
College of Engineering & Science
University of Windsor
Windsor, Ontario
Canada N9B 3P4

1. ABSTRACT

Peroxidase-catalyzed *oxidative coupling*, previously shown to be an effective strategy for removal of phenols from water, has been optimized for analogous reaction and removal of several *anilines*. *Nitroaromatic* compounds, even when they are phenolic or anilino compounds, are outside the scope of *peroxidase* catalysis. However, *zero-valent iron* has been found effective in conversion of aqueous *nitrobenzene* to *aniline*, quickly and quantitatively in the 0.5 to 10 mM concentration range in the absence of oxygen. The sequential application of these two steps, *zero-valent iron* reduction followed by *peroxidase*-catalyzed *oxidative coupling*, to *nitrobenzene* and preliminary results in the development of a continuous flow system for removal of *nitrobenzene* from water are presented.

2. INTRODUCTION

Enzymes such as *peroxidases*, in the presence of *hydrogen peroxide*, and *laccases*, in the presence of oxygen, catalyze the oxidation of a wide variety of phenols, biphenols, *anilines*, *benzidines* and related heteroaromatic compounds. The reactive radicals generated (phenoxyl radicals from phenols, anilinium cation radicals from *anilines*) diffuse from the enzyme's active center into solution and form dimers, trimers, *etc.* The

---

*corresponding author; Tel.: 519-253-3000, ext. 3526; Fax: 519-973-7098; e-mail: taylor@uwindsor.ca

Emerging Technologies in Hazardous Waste Management 8, edited by Tedder and Pohland
Kluwer Academic/Plenum Publishers, New York, 2000.

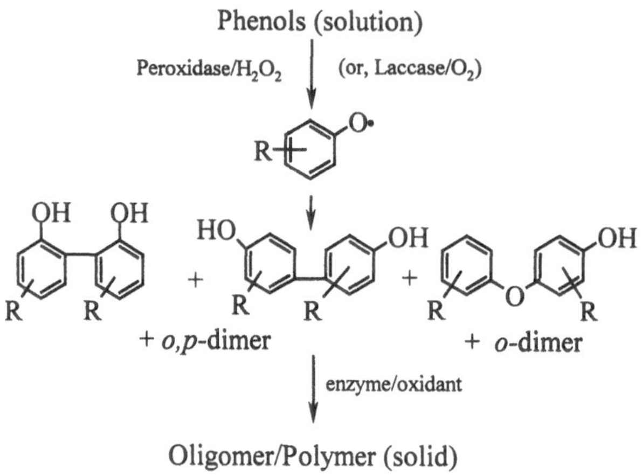
Phenols (solution)
Peroxidase/H_2O_2
(or, Laccase/O_2)
+ o,p-dimer
+ o-dimer
enzyme/oxidant
Oligomer/Polymer (solid)

Scheme 1

polyaromatic compounds grow in size until they reach their solubility limits and then they precipitate (Scheme 1). Waste treatment applications of individual enzymes (as opposed to micro-organisms) have recently been reviewed.[1] We have worked on the optimization of this approach for the *peroxidase*-catalyzed removal of phenols from water, including an economic analysis[2] and a partial product analysis.[3] This *enzyme-based remediation* without degradation, which would allow capture of the organic material in minimally-modified form, has been enabled in recent years as enzymes become available as commodities. *Peroxidase*-catalyzed *polymerization* of *aniline* has been known for some time,[4,5] Scheme 2, and some product analysis has been carried out for *aniline*[4] and *chloroaniline*,[6] Scheme 3, and for anisidine[7] (not shown). However, the systematic devel-

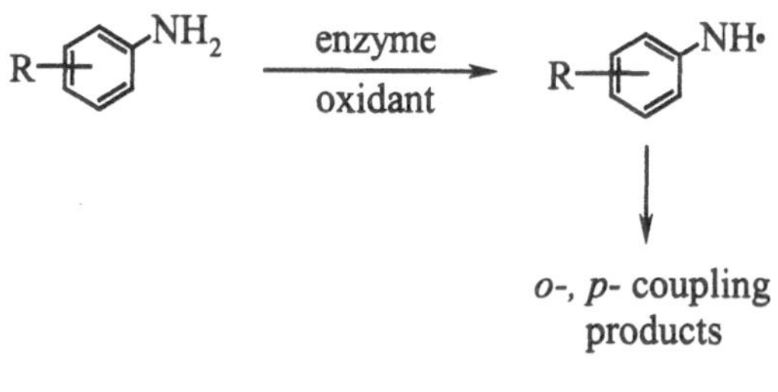
enzyme
oxidant
o-, p- coupling products

Scheme 2

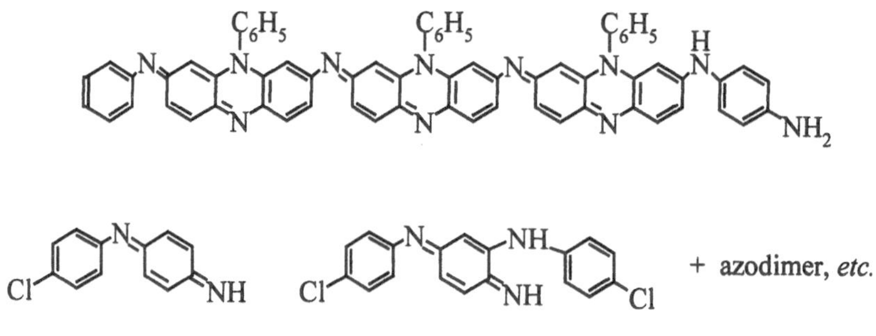
+ azodimer, etc.

Scheme 3

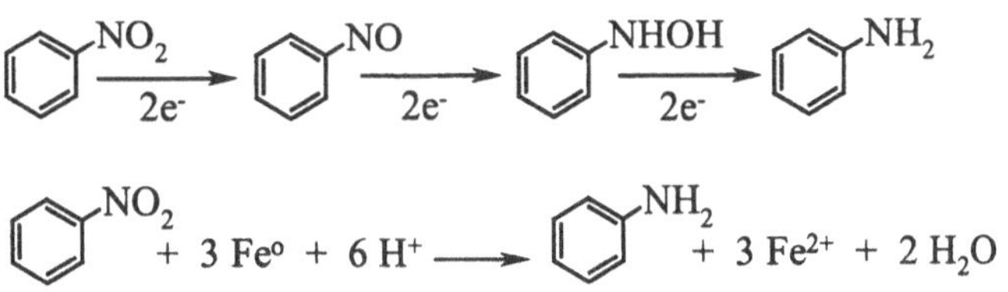

Scheme 4

opment of an *enzyme-based treatment* strategy, such as we and others have done for phenols, has not yet been carried out on *anilines*, except for studies of cross-coupling to humic substances catalyzed by an analogous oxidative enzyme, *laccase*.[8]

Nitroaromatic compounds, even when they are phenolic or anilino compounds, are outside the scope of *peroxidase* catalysis. They are produced in large scale for use as explosives, herbicides, insecticides, dyes, pharmaceuticals, solvents and as intermediates for plastics. Microbial degradation of *nitroaromatic* compounds is known,[9,10] but in general it is difficult and in many cases incomplete, leaving a set of metabolites which are themselves toxic. However, very recent reports[11,12] indicate significant progress in this area, with complete mineralization being achieved in some instances.

The use of *zero-valent metals*, especially iron, to carry out reductive transformations on organic molecules classed as pollutants in the aqueous environment has been introduced as a treatment strategy[13] for halogenated organics,[14,15] *azoaromatics*[16] and *nitroaromatics*.[17] Scheme 4 shows the proposed[17] pathway for *nitrobenzene* reduction to *aniline* and a balanced redox equation for the conversion. With respect to *nitrobenzene*, the combination of *zero-valent iron* reduction of it to *aniline* followed by enzyme-catalyzed coupling reactions was foreseen.[17] The enzymic reaction suggested was of the same type discussed above for phenols and *anilines*, but in this instance it was seen as a coupling to soil organic matter during treatment of groundwater.[18] Groundwater treatment is the main focus for the research groups mentioned above, hence the concentration range studied with *nitrobenzenes* is usually in the 20–50 µM range and in all cases is <0.3 mM. *Zero-valent iron* reduction of nitro-[17] and azo-[16] aromatic compounds is limited by mass transfer to the metal surface. A detailed mechanism is not known for the process.[13] The involvement of surface-bound Fe^{2+}-species is possible but it is known that Fe^{2+} in solution is not capable of *nitrobenzene* reduction.[19] The product of *zero-valent iron* reduction is *aniline*, whether from *nitrobenzene*[17] or *azobenzene*[16]; detectable levels of nitroso- and hydroxylamino- intermediates are found during the reaction but at completion their levels have also diminished to near zero. In any case, the toxic nature of the products formed by *zero-valent metal* reduction has been noted[17] and, therefore, any treatment strategy incorporating this step will also have to incorporate an efficient means of capturing the *anilines* (and any other aromatic nitrogen-containing intermediates).

A removal strategy based on the sequential application of the two transformations, *zero-valent iron* reduction followed by *peroxidase*-catalyzed *oxidative coupling*, to *nitrobenzene* (Scheme 5) and preliminary results in the development of a continuous flow

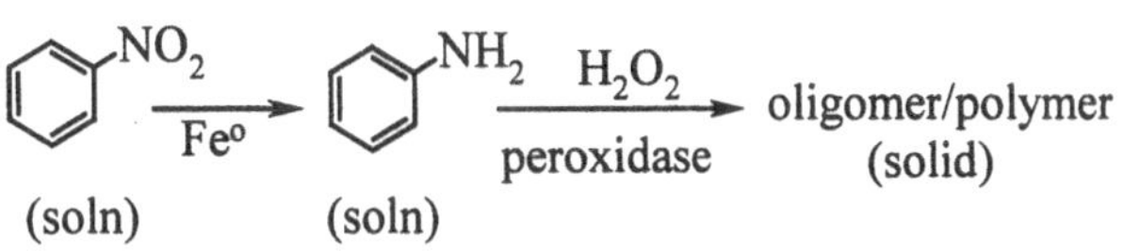

Scheme 5

system for removal of *nitrobenzene* from water are the subject of this report. The concentration range of interest is that which could occur in industrial process streams, the millimolar range for *nitrobenzene* (aqueous solubility *ca.* 20 mM).

3. EXPERIMENTAL

3.1. Materials

British Drug Houses (as BDH Inc., Toronto, ON) was the source of *aniline* (analytical grade), *nitrobenzene*, ferrous sulfate heptahydrate (both ACS Assurance), liquified phenol, 90%, and *hydrogen peroxide*, 30%. *Nitrosobenzene*, 97%, and 4-aminoantipyrine, 98%, were from Aldrich Chemical Co, Milwaukee, WI. Horseradish *peroxidase* (HRP) liquid preparation of purity number 0.37 was a generous gift of Biozyme Labs Ltd., Blaenavon, Wales. The microbial enzyme referred to as ARP was obtained as a concentrated liquid preparation, purity number 0.5, a developmental sample prepared by Novo Nordisk, Bagsvaerd, DK and obtained from Biotech Environmental Inc., Windsor, ON. Two iron preparations were used in this work: initial studies were carried out with a "fine" powder from BDH Inc. (hydrogen-reduced, product number 286024P); subsequent studies were with iron filings, referred to in the paper as "coarse", around 40 mesh, from Aldrich (product number 157–500). Both preparations were acid-washed before use, then neutralized under anaerobic conditions as recommended.[17]

3.2. Enzyme Activity Test

Peroxidase activity tests[2] were based on the rate of *oxidative coupling* of phenol (10 mM) and 4-aminoantipyrine (2.4 mM) in the presence of *hydrogen peroxide* (0.2 mM) in 0.05 M phosphate buffer of pH 7.4 at 25 °C. The reaction rate was monitored at 510 nm in a 1.0 mL reaction volume and converted to μmol of peroxide consumed per minute based on an extinction coefficient of $6,000 \, M^{-1} \cdot cm^{-1}$. Enzyme doses referred to in this paper are given in units (U) per mL of treated solution where one unit of enzyme activity is that amount capable of consuming peroxide at a rate of $1 \, \mu mol \cdot min^{-1}$ in this test.

3.3. TNBS Test for Aniline

In a final volume of 1.0 mL were mixed 0.10 mL of 10 mM tri*nitrobenzene*sulfonic acid sodium salt (TNBS), 0.10 mL of 0.5 M phosphate buffer of pH 6.3 and 0.80 mL of sample plus water. Duplicate samples were allowed to stand for 15 min and then absorbance was measured at 384 nm against a reagent blank. Concentrations were calculated using an extinction coefficient of $12,600 \, M^{-1} \cdot cm^{-1}$.

3.4. Precipitation of Iron Salts

Solutions containing ferrous sulfate were diluted with an equal volume of buffer or sodium hydroxide solution at a concentration twice that of the ferrous sulfate (eg. 1 volume of 20 mM ferrous sulfate plus one volume of 40 mM buffer or NaOH). The suspension that formed was stirred 20–30 min and then centrifuged at $3,000 \times g$ for 30 min. and the supernatant was taken for analysis.

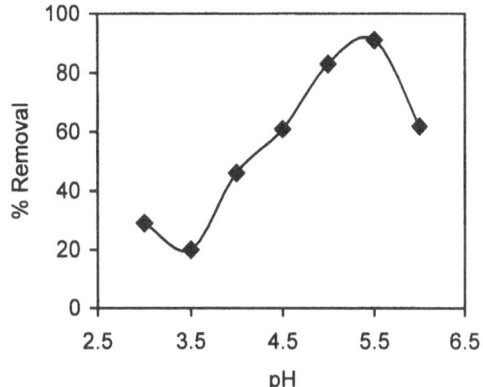

Figure 1. pH-effect on ARP-catalyzed Removal of Aniline. Aniline (1.0 mM), hydrogen peroxide (1.3 mM) and ARP (10 U/mL) in unbuffered solution were adjusted to the pH noted allowed to react for 3 h, then pH adjusted to 7.4, mixture stirred 15 min and then centrifuged 3,000 × g to remove solids; the supernatant was analyzed by TNBS test.

3.5. UV-Vis Measurements

Hewlett-Packard model 8451A and 8452A instruments were used for these measurements; *multi-component analysis* software with the former was used with authentic standards over the wavelength range 230 to 400 nm to determine *nitrobenzene, aniline* and *nitrosobenzene* in duplicate reaction solutions. When those solutions contained ferrous sulfate, the precipitation procedure described above was used before the *multi-component analysis* was performed. Control experiments indicated quantitative recovery of *nitrobenzene* and *nitrosobenzene* through the procedure while *aniline* recovery was 75 and 90% at 20 mM and 10 mM ferrous sulfate, respectively; these factors were used for correction of *aniline* concentrations unless noted.

3.5. General Reaction Conditions

Except as noted, all reactions and tests were carried out at ambient temperature, 22–25 °C. *Zero-valent iron* batch reactor runs were carried out in 30 mL vials sealed by crimp- or screw-cap. The vials were shaken on a wrist shaker (Burrell, Model 75, Pittsburgh, PA) at setting 10. Solutions were made anaerobic where noted by bubbling with nitrogen for 15 min before closing. Settling of iron particles after the reaction was facilitated by standing the vial on a magnet for a few min. Enzyme-catalyzed batch reactor runs were conducted in the same vials equipped with a simple push cap and magnetic stirring bar. After reaction (see details in the legend to Fig. 1) the solids formed were separated by centrifugation and the supernatants were analyzed. All analyses were carried out in duplicate with the averages being reported; in all cases average deviations were smaller than the symbols used in plotting. Experiments were likewise carried out at least twice to confirm reproducibility.

4. RESULTS

Aniline removal from solution through precipitation by *peroxidase*-catalyzed *oxidative coupling* was optimized by analogy with our phenol studies.[2,3] Parameters examined were pH (both for reaction and for settling), enzyme dose, peroxide stoichiometry, influence of the additive polyethylene glycol (PEG) and utility of alum as a coagulant. Initial

Table 1. Minimum Enzyme Doses (U/mL) for ≥95% Removal of Anilines

Enzyme, conditions	Aniline	3-Chloro-Aniline	4-Chloro-Aniline
HRP, buffered, pH 4.5	3	—	—
HRP, unbuffered, pH 4.0	3	—	—
ARP, buffered, pH 4.5	8–10	30	2
ARP, unbuffered, pH 5.5	10	30	2.5

Anilines (1.0 mM) and H_2O_2 (1.0 mM) were stirred in buffer or pH-adjusted water with enzyme for 3 h at room temperature, then pH was adjusted to 7.4 and samples were centrifuged 30 min at 3,000 × g. Buffers used were acetate or phosphate (0.05 M).

studies were with the horseradish enzyme (HRP) while most of the work was with the enzyme from *A. ramosus* (ARP). *Aniline* disappearance from the supernatants was monitored by a colorimetric test (in duplicate) based on derivatization with tri*nitrobenzene*-sulfonate (TNBS). For removal of *aniline* and its 3- and 4-chloro- derivatives from unbuffered solutions at 1.0 mM (around 100 ppm), both enzymes worked best in the pH 4.0–5.5 region (Fig. 1; Table 1), with a peroxide stoichiometry of 1.0 to 1.1, with best settling at pH 7.5 and above and without obvious benefit from either PEG or alum. Enzyme doses are given as the minimum number of units (U) of catalytic activity (which correspond to peroxide turnover of 1 μmol/min in a standard colorimetric assay) capable of catalyzing the removal of >95% of the *aniline*. A typical enzyme dose dependence curve is shown in Fig. 2 and results are given in the table for both enzymes with *aniline* and additionally for ARP with two *chloroaniline* isomers. For comparison, the minimum dose of ARP required[2] to catalyze analogous removal of phenol is 0.6 U/mL.

The *zero-valent iron* reduction of *nitrobenzene* was studied first in shaken batch reactors using a very fine (particle size not specified or measured) powder preparation. Analysis (in duplicate) was by the colorimetric test based on TNBS (above) for the product *aniline* and by direct *multi-component analysis* (MCA) of the ultra-violet (uv) spectra based on *nitrobenzene, nitrosobenzene* and *aniline* standards (*phenylhydroxylamine* standard unavailable). Preliminary results confirmed the literature[17] in demonstrating the need for anaerobic conditions and the improvements gained through acid-washing the iron surface and maintaining anaerobic conditions thereafter. Under these conditions, for example, 30 mL of 0.5 mM *nitrobenzene* was completely converted to *aniline* in under 1 hour at room temperature by 0.5 g of iron as shown in Fig. 3. The same figure

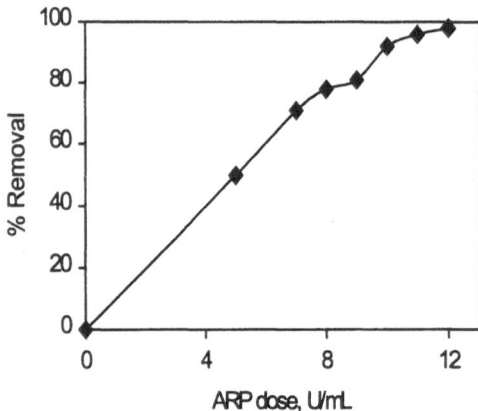

Figure 2. Effect of Enzyme Dose on Aniline Removal. Conditions as for Fig. 1, except at pH 5.5.

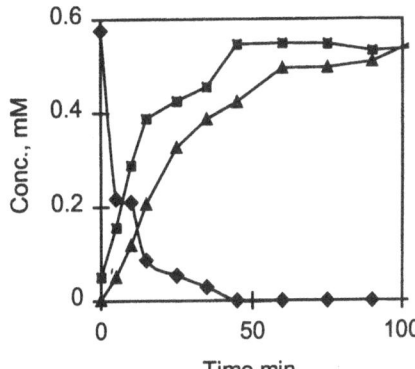

Figure 3. Zero-valent Iron Reduction of Nitrobenzene under Anaerobic Conditions. Nitrobenzene (0.6 mM) was incubated with fine iron (0.5 g); analysis by MCA for nitrobenzene (◆) and aniline (■) and by TNBS-test for aniline (▲).

also shows the correspondence of *aniline* estimates made by the TNBS test and by uv-MCA.

In order to avoid the nuisance of making solutions anaerobic, the study turned to the inclusion of an oxygen scavenger. Sodium sulfite up to 5 mM was ineffective but ferrous sulfate in the 5–20 mM range allowed *nitrobenzene* conversion rates comparable to anaerobic solutions. Figure 4 shows experiments carried out in air-saturated water and 5 mM ferrous sulfate solution while Fig. 5 shows that with prior oxygen outgassing the latter reaction was markedly faster. In fact, comparison of Figs. 3 and 5 might suggest that the rate enhancement in Fig. 5 was due to the anaerobic condition alone. This benefit of ferrous sulfate is seen by comparison of Figs. 6 and 7, which represent identical experiments carried out in air-saturated 20 mM ferrous sulfate solution and oxygen-free water, respectively. Thus, 30 mL of 1.0 mM *nitrobenzene* in 20 mM ferrous sulfate is completely converted to *aniline* in 40 min at room temperature by 0.5 g of fine iron; the same iron could be used over again with the same result, as long as it was maintained in ferrous sulfate solution between uses. Under the same conditions, *nitrosobenzene*, detected at short times in the above reaction (Figs. 6 and 7), was very quickly removed (data not shown). It was also noted that when the reaction shown in Fig. 7 was repeated but with twice the iron dose, conversion to *aniline* was quantitative in the same time frame (data not shown). These reactions involving *zero-valent iron* (occurring around pH 5.0, due to

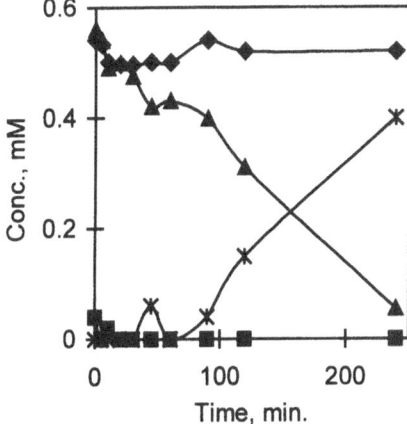

Figure 4. Effect of Ferrous Sulfate (5 mM) in Air-saturated Zero-valent Iron Suspensions. Nitrobenzene (0.55 mM) and fine iron (0.5 g) were incubated in water or in 5 mM FeSO$_4$ without prior outgassing of oxygen; analysis by MCA: nitrobenzene, in water (◆) or FeSO$_4$ (▲), aniline in water (■) or FeSO$_4$ (*).

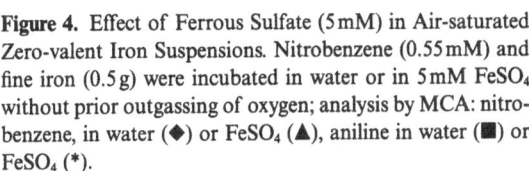

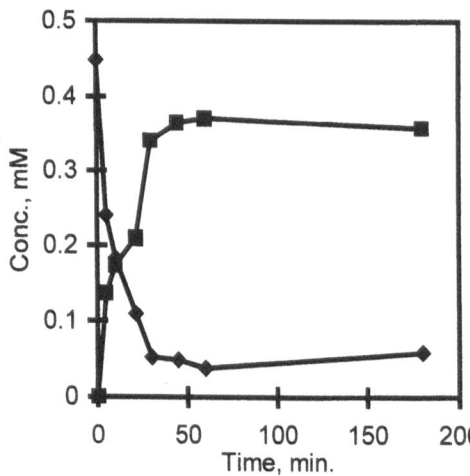

Figure 5. Effect of Ferrous Sulfate (5mM) under Anaerobic Conditions. Conditions analogous to Fig. 4 but in the absence of oxygen; analysis by MCA: nitrobenzene (◆), aniline (■).

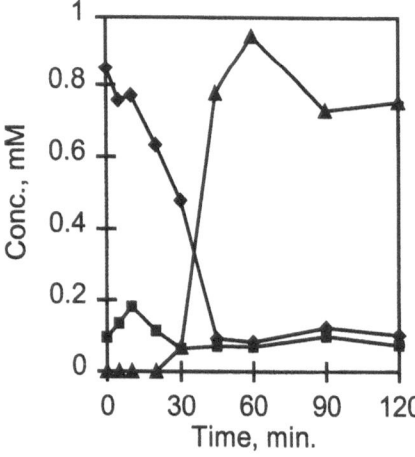

Figure 6. Effect of Ferrous Sulfate (20mM) in Air-saturated Zero-valent Iron Suspensions. Nitrobenzene (0.85mM) and fine iron (0.5g) were incubated in 20mM FeSO$_4$ without prior outgassing of oxygen; analysis by MCA: nitrobenzene (◆), nitrosobenzene (■), aniline (▲).

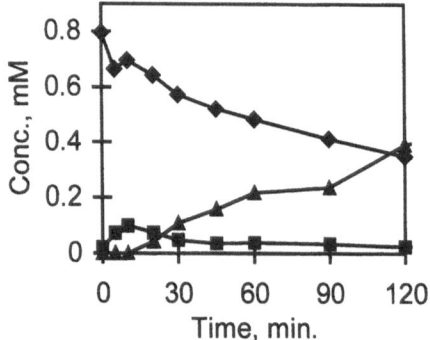

Figure 7. Anaerobic Conditions without Ferrous Sulfate. Conditions as for Fig. 6 but in absence of FeSO$_4$ and with prior outgassing of oxygen; analysis by MCA: nitrobenzene (◆), nitrosobenzene (■), aniline (▲).

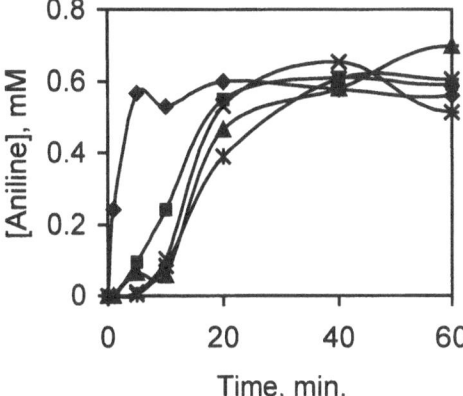

Figure 8. pH Effect on Zero-valent Iron Reduction of Nitrobenzene in Ferrous Sulfate. Nitrobenzene (1.0 mM) and fine iron (1.0 g) were incubated in 20 mM FeSO$_4$ without prior outgassing of oxygen after adjustment of pH to the values noted; analysis by MCA, aniline concentrations not corrected for recovery; pH 2.5 (◆), pH 3 (■), pH 4 (▲), pH 5 (×), pH 5.5 (*).

the ferrous sulfate) could be made faster by lowering pH below 3.0 as seen in Fig. 8 (diamond symbols, pH 2.5).

Column operation of the *zero-valent iron* reduction step was initially investigated with 5 g of the fine iron preparation above, acid-washed and maintained in ferrous sulfate solution. In a 1 × 30 cm column flowing upwards at 2.7 mL/min, this charge of iron expanded 2–3-fold from a settled height of 6 cm (contact time of 2–3 min). Under these conditions, the column effected complete conversion of 1, 2 and 10 mM solutions of *nitrobenzene* (in 20 mM ferrous sulfate). Fig. 9 shows a run for 2.2 mM *nitrobenzene* in which it took about two system volumes to reach a steady state at quantitative conversion. Re-use of this column was easily possible, provided it was maintained in ferrous sulfate solution. However, with use this fine iron charge suffered notable attrition with the result that particles were increasingly "elevated" out of the column.

Subsequent work was carried out with a coarser preparation of iron filings (about 40 mesh). Batch reactor runs showed ferrous sulfate at 5–20 mM to be effective as an oxygen scavenger. For example, with 1.0 mM *nitrobenzene*, 1 g of acid-washed iron required 60 min for >90% conversion to *aniline*, while 2 and 3 g iron completed the conversion in 30 and 15 min, respectively (Fig. 10). Column operation with this iron preparation has been studied with a 50 g charge in a 1.6 × 12 cm bed. Continuous, quantitative conversion of 1.0 mM *nitrobenzene* to *aniline* in 5 or 10 mM ferrous sulfate has been carried out for more than 80 h at 1.1 mL/min (contact time 10–12 min).

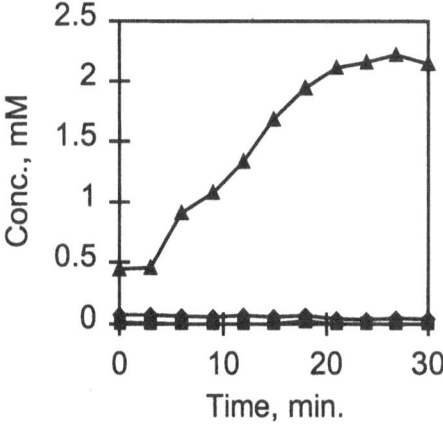

Figure 9. Continuous Conversion of Nitrobenzene in a Zero-valent Iron Column. Nitrobenzene (2.2 mM) in 20 mM FeSO$_4$ was pumped upwards through a bed of fine iron (5 g) in a column (1 × 30 cm) at 2.7 mL/min (bed depth 6 cm settled, 15 cm flowing); analysis by MCA: nitrobenzene (◆), nitrosobenzene (■), aniline (▲).

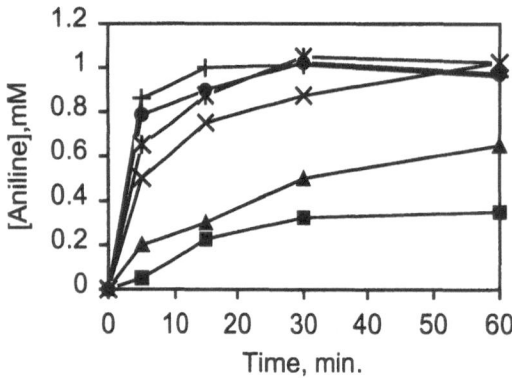

Figure 10. Coarse Iron Dose Effect on Nitrobenzene Reduction. Nitrobenzene (1.0 mM) in 20 mM $FeSO_4$ without prior outgassing of oxygen was incubated with various amounts of coarse iron; analysis by MCA; iron doses (in 30 mL): 0.2 g (■), 0.5 g (▲), 1 g (×), 1.5 g (*), 2 g (•), 3 g (+).

The presence of ferrous salts in the *aniline* solutions produced by the foregoing continuous process precludes direct combination with enzyme and peroxide for the *oxidative coupling* and precipitation of aromatic polymer. Soluble iron species are first precipitated by raising the pH, which promotes oxidation of the ferrous to ferric species, followed by precipitation of the latter. Addition of base (2 molar equivalents relative to $FeSO_4$), mixing (few minutes) and sedimentation (1 h detention) have been combined in a continuous process which furnishes an effluent suitable for enzyme-catalyzed *polymerization*. One side reaction that should be noted is the slow conversion of *nitrobenzene* that occurs after the pH is raised, presumably by ferrous species deposited on the ferric particles that have formed.[19] This is both a benefit, in that it results in a "polishing" of any unreacted nitro compound, and a complication, in that analysis is more complex. Figure 11 shows experiments to test the stability of *nitrobenzene* in ferrous sulfate solution after a nominal two equivalents of buffer or strong base was added (no iron present). Phosphate buffer of pH 6.4 was the normal quenching solution for analytical samples taken from zerovalent reduction reactions reported above. As seen, this means of sample workup leads to negligible additional reduction of the nitro compound. Progressively more basic buffers show greater degrees of conversion culminating in the result for addition of sodium hydroxide, instantaneous and quantitative reduction.

Finally, it has been found in control experiments that *aniline* solutions in 20 mM ferrous sulfate, after subjection to the removal of soluble iron species as above, can be

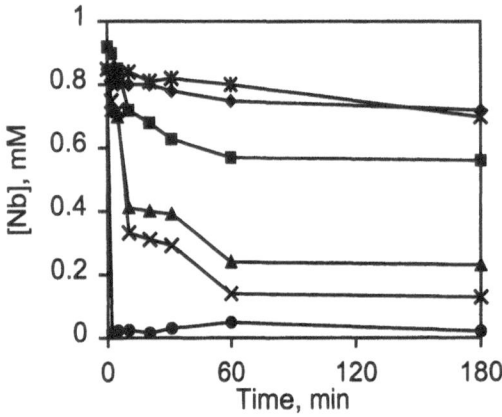

Figure 11. Nitrobenzene Stability During Iron Salt Precipitation. Nitrobenzene (Nb, 0.85 mM) in 20 mM $FeSO_4$ was diluted with an equal volume of buffer or sodium hydroxide solution and stirred; samples taken were centrifuged and analyzed by MCA; diluents (40 mM) added: phosphate pH 6.4 (*), Tris pH 7.2 (♦), Tris pH 8.0 (■), Tris pH 8.5 (▲), Tris pH 9.0 (×), NaOH (•).

treated by the oxidative enzymic process with the same efficiency (with respect to enzyme dose, peroxide stoichiometry and degree of conversion) as noted at the outset for authentic *aniline*.

5. DISCUSSION AND CONCLUSIONS

The ability of *peroxidase* to catalyze the removal of *anilines* from solution through oxidative *polymerization*, analogous to phenols,[2] sets the stage for development of a two-step process for removal of the corresponding *nitroaromatic* compounds. Unfortunately, *aniline polymerization* requires about an order of magnitude more enzyme for the removal than the same molar amount of phenol. As found with the phenols,[2] the minimum doses of enzyme required vary widely with the type and position of substituents on the *aniline*. The *zero-valent iron* step requires anaerobic conditions, but it can be carried out without solution degassing by inclusion of an oxygen scavenger. Ferrous sulfate has been used in this feasibility study but it may not be the most appropriate either for efficacy or for process economics. Current studies are addressing this issue. The present study has shown that the elements of a continuous reductive-oxidative process for the removal of *nitroaromatic* compounds from water are in place.

ACKNOWLEDGMENTS

The Natural Sciences and Engineering Research Council of Canada provided the main financial support. HRM was also supported in part by a training fellowship from the Ministry of Health and Medical Education of The Islamic Republic of Iran. DAM was also supported in part by funds from the Summer Career Placement 1996 (SCP96) program of Human Resource Development Canada. Valuable discussions were held with Piero Tartaro and Anthony Mandarino, SCP97 students, who worked on related studies of the *zero-valent iron* system.

REFERENCES

1. M.D. Aitken (1993), Chem. Eng. J. **52**, B49–B58.
2. K.E. Taylor, L. Al-Kassim, J.K. Bewtra, N. Biswas, and J. Taylor (1996), in *Environmental Biotechnology: Principles and Applications*, M. Moo-Young, W.A. Anderson, A.M. Chakrabarty (editors), Kluwer Academic Publishers, Dordrecht, pages 524–532.
3. J. Yu, K.E. Taylor, H. Zou, N. Biswas, and J.K. Bewtra (1994), Env. Sci. Technol. **28**, 2154–2160.
4. B.C. Saunders *et al.* (1964), in *Peroxidase*, Butterworths, London, pages 10–24.
5. A.M. Klibanov, B.N. Alberti, E.D. Morris, and L.M. Felshin (1980), J. Appl. Biochem. **2**, 414–421.
6. C.-W. Chang and J.A. Bumpus (1993), FEMS Microbiol Lett. **107**, 337–342.
7. D.C. Thompson and T.E. Eling (1991), Chem. Res. Toxicol. **4**, 474–481.
8. K.Tatsumi, A. Feyer, R.D. Minard, and J.-M. Bollag (1994), Env. Sci. Technol. **28**, 210–215.
9. F.K. Higson (1992), Adv. Appl. Microbiol., S.I. Neidleman and A.I. Laskin (Eds) **37**, 1–19.
10. S.B. Haderlein and R.P. Schwarzenbach (1995), in *Biodegradation of Nitroaromatic Compounds*, J.C. Spain (Ed.); Plenum, New York.
11. R. Boopathy and J.F. Manning (1996), Can.J. Microbiol. **42**, 1203–1208.
12. D.L. Widrig, R. Boopathy, and J.F. Manning (1997), Environ. Toxicol. Chem. **16**, 1141–1148.
13. Chem. & Eng. News, July 3/95, pages 19–22.
14. W.S. Orth and R.W. Gillham (1996), Env. Sci. Technol. **30**, 66–71.

15. T.L. Johnson, M.M. Scherer, and P.G. Tratnyek (1996), Env. Sci. Technol. **30**, 2634–2640.
16. E.J. Weber (1996), Env. Sci. Technol. **30**, 716–719.
17. A. Agrawal and P.G. Tratnyek (1996), Env. Sci. Technol. **30**, 153–160.
18. J.-M. Bollag (1992), Env. Sci. Technol. **26**, 1876–1881.
19. J. Klausen, S.P. Trober, S.B. Haderlein, and R.P. Schwarzenbach (1995), Env. Sci. Technol. **29**, 2396–2404.

13

IN SITU FENTON-LIKE *OXIDATION* OF VOLATILE ORGANICS

Laboratory, Pilot and Full-Scale Demonstrations

Richard S. Greenberg[1], Thomas Andrews[1], Prasad K. C. Kakarla[1], and Richard J. Watts[2]

[1]In-Situ Oxidative Technologies, Inc.
Lawrenceville, New Jersey, USA
[2]Department of Civil & Environmental Eng.
Washington State University
Pullman, Washington, USA

1. ABSTRACT

Laboratory, pilot, and full-scale experiments were used to evaluate and optimize the ISOTEC[SM] remedial process at a warehousing facility in Union, N. J. Based on modified Fenton's oxidative chemistry, the *ISOTEC[SM] process* uses a proprietary catalytic agent which delays formation of reactive hydroxyl radicals. This allows adequate dispersion of the hydroxyl radicals, which is an oxidizing agent, throughout a contaminant plume. Groundwater at the site was contaminated with high levels of gasoline and waste oil constituents, principally benzene, toluene, ethylbenzene, xylenes (*BTEX*) and methyl-t-butylether (*MTBE*). Bench scale microcosm studies were used to evaluate the appropriate site-specific stoichiometric relationships between catalyst, stabilizers, and oxidizers; the effect of contaminant type and concentration; and the pH optima. Based on results of the laboratory studies, a pilot-scale study was performed at the site. One injection point for catalyst, stabilizers, and oxidant was installed in the contaminated zone at the site, with one hydraulically connected downgradient well used for monitoring. A single treatment of the reagents in the optimal stoichiometry determined from the laboratory study was injected *in situ* over a period of three days. A 98.5% reduction in volatile organics was observed in the area treated, with the radial extent of treatment estimated to be approximately 20 feet, based on the presence of hydroxyl radicals detected in hydraulically connected areas and at the surface. The full-scale process employed six injection points and three treatment cycles over a three month period. Subsequent to treatment, contaminant levels were either non-detectable or were reduced to below applicable New

Emerging Technologies in Hazardous Waste Management 8, edited by Tedder and Pohland
Kluwer Academic/Plenum Publishers, New York, 2000.

Jersey groundwater standards, with regulatory closure on the site achieved in less than one year.

2. INTRODUCTION

Remediation of groundwater contamination is a chronic environmental problem due primarily to the complexity associated with a subsurface soil-water matrix. In most cases, contamination exists in both soil and aqueous phases dictated by the corresponding distribution coefficient (K_d) and the octanol water partition coefficient (K_{ow}) (Watts 1997). Several factors limit the application of conventional treatment techniques to subsurface contamination. For example, pump and treat and soil vapor extraction (SVE) have high capital and operational costs, large treatment time spans, and are not effective against sorbed contaminants. In situ bioremediation also is plagued by large treatment time spans and is limited by the oxygen supply to subsurface microorganisms and the absence of indigenous microbial strains to degrade the recalcitrant (or biorefractory) contaminants (Thomas and Ward 1989).

Recent years saw a significant increase in the use of catalyzed *hydrogen peroxide* (known as *Fenton's reagent*) for soil and groundwater remediation because of its ability to oxidize a wide variety of contaminants in a short time frame. *Hydroxyl radicals* are the principal oxidant involved in Fenton's process (Haber and Weiss 1934) and are known to react with most contaminants at near diffusion controlled rates i.e. at rates approaching $1 \times 10^{10} M^{-1} s^{-1}$ (Dorfman and Adams 1973). Early researchers successfully transformed benzenes and toluenes into phenols, cresols, and biphenyls using Fenton's reagent (Merz and Waters 1949; Baxendale and Magee 1953). Eisenhauer (1964) suggested oxidation of phenol, an aromatic compound, by hydroxyl radicals into cis, cis-muconic acid, a linear compound, by virtue of ring breakage. More recently, other contaminants such as p-toluenesulfonicacid (Feuerstein *et al.* 1981), formaldehyde (Murphy *et al.* 1989), perchloroethene (Leung *et al.* 1992), and trichloroethene (Ravikumar and Gurol 1994) were oxidized successfully using catalyzed hydrogen peroxide. Furthermore, Stanton and Watts (1995) and Watts *et al.* (1991, 1994) oxidized strongly sorbed contaminants in soils including hexadecane, octachlorodibenzo-p-dioxin (OCDD), and hexachlorobenzene. Two mechanisms were proposed as possible for contaminant oxidation with hydroxyl radicals in a soil-water matrix (Watts *et al.* 1994): (1) desorption of the contaminants followed by contaminant oxidation under normal or aggressive reaction conditions, and (2) contaminant oxidation on the surface of the soils when hydroxyl radicals break the soil-water barrier under aggressive conditions.

Most research in the past was performed at bench scale level although the potential of Fenton's/*Fenton-like reactions* for in situ soil and groundwater remediation was recognized. In the 1990s, growth in commercial usage of Fenton's reagent to achieve full-scale remediation occurred. The process was used in reactors for industrial waste treatment (Bigda 1996) and also for in situ groundwater remediation (ISOTEC[SM] or In-Situ Oxidative Technologies, Inc.).

ISOTEC[SM] is actively involved in the in-situ application of catalyzed hydrogen peroxide for soil and *groundwater remediation*. The ISOTEC[SM] process involves injection of hydrogen peroxide and proprietary iron-based catalysts directly into or around areas of known contamination in the subsurface. A site specific delivery system designed to treat organic contaminants within an area of concern (AOC) is utilized. The injectants react in the subsurface generating hydroxyl radicals which oxidize the organic contaminants to produce carbon dioxide and water.

In the ISOTEC[SM] process, a series of laboratory studies are first performed on representative samples to determine the preliminary treatment quantities and the site specific stoichiometry of reagents needed. Application is later tested in the field during a pilot program to determine process efficiency and extent of treatment, which may vary depending on the site's subsurface characteristics. Based on a successful laboratory study and pilot program, full-scale treatment of the site is performed.

The objective of this paper is to discuss (a) the laboratory studies on a representative contaminated sample for potential application of the ISOTEC[SM] process, and (b) the macroscale field pilot study at a warehousing facility in Union, New Jersey with gasoline and waste oil contamination followed by full-scale application of the ISOTEC[SM] process and case closure.

3. MATERIALS AND ANALYSIS

3.1. Materials

All chemicals used for proprietary laboratory scale ISOTEC[SM] catalyst (chelated iron complex) preparation were reagent grade and purchased from Aldrich (Milwaukee, WI). Stock hydrogen peroxide (35%), sodium thiosulfate, and potassium iodide (reagent grade) were purchased from Fisher Scientific. Ammonium molybdate and starch were purchased from Aldrich (Milwaukee, WI). Distilled Water (Polar Spring) was used for all laboratory reagent preparation and analyses. The hydrogen peroxide field test kit was purchased from Hach (Colorado). Bulk quantities of chemicals and equipment used for full-scale remediation were obtained from local suppliers in New Jersey.

3.2. Analysis

All laboratory analyses was performed by a EPA certified analytical laboratory. Pre- and post-treatment groundwater samples were analyzed for volatile organics (EPA Method 624 + 10), with total iron (Fe) and total organic carbon (TOC) also measured during the post-treatment event. Field sampling parameters, as specified within the *New Jersey Department of Environmental Protection (NJDEP), Field Sampling Procedures Manual (FSPM), dated May, 1992,* were measured during the sampling events. In addition, qualitative tests for hydrogen peroxide and ISOTEC's proprietary method of directly detecting hydroxyl radicals, utilizing a color free radical trap, were conducted during the pilot program and full-scale remediation. The laboratory hydrogen peroxide analysis was performed using a modified thiosulfate technique (FMC 1989). Field monitoring for hydrogen peroxide was performed using the Hach hydrogen peroxide test kit.

4. LABORATORY STUDIES

4.1. Optimization Experiments

In order to optimize the oxidation of gasoline contaminated water, three-level rotatable central composite design experiments (Diamond 1989) were performed. A representative sample contaminated with high concentrations of BTEX, MTBE, and TBA (*t-butyl alcohol*) was used for the experiments. The catalyst (chelated iron complex) and oxidizer (stabilized hydrogen peroxide) concentrations were varied from 0–75 ppm and 0–530 ppm respectively. Response surfaces (curves that represent percent contaminant destruction as a function of oxidizer, catalyst, and stabilizer concentrations) to optimize percent contaminant destruction were generated using the catalyst and oxidizer as independent variables at a fixed stabilizer concentration. Although pH adjustment to higher acidic range was performed initially, the final pH of the treated samples varied according to the concentrations of the reagents added.

4.2. Bench Scale Studies

The groundwater sample which exhibited the highest contaminant levels was collected from MW-2 at the warehousing facility (see site description under "Field Studies") and analyzed for dissolved iron and volatile organic concentrations. The optimum reagent combination obtained from the central composite experiments was used as the basis for the laboratory study. Specific concentrations and volumes of the reagents to be injected in the field were determined based on a series of bench scale studies which tested several catalyst and oxidizer amendments in the optimal pH range of 3–4 (Watts *et al.* 1990). The bench scale studies were conducted in 140 ml vials sealed with aluminum caps containing rubber septums to facilitate easy injection of the reagents with minimum volatilization loss. Samples were placed in vials with enough head space for the predetermined reagent volumes to be injected. A control sample was set up to correct for any volatilization loss. The catalyst mix and oxidizer were injected into each experimental vial as hourly treatment cycles with one treatment corresponding to half the concentration of the optimum combination obtained from the central composite experiments. A total of three treatments were completed with a vial sacrificed after each treatment for volatiles analysis. All experimental vials were left undisturbed at room temperature for approximately 48 hours and periodically monitored for oxidant concentration and pH variations. After determining that the oxidizer was fully consumed, all samples were analyzed for volatile organics. The treatment efficiencies determined from the results of each reagent amendment was further used to arrive at the proprietary reagent combination and treatment cycles specific to the samples.

4.3. Microcosm Studies

The above study only tested the oxidation capability of the reagents for site specific contaminants of interest in water and did not examine the interactive effects of a soil-water matrix. Therefore, additional microcosm studies were performed in glass tanks in an effort to gain insight into a real subsurface environment. Soil collected from an uncontaminated zone of the warehousing facility was used as the media in order to simulate down-gradient flow effects of the injected reagents. Factors such as organic content or permeability were not considered due to the difficulty associated with replicating those specific conditions and in lieu of a forthcoming field pilot study (described under "Field Studies"). Hence, the objective of these experiments was to confirm that the optimum chemical combination determined for the contaminant concentrations in aqueous form was capable of treating a contaminant plume within a soil-water matrix.

The microcosm study was set up in a 10 gallon glass tank [10″W × 12″H × 20″L; see Fig. 1]. The tank was divided into three compartments using two plexiglas panels per-

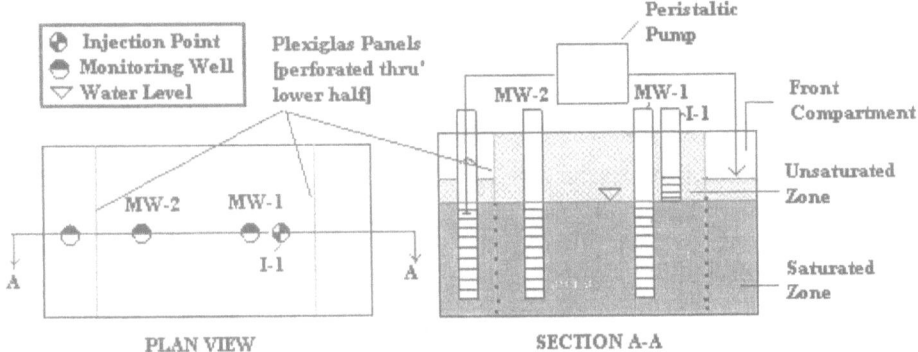

Figure 1. Microcosm Study—Experimental Set Up.

forated through the bottom half such that the central compartment had approximately 80% of the tank capacity and the two extreme compartments occupied the remaining approximately 20% of the tank capacity equally. Nearly 80 lb. of soil was placed in the tank, with the central compartment filled to the top, while the two extreme compartments were filled at half the depth of the tank. The surface of the soil was covered with plastic sheeting to minimize volatilization losses. Three (3)—one and one half inch diameter (1–1/2″ Ø) PVC casings screened through the bottom half of the tanks were installed. Two of the PVC casings were placed in the central compartment to serve as monitoring wells and one was placed in the rear compartment to collect water, which is then pumped back to the front compartment. Nearly 2.7 gal of gasoline spiked water (approximately 20,000 ppb) was introduced into the front compartment and the system was allowed to equilibrate. After about one hour, water from the well in the rear compartment was pumped into the front compartment using a peristaltic pump. The pumping rate was approximately 5 ml/min to simulate the groundwater flow rate. As the separating plexiglas panels were only perforated on the bottom half, the bottom portion of the tank represented a saturated zone. Due to capillary action and percolation, the top portion was also wet but not saturated. Therefore, the top portion of the tank represented the vadose or the unsaturated zone. After the system equilibrated for one day, the catalyst mix followed by hydrogen peroxide were introduced into the upgradient well (I-1). Two treatment cycles of the reagent combination were injected with a two day gap between each treatment (one treatment was equivalent to half the optimum concentrations from the central composite experiments but in the same stoichiometric molar ratio). The water was continually pumped to maintain a steady flow condition in the tank. Sampling was conducted before, during, and after reagent injection to determine the effects of each treatment. The wells were also monitored for depth, hydroxyl radicals (using ISOTEC[SM]'s color free radical trap), hydrogen peroxide, and pH during the treatment.

4.4. Results and Discussion

The results obtained from the central composite design experiments were used to arrive at a regression equation which expresses the percent of contaminant destruction as a function of the molar concentration of catalyst, oxidizer, and stabilizer. The regression equation is given below:

$$D = -80.1 + 2.03C - 0.255S + 0.155H - 0.0289C^2 + 0.0072S^2$$
$$- 0.00028H^2 - 0.0149CS + 0.0011SH + 0.0028CH$$

where,

D = Percent Contaminant Destruction
C = Catalyst Concentration, mM
S = Stabilizer Concentration, mM
H = Oxidizer Concentration, mM

In view of the difficulty associated with delineating the above equation three dimensionally, response surfaces for percent contaminant destruction were plotted at a constant stabilizer concentration of approximately 15 mM as a function of catalyst vs oxidizer (see Fig. 2).

From Fig. 2, it is clear that the contaminant destruction is strongly dependent on the concentrations and molar ratio of oxidizer and catalyst at a fixed stabilizer concentration. While 80–90% contaminant destruction was noticed only for catalyst and oxidizer concentrations greater than 30 mM and 300 mM respectively, higher concentrations (>60 mM catalyst, >475 mM oxidizer) made the reaction too aggressive for all practical purposes. It is evident that the optimum catalyst concentration for 80–90% contaminant destruction lies between 40–50 mM while the optimum oxidizer concentration lies between 300–400 mM in the sample. Along with a stabilizer concentration in the range 12–17 mM, these concentrations formed the basis for further investigation on the site-specific sample.

4.4.1. Bench Scale Studies.   Bench scale studies on samples from MW-2 from the warehousing facility were conducted using the reagent amendments in the optimum ranges obtained from the central composite experiments. One bench scale study treatment was

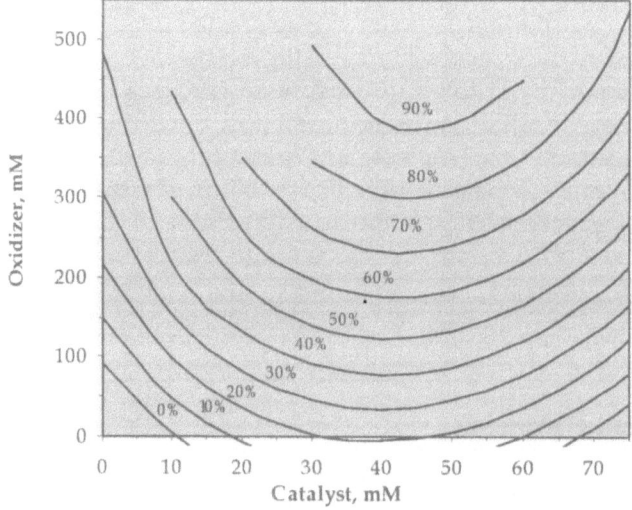

Figure 2. Response surfaces for percent contaminant destruction as a function of oxidizer and catalyst concentrations at a stabilizer concentration of 15 mM.

Table 1. ISOTEC Laboratory Study Results (MW-2)
Warehousing Facility, New Jersey

Contaminant	Original (ppb)	Treated (ppb)
Benzene	4,800	<10
Toluene	12,700	<10
Xylenes, total	13,000	<10
MTBE	2,040	<10
Total VO's	32,540	ND
Total TIC's	20,970	ND
Total VO's & TIC's	53,510	ND

Notes:
(1) ppb—Indicates parts per billion or micrograms per liter (μg/L).
(2) ND—Indicates that the compound was analyzed for but not detected.
(3) TIC's—Tentatively Identified Compounds (or Non-targeted VO's).
(4) <(MDL) = concentration detected at a value below method detection limit (MDL).

approximately equivalent to half the concentration of the optimum combination but in the same molar ratio of catalyst, oxidizer, and stabilizer. The results indicated 100% contaminant destruction after only one treatment (Table 1). This was believed to be due to lower contaminant concentrations in sample MW-2 compared to the representative sample used in the central composite experiments.

4.4.2. Microcosm Study. The effectiveness of the chemical combination was further verified by injecting the reagents into a soil-water matrix set up as a microcosm study. Results of the study indicated an 86% contaminant destruction in I-1 (the up-gradient injection well) and 68% treatment in MW-2 (the down-gradient monitoring well) after the first cycle of injection and 100% treatment in both wells after second cycle of injection. Despite the precautions taken, some volatilization losses may have occurred from the tank. Results of the hydrogen peroxide analysis indicated low to undetectable levels of hydrogen peroxide (in the tank one week after injection. The pH after equilibrium was observed to be in the optimal range as described by Watts *et al.* (1990). ISOTEC's proprietary test for hydroxyl radicals indicated positive results in the wells for up to five days after injection. The results obtained from the microcosm study confirmed the potential of Fenton's process for a field application.

5. FIELD STUDIES

5.1. Site Characteristics

The site is a operating warehousing facility in Union County, New Jersey. The areas of environmental concern at the site include former gasoline, waste oil, and fuel oil underground storage tanks. Groundwater was contaminated with high concentrations of gasoline constituents principally BTEX, MTBE and TBA. The site geology is characterized by soft red shales with interbedded harder sandstones and minor amounts of conglomerate while the site soils were unsorted and unstratified material consisting of pebbles, cobbles, and boulders in a mixture of sand, silt, and clay. A site map is shown in Fig. 3.

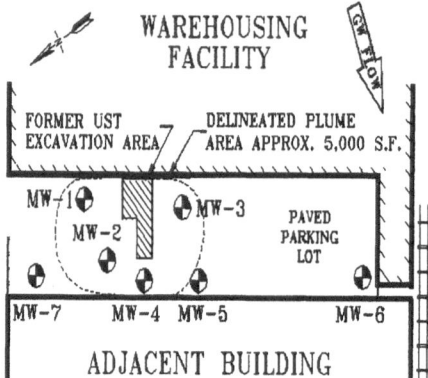

Figure 3. Site Map of the Warehousing Facility.

5.2. Field Pilot Study

Based on ISOTEC's laboratory study results, the New Jersey Department of Environmental Protection (NJDEP) approved ISOTEC's request to proceed with a field pilot program and issued a permit-by-rule authorizing a discharge via a Class V Underground Injection Control (UIC) system. A pilot program was performed to evaluate the effectiveness of the ISOTEC process as an alternative remedial technology to treat organic contaminants identified in the groundwater at the warehousing facility.

The ISOTEC pilot program activities were performed on the specific contaminant plume located side and down-gradient of the former underground storage tank (UST) field. One injection point I-2 was installed (Fig. 4) in the area of highest contamination and monitoring was conducted from a hydraulically connected downgradient well 18 feet away. A single treatment of reagents was completed over a period of 3 days. The initial volume and chemical composition of the treatment injectants were based on contaminant levels, volume of area to be treated, subsurface characteristics, and the specific stoichiometry determined during the laboratory study. Injections were performed in cycles in the sequence of catalysts followed by oxidizer. A site-engineered injection apparatus was used to control flow of oxidizer and the proprietary ISOTEC catalyst into the capillary fringe of the vadose zone under a pressurized condition. The site specific molar ratio of the reagents was maintained at all times during injection.

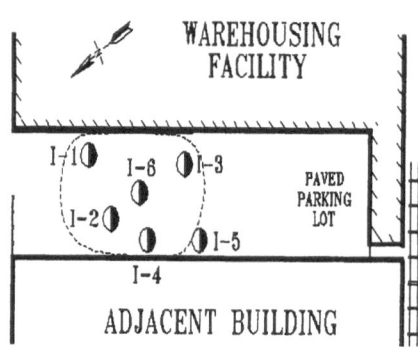

Figure 4. Injection Point Location at the Warehousing Facility.

5.3. Full-Scale Remediation

Full-scale treatment using the ISOTEC process was based on results of the pilot program. Six injection locations (I-1 through I-6) were installed, as determined by the extent of treatment during the pilot study. The location of the injection points are depicted in Fig. 4. The treatment procedures used were similar to that of the pilot program, except that the injections were performed at multiple points. The stoichiometry determined during the laboratory study and used during the pilot program was also used during this treatment. The full-scale treatment activities consisted of three treatment cycles performed over several days within a three month period.

5.4. Site Monitoring

Site monitoring data were collected and reviewed during the pilot program and full-scale remediation to obtain information relative to the treatment process and subsurface characteristics. All the monitoring wells on site were sampled according to the NJDEP sampling protocol and analyzed for volatile organics. The data from samples collected prior to the start of the pilot study served as pre-treatment data.

5.5. Results and Discussion

5.5.1. Pilot Program. The ISOTEC pilot program was designed to treat a specific volume within the source area to be tested (I-2) and to determine the extent of treatment of subsurface organic contaminants at the warehousing facility. Based on the pre-treatment versus post-treatment analytical data, over a 98.5% reduction in organic contamination was noted throughout the delineated plume area (See Fig. 5). The extent of treatment during the pilot program was estimated to be approximately 20 feet, based on the presence of radicals within hydraulically connected areas and at the ground surface over 18 feet away (from I-2) during injection activities into I-2. Hence, the full-scale treatment program was initiated using multiple injection points.

5.5.2. Full-Scale Treatment. The continued ISOTEC treatment application consisted of introducing an oxidizer and the site-specific ISOTEC catalysts into and around the injection points I-1 through I-6. Hydroxyl radical samples were collected during and after injection activities to examine the radial effects of the ISOTEC process, with positive results present at all of the injection points. Pursuant to the NJDEP's permit-by-rule approval, groundwater samples were collected from the existing monitoring wells (MW-1 through MW-7) prior to and after the ISOTEC treatment process. The monitoring well locations are shown in Fig. 3. The pre-treatment versus post-treatment analytical results for the subject site are presented in Table 2.

A review of the data confirms that all contaminants of concern were treated to either non-detect (ND) levels or below applicable regulatory criteria. A comparison of the gradual reduction in total concentration of the volatile organics for all the wells has been plotted in Fig. 6.

The warehousing facility case had been under regulatory oversight since 1991. Following review of the post-treatment sampling results, which indicated contaminant concentrations remaining below regulatory levels four months after the final treatment application, NJDEP closed the case with "No Further Action Required" declared at the

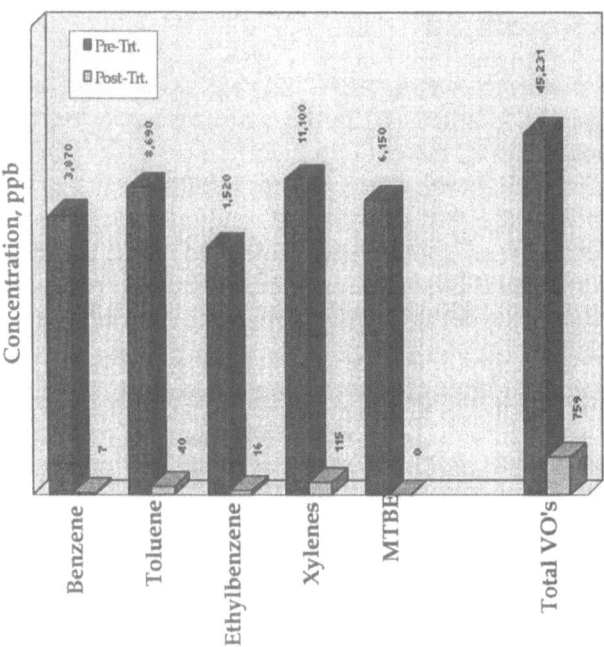

Figure 5. Pre- vs post-treatment results of pilot study at the warehousing facility. The bars represent the BTEX + MTBE concentrations. Post-treatment samples collected approximately four weeks after application.

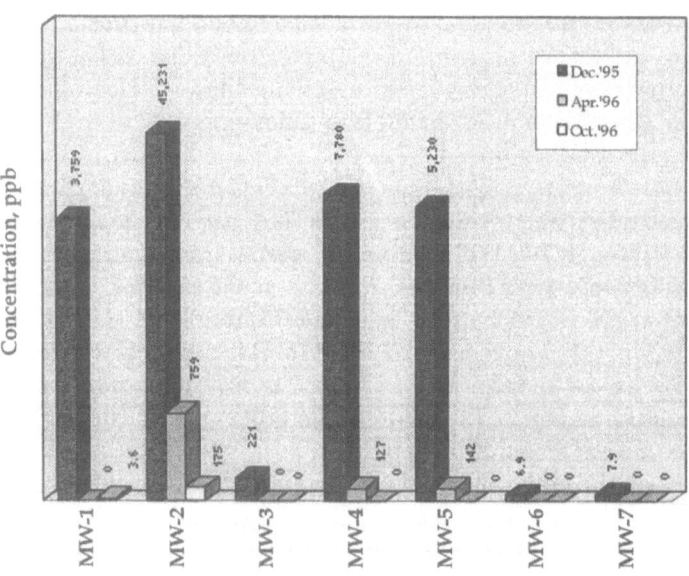

Figure 6. Results of full-scale remediation of groundwater at the warehousing facility. The bars represent the total volatile concentrations prior to application, approximately four weeks after initial application, and four months after final application, respectively.

Table 2. Results after Full-scale Remediation of the Warehousing Facility

Well #	Location/ Characteristics	Contaminants (NJ GWQ criteria, ppb)	Pre-Trt (Dec-95) (ppb)	Post-Trt-1 (Apr-96) (ppb)	Post-Trt-2 (Oct-96) (ppb)
MW-1	Down to side-gradient of the backfilled UST excavation.	Benzene (1)	880	<1	<1
		Toluene (1,000)	481	<1	<1
		Ethylbenzene (700)	100	<1	<1
		Xylenes (40)	639	<1	3.63
		MTBE (70)	1,090	<1	<1
		Non-targets	569	ND	ND
MW-2	Hydrogeologically down-gradient of the former UST field. Historically revealed the highest contaminant concentrations at the site.	Benzene (1)	3,870	6.85	<1
		Toluene (1,000)	8,690	40.4	<1
		Ethylbenzene (700)	1,520	15.5	5.11
		Xylenes (40)	11,100	115	19.6
		MTBE (70)	6,150	<1	<1
		Non-targets	13,020	581.1	149.8
MW-3	Side-gradient of the former UST field.	Benzene (1)	6.66	<1	<1
		Toluene (1,000)	<2	<1	<1
		Ethylbenzene (700)	<2	<1	<1
		Xylenes (40)	4.56	<1	<1
		MTBE (70)	200	<1	<1
		Non-targets	ND	ND	ND
MW-4	Down-gradient to side-gradient of the former UST field.	Benzene (1)	541	<1	<1
		Toluene (1,000)	25.8	<1	<1
		Ethylbenzene (700)	64.2	<1	<1
		Xylenes (40)	114	<1	<1
		MTBE (70)	6,400	<1	<1
		Non-targets	635.4	127	ND
MW-5	Side-gradient of the former UST field.	Benzene (1)	<50	<1	<1
		Toluene (1,000)	<50	<1	<1
		Ethylbenzene (700)	<50	<1	<1
		Xylenes (40)	<50	<1	<1
		MTBE (70)	5,230	<1	<1
		Non-targets	ND	142	ND
MW-6	Up-gradient to side-gradient of the former UST field. Historically revealed low concentrations at the site.	Benzene (1)	<1	<1	<1
		Toluene (1,000)	<1	<1	<1
		Ethylbenzene (700)	<1	<1	<1
		Xylenes (40)	<1	<1	<1
		MTBE (70)	6.89	<1	<1
		Non-targets	ND	ND	ND
MW-7	Down-gradient of the former UST field. Historically revealed the lowest contaminant concentrations at the site.	Benzene (1)	<1	<1	<1
		Toluene (1,000)	<1	<1	<1
		Ethylbenzene (700)	<1	<1	<1
		Xylenes (40)	<1	<1	<1
		MTBE (70)	6.78	<1	<1
		Non-targets	ND	ND	ND

Notes:
(1) Post Trt-1 is 4 weeks after initial application was completed and Post Trt-2 is 4 months after final treatment application was completed.
(2) ND = analyzed for but not detected.
(3) <(MDL) = concentration detected at a value below method detection limit (MDL).
(4) ppb = parts per billion or micrograms/liter (μg/L).
(5) MTBE = Methyl-t-butylether.
(6) Non-targets are compounds which are not on the EPA's targeted list of volatiles but have been tentatively detected in the forward library search.

site. ISOTEC's treatment activities were completed over several days within a six month period, with the case closed in under a year from start to finish.

6. SUMMARY AND CONCLUSIONS

The in situ oxidation of gasoline and waste oil contaminated groundwater was conducted at a warehousing facility in New Jersey using catalyzed hydrogen peroxide (ISOTEC[SM] process). The laboratory studies used three level central composite experiments followed by bench scale studies on representative aqueous samples to identify a chemical combination that was best suited for contaminant destruction at the subject site. The chemical combination was further verified in microcosms set up as a soil-water matrix in glass tanks. The microcosm study achieved 100% destruction of BTEX and MTBE.

Field studies were conducted at the subject site with groundwater contamination. A pilot study was initially performed with a significant contaminant destruction (98.5%) observed. In addition, hydroxyl radicals were detected in a hydraulically connected downgradient well indicating the radial effect and the potential for full-scale remediation. Full-scale remediation was conducted using similar procedures as the pilot study with multiple injection points utilized to cover the entire contaminant plume. Following three treatments and review of monthly monitoring data which indicated contaminant levels at either non-detect (ND) or below applicable regulatory criteria for four months thereafter, the subject site was closed by NJDEP declaring a "No Further Action Required" on the site.

REFERENCES

Baxendale, J.H. and Magee, J. (1953) "The Oxidation of Benzene by Hydrogen Peroxide and Iron Salts." *Disc. Faraday Soc.*, 14, 160–167.

Bigda, R.J. (1996) "Fenton's Chemistry: An Effective Advanced Oxidation Process." *Environ. Technol.*, May/June 1996, 34–39.

Diamond, W.J. (1989) "Practical Experiment Designs for Engineers and Scientists." 2nd Edn., Van Nostrand Reinhold, NY.

Dorfman, L.M. and Adams, G.E. (1973) "Reactivity of the Hydroxyl Radical". *National Bureau of Standards*, Report # NSRDS-NBS-46.

Eisenhauer, H.R. (1964) "Oxidation of Phenolic Wastes." *JWPCF*, 36, 1116–1128.

Feuerstein, W., Gilbert, E., and Eberle, S.H. (1981) "Model Experiments for the Oxidation of Aromatic Compounds by Hydrogen Peroxide in Wastewater Treatment." *Vom Wasser*, 56, 35–54.

FMC Corporation (1989) Technical Bulletin: *Hydrogen Peroxide*.

Haber, F. and Weiss, J.J. (1934) "The Catalytic Decomposition of Hydrogen Peroxide by Iron Salts". *Proc. Roy. Soc. London, Ser. A*, 147, 332–351.

Leung, S.W., Watts, R.J., and Miller, G.C. (1992) "Degradation of Perchloroethylene by Fenton's Reagent: Speciation and Pathway." *J. Environ. Qual.*, 21, 377–381.

Mertz, J.H. and Waters, W. (1949) *J. Chem. Soc.*, 2427.

Murphy, P., Murphy, W.J., Boegli, M., Price, K., and Moody, C.D. (1989) "A Fenton-like Reaction to Neutralize Formaldehyde Waste Solutions." *Environ. Sci. Technol.*, 23, 166–169.

Ravikumar, J.X. and Gurol, M.D. (1994) "Chemical Oxidation of Chlorinated Organics by Hydrogen Peroxide in the Presence of Sand." *Environ. Sci. Technol.*, 28, 394–400.

Thomas, J.M. and Ward, C.H. (1989) "In situ Biorestoration of Organic Contaminants in the Subsurface". *Environ. Sci. Technol.*, 23, 760–766.

Watts, R.J., Udell, M.D., Rauch, P.A., and Leung, S.W. (1990) "Treatment of Pentachlorophenol Contaminated Soils using Fenton's Reagent." *Haz. Waste Haz. Mater.*, 7, 335–345.

Watts, R.J., Smith, B.R., and Miller, G.C. (1991) "Treatment of Octachlorodibenzo-p-dioxin (OCDD) in Surface Soils Using Catalyzed Hydrogen Peroxide." *Chemosphere.*, 23, 949–956.

Watts, R.J., Kong, S., Dippre, M., and Barnes, W.T. (1994) "Oxidation of Sorbed Hexachlorobenzene in Soils Using Catalyzed Hydrogen Peroxide." *J. Haz. Mat.*, 39, 33–47.

Watts, R.J. (1997) "Hazardous Wastes: Sources, Pathways, Receptors". John Wiley & Sons, Chapter 5, Publication scheduled for Fall 1997.

14

DEVELOPMENT OF AN ENHANCED OZONE-HYDROGEN PEROXIDE ADVANCED OXIDATION PROCESS

Kevin C. Bower, Stephen Duirk, and Christopher M. Miller*

Department of Civil Engineering
210 Auburn Science and Engineering Center
University of Akron
Akron, Ohio 44325-3905

ABSTRACT

Advanced chemical oxidation processes (AOP's), particularly processes using *ozone*, involve a rapidly growing area of environmental research and applications. A need exists, however, for improvements to AOP's to make them less costly and more effective. This paper evaluated the addition of a fixed bed of *sand* to the conventional H_2O_2-O_3 process as a function of reactor configuration, pH, and sand type. The West Liberty sand consisting of the greatest concentrations of iron and manganese yielded the best results, showing 15% more phenol removal than Muscatine sand and 28% more removal than no sand (i.e. H_2O_2-O_3). Phenol degradation product formation and disappearance rates in comparison to the direct ozonation and O_3-H_2O_2 processes indicate metal-oxide sand surface-catalyzed H_2O_2 decomposition improves the conventional O_3-H_2O_2 advanced oxidation process at pH 7. Analysis of results suggests the enhancement is a result of increased formation of essential oxygen radical intermediates (i.e. hydroxyl radical). Direct ozonation, however, was equally effective at phenol removal at pH 8.9 because the conjugate base of *phenol* (i.e. phenate ion) has a direct ozonation reaction rate constant six orders of magnitude greater than phenol.

1. INTRODUCTION AND BACKGROUND

Hazardous organic chemicals in drinking water supplies continue to be an environmental threat. One class of technologies to clean up contaminated water includes

---

*Corresponding author Fax: (330) 972-6020; e-mail address: cmmiller@uakron.edu

Emerging Technologies in Hazardous Waste Management 8, edited by Tedder and Pohland
Kluwer Academic/Plenum Publishers, New York, 2000.

advanced oxidation processes (AOP). These processes involve the application of strong oxidants such as ozone (O_3) and hydrogen peroxide (H_2O_2). The oxidants directly react with organic contaminants and degrade them,[3] or the oxidants themselves react under certain conditions to form hydroxyl radical ($\cdot OH$), a highly reactive species capable of degrading organics.[2] The combined application of ozone and hydrogen peroxide has been shown particularly effective.[4] Depending on contaminant levels, water quality, and the required ozone dose, however, this process can be energy and chemical intensive. Attempts to improve the ozone-hydrogen peroxide process have primarily used ultraviolet light (UV) to initiate reactions.[1] This has shown promise, but UV requires additional energy input and engineering considerations (i.e. equipment and electrical requirements).

The chemical reactions and kinetics associated with the O_3-H_2O_2 process are well known. A summary of the principal reactions is shown in Table 1, including phenol degradation by direct ozonation and hydroxyl radical reaction.[5,12] The initiation reactions (reaction 1 and 2) are both solution pH driven due to equilibrium speciation of hydroxide ion (OH^-) and hydroperoxide (HO_2^-) from hydrogen peroxide. Kinetics of the initiation reactions (1–2) are relatively slow compared to the propagation (4–6) and degradation (7–8) reactions. For example, reaction 2 has a rate constant of $70\,M^{-1}s^{-1}$, eight orders of magnitude slower than reactions involving hydroxyl radical.[4] It should also be understood that superoxide formation is pH dependent because perhydroxyl radical generated in reaction 2 is in equilibrium ($pK_a = 4.8$) with superoxide (reaction 3). Therefore, at pH greater than six, almost all perhydroxyl radical generation results in additional superoxide formation.

Several studies have indicated metal oxide reaction with H_2O_2 leads to the formation of perhydroxyl radical and superoxide[7,10] according to:

$$S^+ + H_2O_2 \rightarrow S + H^+ + \cdot HO_2 \tag{1}$$

$$\cdot HO_2 \leftrightarrow H^+ + O_2^- \tag{2}$$

Based on their results and the work of Miller and Valentine,[9] we propose metal-oxide surface-catalyzed H_2O_2 decomposition can enhance the conventional O_3-H_2O_2 advanced oxidation process by increased formation of perhydroxyl radical ($\cdot HO_2$) and superoxide anion (O_2^-). The fundamental premise for the enhanced process lies in the formation of superoxide. Specifically, if the initiation reactions leading to superoxide generation are bypassed, ozone could react directly with superoxide and ultimately form hydroxyl radical

Table 1. Ozone-H_2O_2 Mechanism Including Phenol Degradation (Staehelin and Hoigne, 1982; Gurol and Singer, 1983)

No.	Reaction
1	$HO_2^- + O_3 \rightarrow O_3^- + \cdot HO_2$
2	$OH^- + O_3 \rightarrow O_2^- + \cdot HO_2$
3	$\cdot HO_2 \leftrightarrow H^+ + O_2^-$
4	$O_2^- + O_3 \rightarrow O_3^- + O_2$
5	$O_3^- + H^+ \rightarrow HO_3$
6	$HO_3 \rightarrow \cdot OH + O_2$
7	$O_3 + Phenol \rightarrow Products$
8	$\cdot OH + Phenol \rightarrow Products$

much faster and more efficiently. Enhanced organic contaminant degradation will not occur, however, unless the proper metal-oxides are used and the contact time between the metal-oxide and hydrogen peroxide are optimized. The objective of this paper is to evaluate the addition of a fixed bed of sand to the conventional H_2O_2-O_3 process on contaminant degradation as a function of reactor configuration, pH, and sand type.

2. EXPERIMENTAL

2.1. Materials and Chemicals

The water used in all of the experiments was deionized with a Barnstead NANOpure water system. Phenol (99% purity), catechol (99% purity), hydroquinone (99% purity), and resorcinol (99% purity) were purchased from Sigma Chemical Company (St. Louis, MO). Phenol was selected as the model contaminant because of well-characterized reactions with ozone and hydroxyl radical leading to formation of hydroquinone, catechol, and resorcinol.[6] Hydrogen peroxide (30% H_2O_2) was purchased from Fisher Scientific (Pittsburgh, PA). Ozone was generated using an Ozotech (Yreka, CA) generator model OZ4BTU with an air power prep module model 2450 (also manufactured by Ozotech). Two sand materials were examined in this study. Characteristics of the two materials are presented in Table 2. A commercially available filter sand (hereafter referred to as Muscatine), with an effective size of 0.49 mm, was obtained from the Northern Gravel Company (Muscatine, IA) and used filter sand with an effective size of 0.50 mm, was obtained from the West Liberty, IA drinking water treatment plant (hereafter referred to as West Liberty).

2.2. Analytical Methods

Phenol was analyzed using a Shimadzu high-pressure liquid chromatograph (HPLC) with a Supelco LC-8 column (15 cm × 4.6 mm ID). The mobile phase consisted of 30% HPLC grade acetonitrile and 70% deionized water at a flow rate of 0.8 ml/minute. The phenol peak retention time under these conditions was approximately 6.6 minutes. The UV detector was set at 280 nm. Phenol degradation products were also measured under these conditions with peak retention times of approximately 4.3 (catechol), 3.3 (hydroquinone), and 3.7 (resorcinol) minutes, respectfully. Solution pH was measured with a Cole Parmer (Chicago, IL) pH meter after appropriate calibration. Aqueous ozone concentration was measured using the indigo method.[8]

2.3. Reactor Systems

Two reactor configurations were investigated (Fig. 1). The first system was a continuously stirred tank reactor (CSTR) consisting of a 500 ml Erlenmeyer flask covered with aluminum foil in which two separate inputs were mixed. The two inputs consisted of an aqueous ozone solution (made by pumping ozone gas into a 6″ diameter capped PVC pipe packed with plastic media and pH 4 buffered water) and a H_2O_2 and phenol solution. The H_2O_2 and phenol solution was pumped across a sand column (hydraulic retention time = 0.2 minutes) before mixing in the CSTR. All solutions were pumped with a Cole Palmer (Chicago, IL) peristaltic pump. Mixing in the CSTR was achieved using a Thermolyne™ stir plate at setting 2 (Scale 1–10). A constant volume was maintained

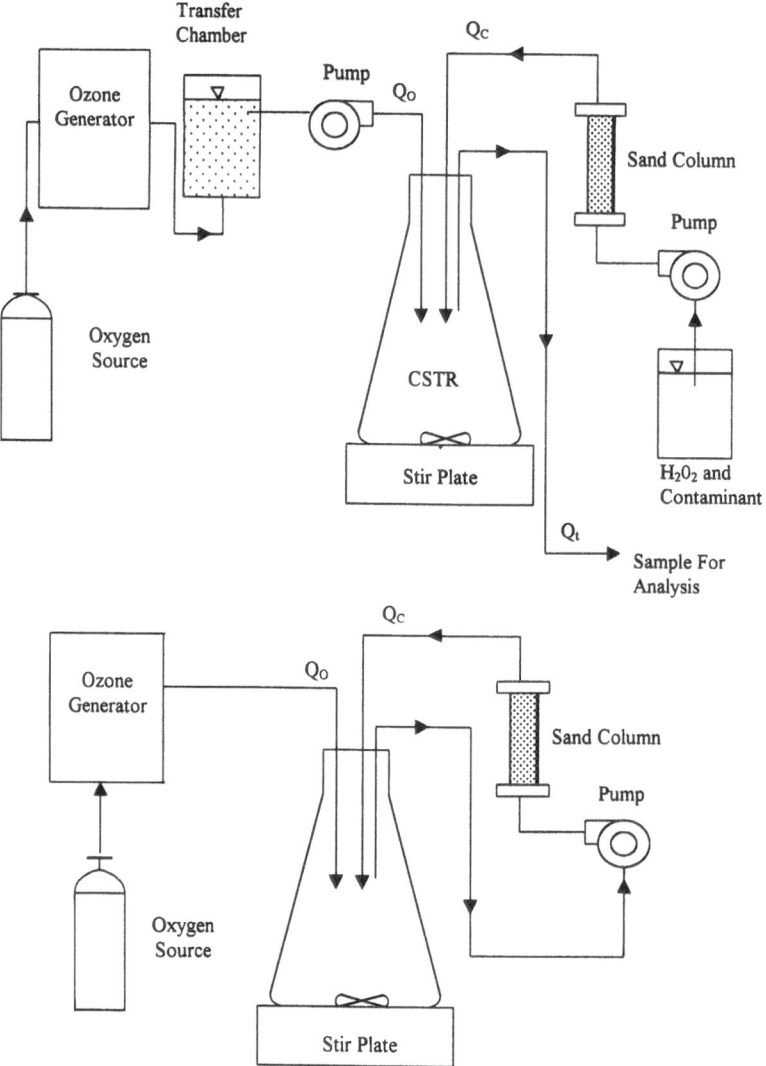

Figure 1. Reactor configurations: CSTR (top), Semi-batch with recycle (bottom).

in the CSTR by pumping out of the CSTR at the same rate of solution addition into it. The second reactor configuration was a semi-batch with recycle (Fig. 1). Ozone gas was continuously supplied to the reactor through an aluminum silicate cylindrical diffusion stone. Solutions were pumped in a recycled configuration through the sand column (15 cm × 1.5 cm) and mixed using a Thermolyne™ stir plate at setting 2 (Scale 1–10). Direct reaction of hydrogen peroxide with phenol and volatilization/stripping were minimal, with less than 2% loss measured in control experiments.

3. RESULTS AND DISCUSSION

Figure 2 shows the effects of sand addition on phenol degradation in the CSTR reactor configuration. West Liberty sand showed the greatest phenol loss (35%), 15% greater than Muscatine and 28% greater than the conventional O_3-H_2O_2 (i.e. no sand)

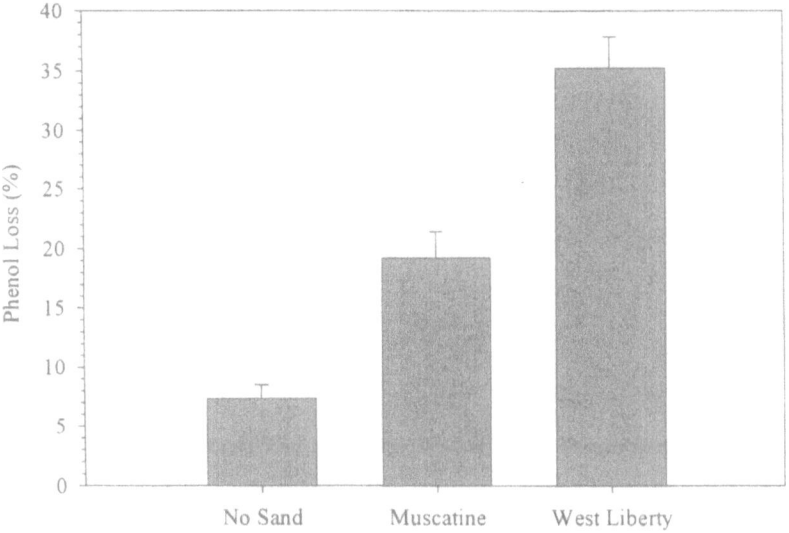

Figure 2. Phenol degradation in CSTR with and without sand addition. CSTR hydraulic residence time = 20 minutes, $[Phenol]_o$ = 10 mg/L, $[H_2O_2]_o$ = 25 mg/L, $[O_3]_o$ = 1.5 mg/L, and pH = 7.0. Reported values are the mean ± one standard deviation (n = 3).

process. Clearly, there is enhanced removal due to the addition of the sand (i.e. exposure of H_2O_2 to sand before mixing with O_3). The observed difference between West Liberty and Muscatine is likely related to their different material properties (Table 2), as West Liberty has greater concentrations of both iron and manganese. It is not possible to distinguish yet what iron, manganese, or a combination thereof, is the optimum for phenol removal. Based on West Liberty sand showing the greatest phenol loss in the CSTR, all remaining experiments were performed using West Liberty sand.

Phenol degradation was also measured in a semi-batch recycle reactor configuration (Fig. 1) under O_3 only (Fig. 3), O_3-H_2O_2 (Fig. 4), and O_3-H_2O_2-West Liberty sand (Fig. 5) conditions. Figure 3 shows phenol degradation by direct ozonation at pH 7 as a function of time and the corresponding formation of catechol, hydroquinone, and resorcinol. Greater than 90% of the initial phenol concentration was degraded in 19 minutes. As phenol degrades, there is a rapid increase and then decline in catechol, hydroquinone, and resorcinol concentrations. The peak hydroquinone concentration (0.014 mM) occurs after 19 minutes and the peak catechol concentration (0.008 mM) occurs after 11 minutes. The peak concentrations of hydroquinone and catechol correspond to 14 and 8% of the

Table 2. Material Characterization

Characteristic	Muscatine	West Liberty
*Organic Carbon	<0.1%	<0.1%
+Surface Area, m²/g	1.04	1.21
Total Iron, mg/kg	2,400 ± 105	4,000 ± 220
Total Manganese, mg/kg	250 ± 25	330 ± 35
pH	7.4	7.1

*Analysis performed by Minnesota Valley Testing Laboratories (Nevada, IA).
+Analysis (N₂-BET isotherm) performed by Porous Materials, Inc. (Ithaca, NY).

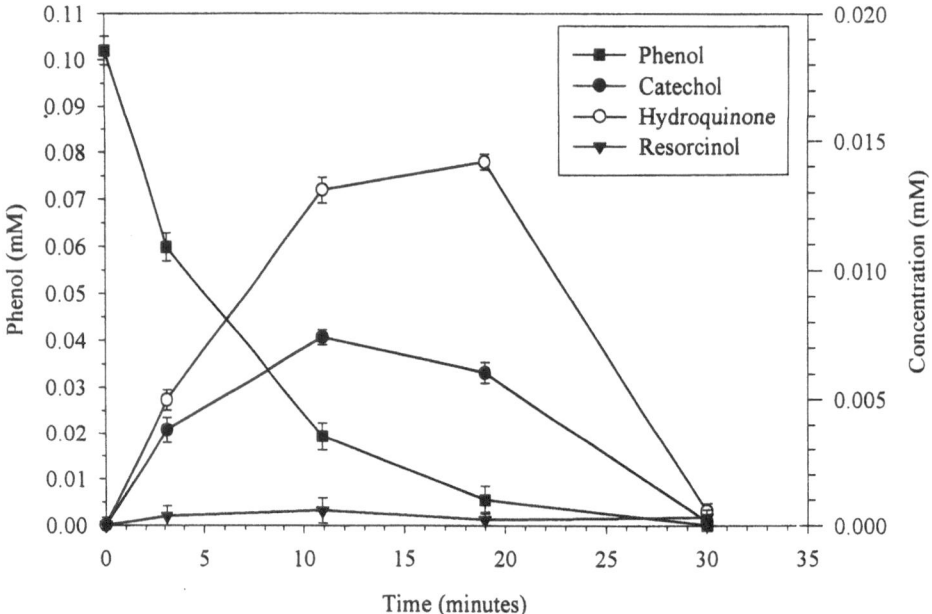

Figure 3. Phenol degradation by direct ozonation in the semi-batch recycle apparatus as a function of time. $[Phenol]_o = 9.7 \, mg/L$, $[O_3]_o = 1.5 \, mg/L$, and pH = 7.0. Pump flow rate = 20 ml/minute. Reported values are the mean ± one standard deviation (n = 3).

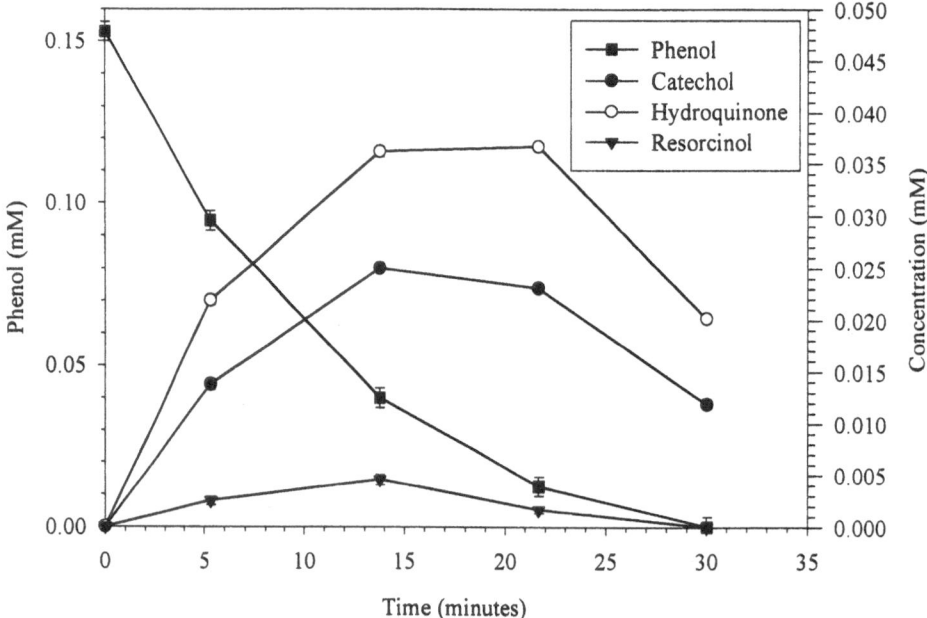

Figure 4. Phenol degradation by ozone and hydrogen peroxide in semi-batch apparatus as a function of time. $[Phenol]_o = 14.6 \, mg/l$, $[H_2O_2]_o = 100 \, mg/l$, $[O_3]_o = 1.5 \, mg/l$, and pH = 7.0. Pump flow rate = 20 ml/minute. Reported values are the mean ± one standard deviation (n = 3).

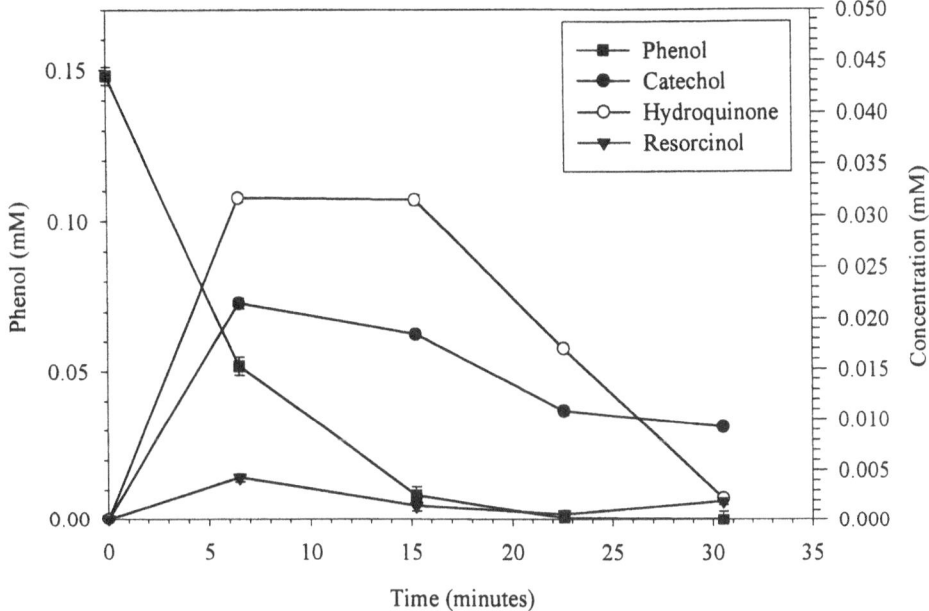

Figure 5. Phenol degradation by H_2O_2-O_3-West Liberty sand in semi-batch recycle apparatus as a function of time. [Phenol]$_o$ = 14.6 mg/l, [H_2O_2]$_o$ = 100 mg/l, [O_3]$_o$ = 1.5 mg/l, and pH = 7.0. Pump flow rate = 20 ml/minute. Reported values are the mean ± one standard deviation (n = 3).

initial phenol respectively. During oxidation of phenol by hydroxyl radical generated in a pH driven ozone decomposition mechanism, hydroquinone and catechol typically account for greater than 95% (mass basis) of the primary (i.e. initial) product distribution.[6] Based on the measured product distribution (Fig. 3), therefore, ring-cleavage by direct ozonation must also be contributing to a significant portion of phenol degradation.[5]

Significant phenol degradation was also observed in the semi-batch recycle reactor by the O_3-H_2O_2 process at pH 7 (Fig. 4). Similar to direct ozonation, the primary products were hydroquinone, resorcinol, and catechol. Peak concentration times for catechol and hydroquinone were similar to direct ozonation, but their concentrations were much greater. Specifically, the peak hydroquinone concentration (0.037 mM) occurred between 14 and 21 minutes, and the peak catechol concentration (0.026 mM) was measured after approximately 14 minutes. A portion of the higher measured peak concentration of hydroquinone and catechol can be attributed to the larger initial phenol concentration (14.6 vs. 9.6 mg/l), but the peak concentrations are approximately 2–3 times greater than their counterparts during direct ozonation. Therefore, it is the combination of direct ozonation and enhanced hydroxyl radical production by the hydrogen peroxide driven ozone decomposition mechanism (Table 1) that causes the increased rate of product formation shown in Fig. 4. The overall rate of phenol degradation, however, is similar in the two systems. This could be due to parallel reactions involving competition for ozone by hydrogen peroxide and phenol (i.e. reduced phenol degradation by direct ozonation) and hydroxyl radical formation from the O_3-H_2O_2 process (i.e. increased phenol degradation by hydroxyl radical).

Addition of West Liberty sand to the O_3-H_2O_2 process at pH 7 also resulted in substantial phenol degradation (Fig. 5). Greater than 93% of the initial phenol concentration was degraded in 15 minutes. By comparison, the O_3-H_2O_2 process took approximately 7 minutes longer to achieve the same removal. Similar to the O_3-H_2O_2 process, the primary products were hydroquinone, resorcinol, and catechol, however, their rate of formation and disappearance were more rapid in the presence of the sand. Hydroquinone reached a peak concentration of 0.034 mM after 7 minutes and was almost nondetectable after 30 minutes. The peak catechol concentration was 0.023 mM, achieved after approximately 7 minutes.

Figure 6 is a comparison of the overall removal of phenol and its observed degradation products in the conventional O_3-H_2O_2 advanced oxidation process to the O_3-H_2O_2-West Liberty sand process at pH 7. Clearly, metal-oxide sand surface-catalyzed H_2O_2 decomposition improves the conventional O_3-H_2O_2 advanced oxidation process at pH 7. Figure 7 shows a similar comparison between direct ozonation and the sand-enhanced process at pH 8.9. Direct ozonation, however, is just as effective as the sand-enhanced process at removing phenol. This can be explained by the disassociation behavior of phenol,[11] as phenol (pKa = 9.82) and its conjugate base phenate ion have significantly different reaction rate constants with ozone. The importance of this regarding phenol degradation by ozone is that phenate ion has a reaction rate constant with ozone that is six orders of magnitude greater than phenol.[8] Therefore, direct ozonation of phenol, and in particular phenate ion, is likely the predominant degradation mechanism at pH 8.9. Finally, with the only variation between the CSTR and semi-batch with recycle reactor configurations being the initial H_2O_2 concentration, the semi-batch reactor (Fig. 6) appears to be more effective (i.e. faster and more complete phenol degradation) than the CSTR configuration (Fig. 2). This is likely due to the constant input of ozone gas and continuous generation of superoxide resulting from recycling hydrogen peroxide across

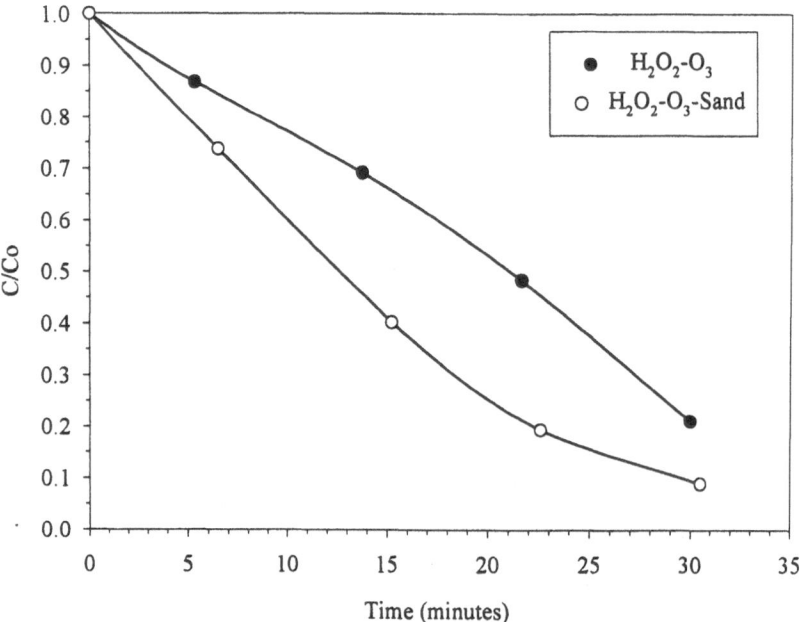

Figure 6. Comparison between H_2O_2-O_3 and H_2O_2-O_3-West Liberty sand process removal of phenol and its degradation products at pH 7.0. [Phenol]$_o$ = 14.6 mg/L, [H_2O_2]$_o$ = 100 mg/L, and [O_3]$_o$ = 1.5 mg/L. Pump flow rate = 20 ml/minute.

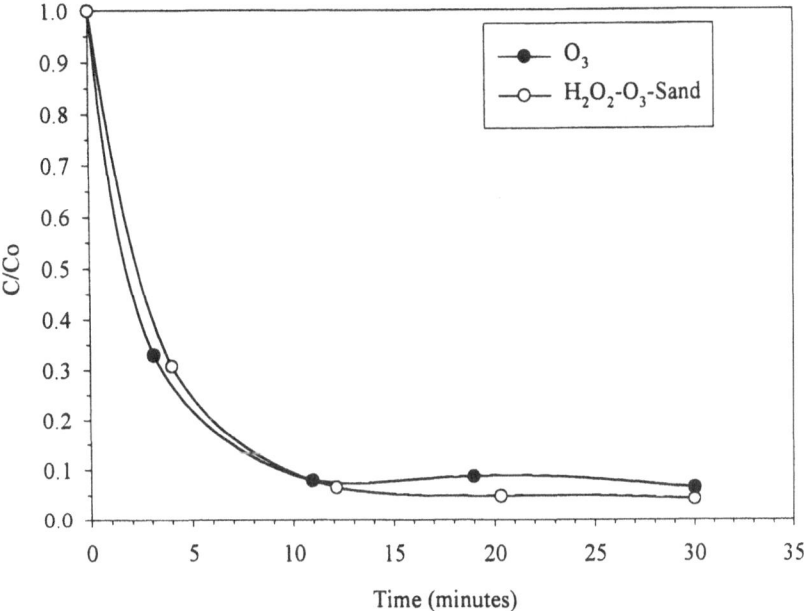

Figure 7. Comparison between O_3 and H_2O_2-O_3-West Liberty sand process removal of phenol and its degradation products at pH 8.9. [Phenol]$_o$ = 14.6 mg/L, [H_2O_2]$_o$ = 100 mg/L, and [O_3]$_o$ = 1.5 mg/L. Pump flow rate = 20 ml/minute.

the sand bed. Additional experiments are needed, however, to further characterize the effect of pH, H_2O_2, and individual metal oxide sand coatings.

ACKNOWLEDGMENTS

This research was supported by the University of Akron Faculty Research Program, the U.S. Army 88th Regional Support Command, GPD Associates (Akron, Ohio), and the Ohio Board of Regents Research Grant. No endorsement by granting agencies should be inferred. Derek Losh assisted with constructing the semi-batch with recycle reactor configuration.

REFERENCES

1. Beltran, F., Encinar, J., and Gonzlez, J., "Industrial Wastewater Advanced Oxidation: Part 2. Ozone Combined With Hydrogen Peroxide of UV Radiation," *Water Research* **1997**, *31*, 2415–2428.
2. Buxton, G.V., Greenstock, C.L., Helman, W.P., and Ross, A.B., "Critical review of rate constants for reactions of hydrated electrons, hydrogen atoms, and hydroxyl radicals (•OH/•O–) in aqueous solution," *Journal of Physical Chemistry Reference Data* **1988**, *17*, 513–886.
3. Glaze, W.H., "Drinking-water treatment with ozone," *Environmental Science and Technology* **1987**, *21*, 224–230.
4. Glaze, W.H. and Kang, J., "Advanced oxidation processes for treating groundwater contaminated with TCE and PCE: laboratory studies," *Journal American Water Works Association* **1988**, *80*, 57–63.
5. Gurol, M.D. and Singer, P.C., "Dynamics of the ozonation of Phenol—II," *Water Research* **1983**, *17*, 1173–1181.

6. Huang, C.R. and Shu, H.Y., "The reaction kinetics, decomposition pathways and intermediate formations of phenol in ozonation, UV/O3 and UV/H2O2 processes," *Journal of Hazardous Materials* **1995**, *41*, 47–64.

7. Kitajima, N., Fukuzumi, S., and Ono, Y., "Formation of superoxide ion during the decomposition of hydrogen peroxide on supported metal oxides," *The Journal of Physical Chemistry* **1978**, *82*, 1505–1509.

8. Langlais, B., Reckhow, D.A., and Brink, D.Ed. 1991. *Ozone in Water Treatment*; Lewis Publishers: Chelsea, pp. 569.

9. Miller, C.M. and Valentine, R.L., "Hydrogen peroxide decomposition and quinoline degradation in the presence of aquifer material," *Water Research* **1995**, *29*, 2353–2359.

10. Ono, Y., Matsumura, T., Kitajima, N., and Fukuzumi, S., "Formation of superoxide ion during the decomposition of hydrogen peroxide on supported metals," *Journal of Physical Chemistry* **1977**, *81*, 1307–1311.

11. Schwarzenbach, R.P., Gschwend, P.M., and Imboden, D.M. 1993. *Environmental Organic Chemistry*; Wiley-Interscience: New York.

12. Staehelin, J. and Hoigne, J., "Decomposition of ozone in water: rate of initiation by hydroxide ions and hydrogen peroxide," *Environmental Science and Technology* **1982**, *16*, 676–681.

15

POLYETHYLENE ENCAPSULATION OF DEPLETED URANIUM TRIOXIDE[1]

J. W. Adams, P. R. Lageraaen, P. D. Kalb, and B. R. Patel

Brookhaven National Laboratory
Department of Advanced Technology
Environmental & Waste Management Group
Building 830
PO Box 5,000
Upton, New York 11973-5000

1. ABSTRACT

Depleted uranium, in the form of uranium trioxide (UO_3) powder, was encapsulated in molten polyethylene forming a stable, dense composite henceforth known as DUPoly (patent pending). Materials were fed by calibrated volumetric feeders to a single screw extruder where they were heated and mixed to form a homogeneous molten extrudate. Oxide loadings as high as 90 weight percent (wt%) UO_3 were successfully processed, yielding a maximum product density of $4.2 \, g/cm^3$. Performance testing included compressive strength, water immersion, and leach testing. Compressive strengths of samples with 50–90 wt% UO_3 were nearly constant, with a mean value exceeding 15 MPa (2,200 psi). Leach rates, which increased as a function of sample waste loading, were less than 1.1% after 11 days at ambient temperature for samples containing 90 wt% UO_3. Ninety day water immersion tests showed sensitivity to "batch" processed UO_3 for samples containing >85 wt% of the oxide. Considering that UO_3 should be insoluble in water, these results indicate the probable presence of other, more soluble uranium compounds. Samples containing UO_3 produced by a "continuous" process showed no deterioration at up to 90 wt% waste loadings.

2. INTRODUCTION

U.S. Department of Energy (DOE) facilities maintain large inventories of depleted uranium (DU). Novel applications are currently being sought to convert these materials

---

[1] Work performed under the auspices of the U.S. Department of Energy

Emerging Technologies in Hazardous Waste Management 8, edited by Tedder and Pohland
Kluwer Academic/Plenum Publishers, New York, 2000.

to stable, useful secondary products. Uses that provide a positive benefit to society while allowing potential recovery or extraction of the uranium are desirable, but techniques for stabilization of DU for long-term storage or disposal are also being evaluated. Potential applications will likely exploit the high density, shielding effectiveness and nuclear applicability of these materials. This study, in particular, was initiated to investigate the feasibility of processing DU (e.g., UO_3 powder) by polyethylene microencapsulation, to mitigate potential health effects and produce useful radiation shielding and other products.

Natural uranium ore in the form of U_3O_8 contains about 0.7 wt% of the fissionable isotope ^{235}U, with the remainder of uranium present as ^{238}U. Reactor fuel is produced from the ore by converting it to uranium hexafluoride gas and enriching the proportion of ^{235}U to around 3.5%, leaving the remaining portion depleted in ^{235}U. This residual material, with ^{235}U concentrations at around 0.25%, is known as depleted uranium. Approximately 560,000 metric tons of DU in the form of UF_6, containing an equivalent mass of 379,000 metric tons of uranium, are stored at the DOE Paducah, Portsmouth, and Oak Ridge Gaseous Diffusion plants. Some of the UF_6 has been converted to uranium trioxide (UO_3); about 20,000 metric tons of DU are currently stored at the Savannah River Site. UO_3 from Savannah River was used in this preliminary investigation.

Alternatives for management of the DU inventory under consideration by the U.S. DOE include: (1) continue current management plan (no action); (2) revise current practices for long-term storage as either UF_6 or in an oxide form; (3) use of DU in shielding or high density applications; (4) disposal of DU[1]. Since uranium and uranium oxides are considered valuable resources, use of the material (option 3) is most attractive. As DUPoly will attenuate neutron as well as gamma radiation, there is currently significant interest for its use in spent fuel storage and transport casks and low-level waste storage, transport and disposal packages. The high density and workability of the material make it appealing for use as a ballast in nautical and aeronautical applications.

Treatment of DU materials by polyethylene encapsulation is a desirable option because of the immediate availability of the technology and proven record to effectively and efficiently process similar powder and granular materials. In addition, the process is very flexible. Polyethylene products can be heated and reworked if future needs change. DU can potentially be retrieved from DUPoly by chemical and/or thermal processing if needed as a resource in the future. Over the last twelve years, Brookhaven National Laboratory (BNL) has extensively developed the polyethylene encapsulation extrusion process for low-level radioactive, hazardous, and mixed wastes.[2-9] During processing, filler materials (e.g., waste) are mechanically mixed into the molten polyethylene binder, producing a workable homogeneous product. The process is not susceptible to chemical interactions between the waste and binder, enabling a wide range of acceptable waste types, high waste loadings, and technically simple processing under heterogeneous waste conditions. The process has evolved from proof-of-principle, through bench-scale development and testing, to full-scale technology demonstration and technology transfer.

3. EQUIPMENT AND PROCEDURES

Representative samples of depleted UO_3 powder were obtained from Westinghouse Savannah River Company (WSRC). Savannah River alone maintains over 22 million kg (50 million pounds) of depleted UO_3 stored in 55-gallon drums. This inventory consists of material of two distinct lots corresponding to two different processes used to prepare the oxide, batch and continuous. Over 99% of this inventory was produced with the batch process. The remaining 1% of the inventory, produced by a newer continuous process, is chemically identical but is characterized by a slightly larger particle size. Extrusion process runs described below were carried out using the batch process material.

The UO_3 materials shipped from WSRC were characterized in a recent report by Carolina Metals, Inc.[10] The drummed material was described as a 200 mesh (74 μm average particle size), 96.5% uranium trioxide with trace impurities of aluminum, iron, phosphorous, sodium, silicon, chromium, and nickel. The material has a bulk density range of about 2.5 g/cm^3 (uncompacted) to about 3.6 g/cm^3 (compacted). The ^{235}U content was assayed at approximately 0.2% and the plutonium content at 3 ppb. Gross gamma was 53,100 dpm per gram of uranium.

A 32-mm (1.25-inch diameter) single-screw, non-vented, Killion extruder was used for processability testing as shown in Fig. 1. The extruder is equipped with a basic metering screw, three heating/cooling barrel zones, and an individually heated die. The polyethylene and the DU powder were metered to the extruder through AccuRate, 300 Series, volumetric feeders. These feeders are designed to provide a constant volume output at a given operating setting (varies as a percentage from zero to 100% output). Due to differing densities of feed materials, calibration of the feeders was required with each material. The resulting data provided feeder output in g/min versus feeder speed setting.

For a given process run, a number of different samples were taken in replicate (typically ten for statistical assurance) to monitor the process and characterize the extruder output. *Rate samples* were one minute collections of the extruder output to determine process consistency over time. Low variation between replicate rate samples indicated the output was continuous and that the material was successfully processed at a given DU loading. *Grab samples* were small (approximately 3–10 g) specimens of the output product taken for pycnometer density measurement. Low variation between replicate grab samples indicated that the DU material fed well and was consistently well mixed with the polyethylene as it was processed in the extruder. *2 × 4 samples*, nominal 5.1-cm (2.0 in) diameter by 10.2-cm (4.0 in) tall right cylindrical specimens, were fabricated for compressive strength[11] and immersion testing as shown in Fig. 2. Specimens were formed in pre-heated brass molds, compressed with up to 0.2 MPa (25 psi) pressure. *ALT samples*, for use in Accelerated Leach Testing,[12] were nominal 2.5-cm (1.0 in) diameter by 2.5-cm (1.0 in) tall right cylinders fabricated in individual Teflon molds. These samples were also compression molded with up to 1.7 MPa (250 psi) pressure. *Disk samples*, for use in future attenuation studies, were formed in glass petri dishes with <0.1 MPa (10 psi) pressure. Disks were nominally 11.7-cm (4.5 in) diameter and were fabricated at varying thicknesses, up to approximately 2.5 cm (1.0 in).

4. PROCESS RESULTS

Processability testing with UO_3 was initially conducted at a loading of 50 wt%. This loading was selected based on previous microencapsulation experience. Process results at 50 wt% were not immediately successful because the extrudate contained trapped gases, producing voids in samples during cooling. DU powders were oven-dried overnight at the maximum process temperature (160 °C) to ensure removal of any excess moisture prior to extrusion.

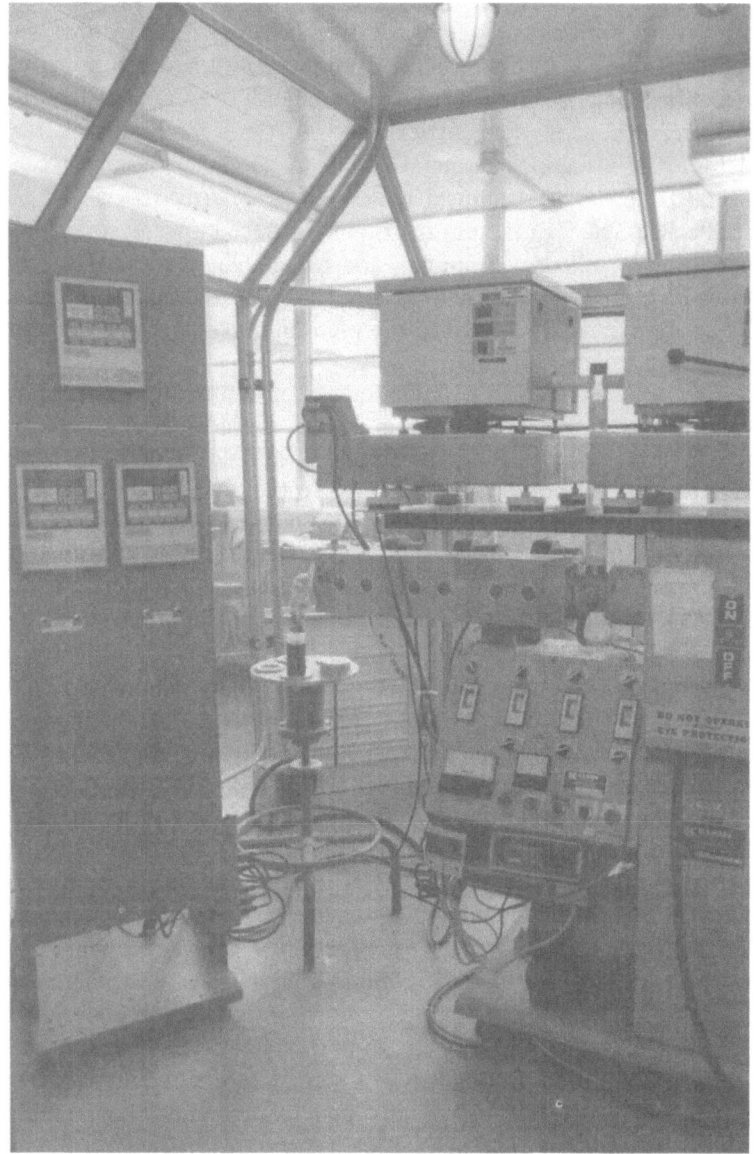

Figure 1. 32 mm Single-screw Extruder Used for Processibility Testing.

Processing with dried DU produced excellent results. DU was successfully processed at loadings of 50, 60, 70, 75, 80, 85, and 90 wt%. As DU loading increased, the extrudate became more viscous and dry, with a rough surface texture due to the decreasing amounts of polyethylene available to encapsulate and lubricate the DU particles. However, even at 90 wt%, the DU was readily processible, and DUPoly samples could be successfully cast with the aid of compression molding. At 95 wt% DU, the material plugged causing output to cease and die pressures to rise above their alarm set point (25 MPa). At this loading, there was insufficient polyethylene to mix, wet and convey the DU through the extruder barrel. DU flow was stopped immediately after noting the plugged condition. The clog was voided within several minutes of stopping the DU flow,

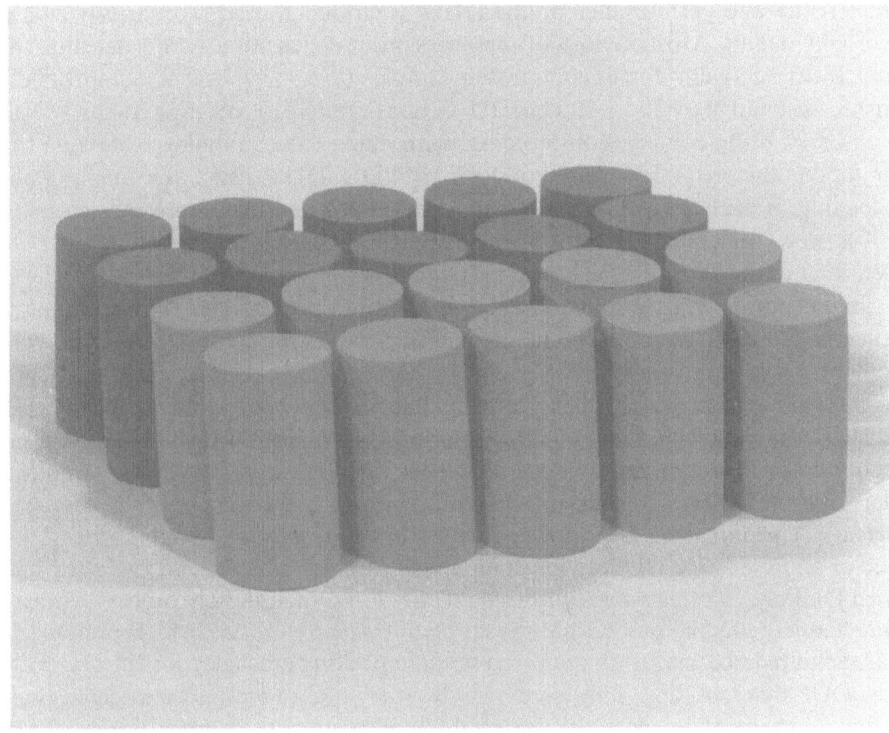

Figure 2. DUPoly "2x4" Cylinders Containing 70 Wt% Batch-process (Front) and Continuous-process (Rear) Depleted UO$_3$.

introducing pure polyethylene to the screw. Motor load, as measured by current draw, rose slightly during this episode, but remained within acceptable limits.

The overall success encountered during processing was evidenced from *rate* and *grab* sample statistics. The low deviation and percent error between ten replicate samples taken at each DU loading, shown in Table 1, indicate that the UO$_3$ powders processed continuously and consistently. These runs were conducted at screw speeds of either 60 or 65 rpm and at combined (DU + binder) feed rates of 100 to 120 g/min. Feed rates were increased proportional to DU loading, as the actual volume of feed material decreased with increased DU loadings.

5. PRODUCT CHARACTERIZATION

The DUPoly product was characterized by density measurement, compressive strength testing, and leach and immersion testing in deionized water. Physical

Table 1. Process Rate Samples (g/min)

	DU Loading					
	50 wt%	60 wt%	70 wt%	75 wt%	80 wt%	85 wt%
Mean	114.23	109.23	111.69	117.78	125.63	124.13
Std.Dev.	3.45	2.71	3.37	1.48	2.27	2.87
2 sigma	2.47	1.94	2.41	1.06	1.62	2.05
% Error	2.16	1.76	2.16	0.90	1.29	1.65

characteristics and performance of the DUPoly product varied significantly as a function of DU loading. Most obvious of these was product density. DUPoly densities ranged from 1.38 to 3.93 g/cm^3 for uncompressed samples (*disk*, *2x4*, and uncompressed *ALT* forms) for the range of 50 to 90 wt% DU. A density increase of approximately 10–15% was observed using compression molding, with mean values ranging from 1.62 to 4.25 g/cm^3 for compressed *ALT* forms at 50 to 90 wt% DU. DU density, as a function of wt% DU loading, is depicted in Fig. 3 for both compressed and uncompressed samples. DU densities were calculated by multiplying the mean DUPoly density times the weight percent DU in the DUPoly, for a given DU loading.

Initial process runs were conducted using batch process UO$_3$. Process runs using continuous process UO$_3$ produced nearly identical values for compressed forms, whereas uncompressed sample densities were slightly higher than corresponding batch process samples. This artifact was probably attributable to improved molding technique with subsequent runs, allowing fewer voids to be trapped in the product while filling the molds. For both batch and continuous process DUPoly, DU densities for 90 wt% samples were higher than the reported density of a vibration compacted sample of the dry powder (3.5 g/cm^3). Uncompacted UO$_3$ powder, which has a density of about 2.5 g/cm^3, is surpassed at about 80 wt% DUPoly for compressed samples and about 85 wt% for uncompressed DUPoly. In other words, at these DU loadings, the DUPoly process represents a volume reduction compared with disposal of untreated UO$_3$. Such high product densities are achieved because of an increased volume packing efficiency for the DU particles during DUPoly processing. This effect may be attributed to reduced particle agglomeration due to drying of the particles during thermal treatment, comminution of the particles due to mechanical abrasion during processing, or compressive forces exerted during forming.

DUPoly *2 × 4* samples were compression tested in accordance with ASTM D-695, "Standard Method of Test for Compressive Properties of Rigid Plastics."[11] Testing was done using a Soiltest hydraulic compression tester at an unloaded crosshead deflection

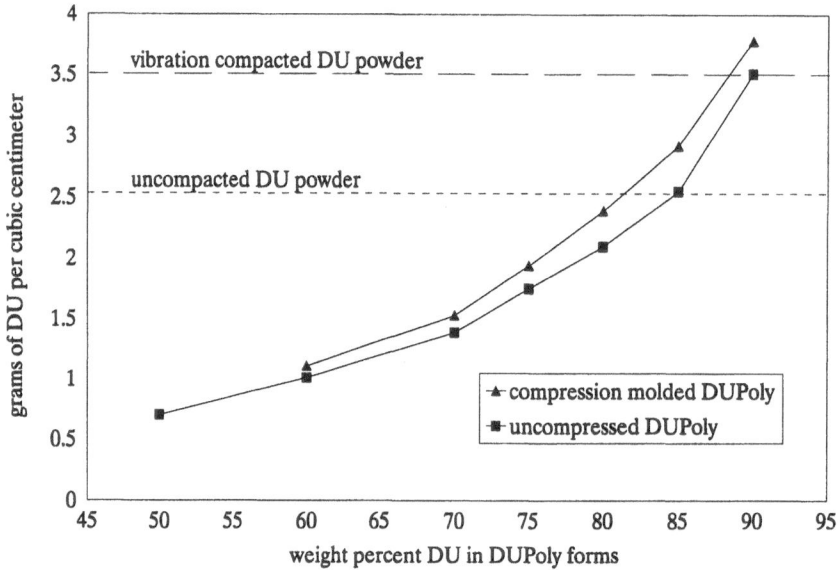

Figure 3. Grams DU per cm^3 DUPoly versus Weight Percent DU in DUPoly Forms.

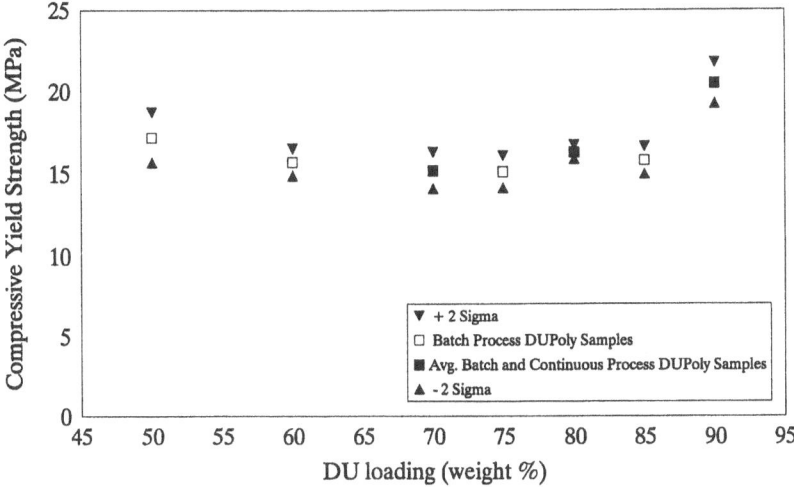

Figure 4. Compressive Yield Strength versus DU Loading.

rate of 1.3 ± 0.3 mm (0.05 ± 0.01 in.)/min. Crosshead speed and total deflection were monitored using a dial gauge and lab timer. Load and deformation were recorded at 60 second intervals. Compressive yield strength is plotted against DU loading in Fig. 4. At least 8 samples were tested at each DU loading.

With batch and continuous process DUPoly data averaged together (filled squares), maximum yield strength is relatively constant between 50 and 85 wt% DU considering the range of measurement error. At 90 wt%, a statistically significant increase was noted, probably due to particle-to-particle contact of the DU in the matrix, with barely enough polyethylene present to fill void spaces. This fact is reflected in the percent deformation at yield, reduced from approximately 26% for 50 wt% DUPoly samples to only 7% for 90 wt% DUPoly samples.

DUPoly forms containing 50, 70 and 90 wt% batch process UO_3 were leach tested in accordance with ASTM C1308[12]. This Accelerated Leach Test (ALT), developed at BNL, was devised to enable prediction of a sample leach rate providing data fits a diffusion-controlled model. In theory, the effective diffusion coefficient (D_e) is temperature dependent according to the Arrhenius equation: $D_e = A\exp(-E_a/RT)$, where E_a is the activation energy. Due to limited scope, all leach testing was done at ambient temperature. Leachates were analyzed by inductively coupled plasma (ICP) spectroscopy for their total uranium metal concentration.

Leach testing of batch process DUPoly forms produced cumulative uranium releases of approximately 1.1% for 90 wt% DU and approximately 0.07% for both 50 and 70 wt% DU samples, after 11 days (Fig. 5). These results are typical for high loadings of soluble salts microencapsulated in polyethylene. However, considering the insolubility of uranium trioxide in water,[13] these data indicate the probable presence of other, more soluble uranium compounds. While the UO_3 is reportedly 96.5% pure (82.25–78.47% total U), it is likely that other hygroscopic uranium compounds are present and unaccounted for in the DU. Although identification of these salts was beyond the scope of this effort, all uranium halides are very soluble, as are uranium and uranyl sulfates and nitrates. The high solubility of the as-received batch DU was further evidenced in that a source term leach sample (50 g batch process DU in 3,000 ml water)

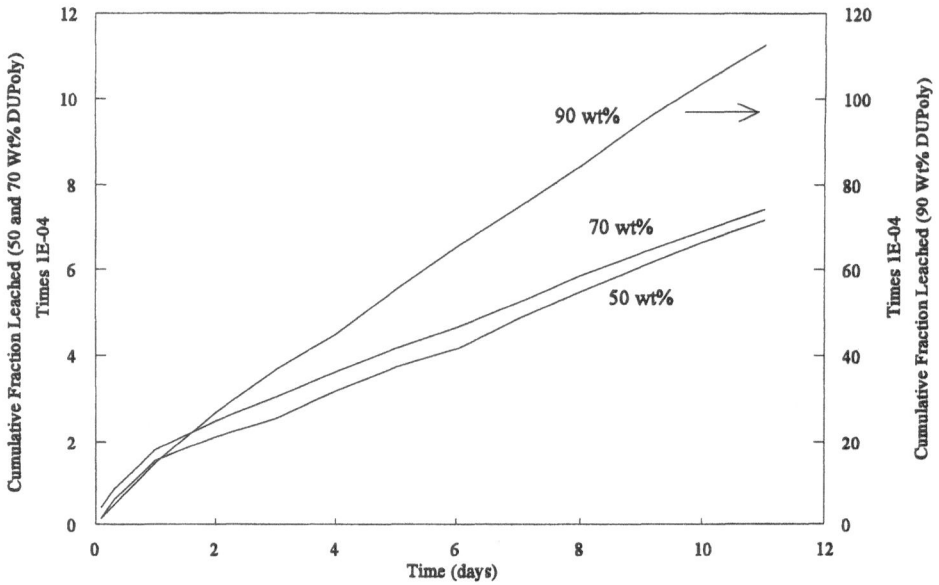

Figure 5. ALT Leach Results for Batch Process DUPoly Samples.

saturated within the first (2 hr.) leach interval. Continuous process DUPoly samples were not tested.

Water immersion testing was performed using one *2 × 4* and one *ALT* form of each DU type and waste loading. Samples were immersed in distilled water to determine possible deleterious effects of a saturated environment. Three or four similar samples were grouped together in a single polyethylene container, with a water/sample ratio of 1,000 ml per sample for *2 × 4* samples and 200 ml per sample for *ALT's*. The test, done at ambient temperature, was a 90-day static immersion after which time the sample weights and volumes were re-measured. Samples remaining intact on completion of the test were compression tested to determine potential for non-visible degradation.

After 90 days, visible degradation was only evident on samples containing 85 and 90 wt% batch process UO_3. Samples containing 80 wt% or less batch process UO_3 were unchanged, as were all samples containing up to 90 wt% continuous process UO_3. The 90 wt% batch process samples began showing signs of cracking around the top and bottom perimeter within the first week of immersion. Cracks in the 85 wt% batch process samples were not noticed until the third month of the test. After 90 days, 90 wt% batch process samples were severely deteriorated, with cracks running lengthwise penetrating nearly to the center of both samples. 85 wt% batch process samples contained only three or four minor cracks (<1 cm) along the sample sides. Since the only DUPoly samples that degraded under water immersion testing contained high concentrations of batch processed DU, this phenomena is thought to be related to the presence of soluble impurities, as discussed above.

Post-immersion compressive strengths of 50, 60, 70, 75, 80 and 85 wt% batch process UO_3 samples were 16.9, 17.0, 9.6, 16.5, 13.6, and 9.2 MPa (2450, 2460, 1390, 2390, 1980, and 1340 psi), respectively. Post-immersion compressive strengths of 70, 80 and 90 wt% continuous process UO_3 samples were 18.5, 16.8, and 18.2 MPa (2,680, 2,440, and 2,640 psi), respectively. Except for the 70 and 85 wt% batch process UO_3 samples (28 and 41% decrease in strength, respectively), post-immersion compressive strengths were statistically equivalent to non-immersed samples.

6. CONCLUSIONS

Depleted UO_3 powders, representative of the depleted uranium inventory at the Savannah River Site, were successfully stabilized by polyethylene encapsulation using BNL's single-screw extrusion process. Material corresponding to two different processes used to prepare the oxide, batch and continuous, were processed. Loadings as high as 90 wt% DU were successfully achieved for both lots. A maximum product density of 4.2 g/cm^3 was obtained. Additional process improvements are under consideration which are expected to lead to DUPoly product densities as high as 7.2 g/cm^3.

Waste form performance testing of microencapsulated UO_3 included compressive strength, water immersion and leach testing. Compression test results were in keeping with measurements made with other waste materials encapsulated in polyethylene, typically around 14 MPa (~2,000 psi). Leach rates, which varied according to UO_3 loading, were similarly in line with polyethylene waste forms bearing high salt waste loadings: approximately 1% after 11 days for samples containing 90 wt% UO_3. Considering the insolubility of uranium trioxide, however, these leach data indicate the probable presence of other, more soluble uranium compounds present in the batch process UO_3 powder. The continuous process DUPoly material was not tested. Ninety-day water immersion tests concluded that water absorption was inconsequential except for batch process DUPoly samples at very high (>85 wt%) waste loadings. Sample degradation in batch process DUPoly corresponds to the increased leach rate observed during ALT testing, probably resulting from hygroscopic impurities. In contrast, continuous process DU showed no evidence of swelling/cracking during 90-day immersion testing even at the highest DU loading of 90 wt%. Therefore, continuous process DU provides a significantly more stable and durable product.

REFERENCES

1. U.S. DOE, "Alternative Strategies for the Long-Term Management and Use of Depleted Uranium Hexafluoride, Notice of Intent," 61 Fed Reg 2239, January 25, 1996.
2. Kalb, P.D. and P. Colombo, "Polyethylene Solidification of Low-Level Wastes, Topical Report," BNL-51867, Brookhaven National Laboratory, Upton, NY, October 1984.
3. Kalb, P.D., J.H. Heiser, and P. Colombo, "Polyethylene Encapsulation of Nitrate Salt Wastes: Waste Form Stability, Process Scale-Up, and Economics," BNL-52293, Brookhaven National Laboratory, Upton NY, July 1991.
4. Kalb, P.D. and M. Fuhrmann, "Polyethylene Encapsulation of Single-Shell Tank Wastes," BNL-52365, Brookhaven National Laboratory, Upton, NY, September 1992.
5. Kalb, P.D., J.H. Heiser, and P. Colombo, "Long-Term Durability of Polyethylene for Encapsulation of Low-Level Radioactive, Hazardous, and Mixed Wastes," Emerging Technologies in Hazardous Waste Management III, D.W. Tedder and R.G. Pohland, ed., American Chemical Society Symposium Series 518, Washington, D.C., 1993.
6. Kalb, P.D., J.W. Adams, M. Meyer, and H. Holmes Burns, "Thermoplastic Encapsulation Treatability Study for a Mixed Waste Incinerator Off-Gas Scrubbing Solution," Third International Symposium on Stabilization/Solidification of Hazardous, Radioactive, and Mixed Wastes, November 1993.
7. Adams, J.W. and P.D. Kalb, "Thermoplastic Stabilization Treatability Study for a Chloride, Sulfate and Nitrate Salts Mixed Waste Surrogate," Emerging Technologies in Hazardous Waste Management VI, American Chemical Society, D.W. Tedder and R.G. Pohland, ed., in press.
8. Kalb, P.D. and P.R. Lageraaen, "Full-Scale Technology Demonstration of a Polyethylene Encapsulation Process for Radioactive, Hazardous, and Mixed Wastes," *Journal of Environmental Science and Health*, in press.
9. Lageraaen, P.R., B.R. Patel, P.D. Kalb, and J.W. Adams, "Treatability Studies for Polyethylene Encapsulation of INEL Low-Level Mixed Wastes," BNL-62620, Brookhaven National Laboratory, Upton, NY, October 1995.

10. Brady, S.D., D.A. Krough, and R.P. Laporte, "Depleted Uranium Trioxide (UO₃) Characterization and Storage," 94AB62387P-00F, Carolina Metals, Inc., January 1995.

11. ASTM, "Standard Method of Test for Compressive Properties of Rigid Plastics," D695, American Society for Testing and Materials, Philadelphia, PA, 1969.

12. ASTM, "Accelerated Leach Test for Diffusive Releases from Solidified Waste and a Computer Program to Model Diffusive, Fractional Leaching from Cylindrical Waste Forms," C1308, American Society for Testing and Materials, Philadelphia, PA, 1995.

13. *CRC Handbook of Chemistry and Physics*, ed. R.C. Weast and M.J. Astle, CRC Press, Inc, Boca Raton, FL, 1981.

16

TREATABILITY STUDY ON THE USE OF BY-PRODUCT SULFUR IN KAZAKHSTAN FOR THE STABILIZATION OF HAZARDOUS AND RADIOACTIVE WASTES[*]

P. D. Kalb, L. W. Milian[1], S. P. Yim[2], R. S. Dyer, and W. R. Michaud[3]

[1]Brookhaven National Laboratory
Department of Advanced Technology
Environmental & Waste Management Group
Building 830
PO Box 5000
Upton, New York 11973-5000
[2]Korean Atomic Energy Research Institute
Nuclear Environment Management Center
150 Dukjin-dong
Yusong-gu, Taejon 305-600
Korea
[3]U.S. Environmental Protection Agency
Office of International Activities
Ronald Reagan Building
1300 Pennsylvania Avenue
Washington, DC 20004

1. ABSTRACT

The Republic of Kazakhstan generates significant quantities of excess elemental sulfur from the production and refining of petroleum reserves. In addition, the country also produces hazardous and radioactive wastes which require treatment/stabilization. In an effort to find secondary uses for the elemental sulfur, simultaneously produce a material which could be used to encapsulate, and reduce the dispersion of harmful contaminants into the environment, BNL evaluated the use of the sulfur polymer cement (SPC)

---

[*]Work performed under the auspices of the U.S. Department of Energy

Emerging Technologies in Hazardous Waste Management 8, edited by Tedder and Pohland
Kluwer Academic/Plenum Publishers, New York, 2000.

produced from by-product sulfur in Kazakhstan. This thermoplastic binder material forms a durable waste form with low leaching properties and is compatible with a wide range of waste types. Several hundred kilograms of Kazakhstan sulfur were shipped to the U.S. and converted to SPC (by reaction with 5 weight percent (wt%) organic modifiers for use in this study. A phosphogypsum sand waste generated in Kazakhstan during the purification of phosphate fertilizer was selected for treatment. Waste loadings of 40 wt% were easily achieved. Waste form performance testing included compressive strength, water immersion, and Accelerated Leach Testing.

2. INTRODUCTION

Chevron Oil Corp., together with the Kazakhstan Ministry of Oil and Gas, has formed a partnership (Tengizchevroil) to explore, develop, and market oil and gas reserves in the Tengiz region of Kazakhstan on the northern shore of the Caspian sea. The growing oil industry in Kazakhstan is already one of their major industries. However, due to limited resources and the required infrastructure to manage resulting toxic and hazardous wastes, there is a growing potential for environmental consequences arising from oil production, refining operations and other environmentally sensitive industries that will need to be addressed. This project, sponsored by the U.S. Environmental Protection Agency Office of International Activities, is investigating potential environmental solutions to foster both commercial and environmental sustainable development in Kazakhstan.

Large quantities of by-product sulfur are currently generated by the cleanup of hydrogen sulfide in the production of petroleum and natural gas at the Tengizchevroil fields in Kazakhstan. Currently about 1,200 metric tons/day of sulfur are generated from processing 60,000 barrels of oil/day, but oil production is expected to grow rapidly. The by-product sulfur has little commercial or social benefit and presently, much of it is disposed as waste. Tengizchevroil has obtained special permission from the Kazakhstan Ministry of the Environment to "block and store" the sulfur by-product, a process in which the molten sulfur is cooled into large solid blocks for storage. But as the volume of sulfur residue increases with increased oil and gas production, this practice may no longer be sound or acceptable. In addition, hazardous oil and gas residuals (e.g., incinerator ash, blowdown solutions), as well as toxic and hazardous wastes generated by other past, currently operating and emerging industries throughout Kazakhstan are produced and require treatment prior to disposal.

This paper evaluates the feasibility of using sulfur polymer (produced from by-product sulfur in the Tengiz region of Kazakhstan) to encapsulate hazardous and radioactive wastes generated on site, at other sites in Kazakhstan, and elsewhere in eastern Europe. Phosphogypsum sand waste generated in Kazakhstan from the production of fertilizers was used for this treatability study. This waste was characterized, encapsulated in sulfur polymer and subjected to selected standardized performance tests (e.g., NRC, ASTM) to evaluate mechanical integrity, durability, and leaching properties.

3. SULFUR POLYMER ENCAPSULATION

Using techniques developed by the U.S. Bureau of Mines[1,2] by-product sulfur can be successfully converted into a stable, durable alternative to concrete with numerous environmental and commercial applications including stabilization of toxic and hazardous wastes. The sulfur is reacted with an organic oligomer (e.g., dicyclopentadiene) and other polymers to form a thermoplastic material known as sulfur polymer or SPC with mechanical properties and chemical durability greater than conventional cement products. Its strength and resistance to harsh chemical environments makes sulfur polymer useful for encapsulation of radioactive, hazardous and mixed wastes, as well as for general construction, paving, piping, and coatings for tanks and pads. Since the processing temperature of SPC is relatively low (melting temperature of 120 °C) compared with thermal processes such as vitrification (process temperatures in excess of 1,200 °C), volatization of contaminants is minimized or eliminated. The commercial cost of sulfur polymer produced in the U.S. is about $0.12/lb., but due to the large inventory of waste sulfur and low operating expenses, the anticipated cost in Kazakhstan would be lower.

The application of sulfur polymer for encapsulating radioactive, hazardous, and mixed wastes was developed over the last 10 years at the BNL Environmental and Waste Technology Center under sponsorship of the U.S. DOE.[3-9] The process uses a heated double planetary orbital mixer to heat and melt the waste and sulfur polymer binder to form a homogeneous molten mixture. The mixture is then poured into suitable molds for cooling to a solid monolithic final waste form. Contrary to conventional cement processes, the sulfur polymer encapsulation process does not rely on a chemical reaction for hardening, so it is more compatible with a wide range of waste types and can accept greater volumes of waste. Thus, overall waste volumes for storage and disposal are reduced and treatment costs are significantly decreased.

The process has been successfully applied to a wide range of waste types including evaporator concentrates, ash, and sludges. Improved waste loadings have been achieved while still exceeding waste form performance standards specified by the U.S. Nuclear Regulatory Commission (NRC) and U.S. Environmental Protection Agency (EPA). A process flow diagram of the sulfur polymer microencapsulation process is shown in Fig. 1.

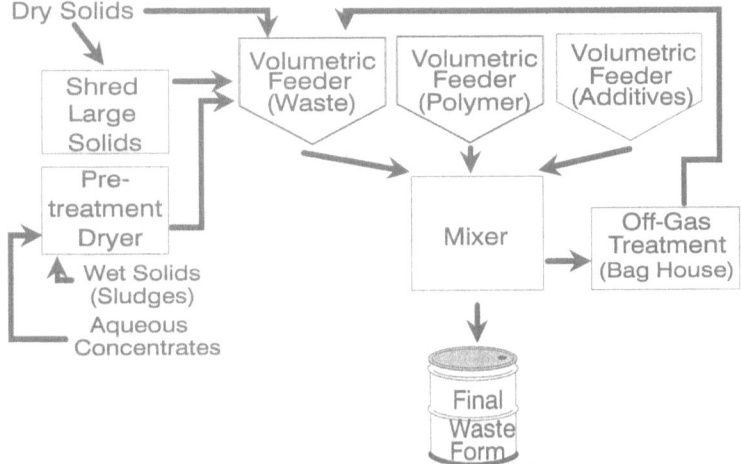

Figure 1. Sulfur Polymer Encapsulation Process Flow Diagram.

4. KAZAKHSTAN SULFUR

Twelve drums containing approximately 44 kg each of by-product elemental sulfur were shipped to the U.S. by Tengizchevroil in Kazakhstan. Six drums of the Kazakhstan sulfur were shipped to McBee and Associates (Lebanon, OR) for conversion to SPC. The material was reacted in a 15-gallon stainless steel, steam-jacketed reactor at a temperature of 140 °C. The molten sulfur was reacted with a mixture of 2.5 wt% polyester grade dicyclopentadiene and 2.5 wt% of a proprietary reactive polymer manufactured by Exxon Chemical Co. The reaction was conducted for about 4 hours until a viscosity of 50 centipoise, measured at 135 °C was achieved. The molten SPC material was then cooled and solidified in 5 gallon steel containers for shipment. Unlike commercially available SPC which is marketed in flaked or granular form, the SPC was received in large blocks from which smaller quantities were chipped off as needed for processing.

Small representative samples of the source sulfur and the SPC produced from Kazakhstan sulfur were analyzed for purity by ACTLABS, Inc., using ICP, INAA, ICP/MS, and XRF techniques. Results are summarized in Table 1. Comparison of carbon content before (<3 ppm) and after conversion to SPC (5 wt%) indicates that the organic polymer modification was accomplished successfully. The source sulfur and the SPC contained a small quantity of chromium which is defined by the EPA as a toxic metal. However, leach testing of these materials according to the EPA Toxicity Characteristic Leaching Procedure (TCLP)[10] resulted in concentrations of chromium (0.05 ppm) below the allowable limit (5 ppm) that define a characteristic hazardous waste.

5. CHARACTERISTICS OF PHOSPHOGYPSUM SAND WASTE

Phosphogypsum sand was selected for encapsulation in SPC because of its availability and potential hazards associated with the by-product material. In addition to being relatively acidic and containing small amounts of fluoride, the material has been determined to be slightly radioactive. Analyses performed on phoshogypsum stockpiled in Florida have shown radium contents between 20–30 pCi/g. This compares with natural gypsum where the radium content of most soils and rocks is <2 pCi/g. The daughter products of radium decay are of equal concern (Radon gas, Pb-210, and Po-210). As a result, the large-scale uses (road construction and agriculture) of phosphogypsum has in part, been prohibited by the EPA because of the elevated levels of these radionuclides.

Analyses of phosphogypsum sand waste received from Kazakstan were conducted to determine density, moisture content, acidity in water, particle size, elemental composition, and radionuclides. Density was measured at room temperature using a Multi Pycnometer MVP-1 (Quantachrome Co.). Moisture content was determined at 105 °C using a Moisture Analyzer MA30 (Sartorius Co.). Acidity in water was obtained by soaking 1 g of the sand into the 10 ml of demineralized water and measuring the pH of the supernatant water after 24 hours. Particle size distribution was analyzed using ASTM standard sieves according to ASTM C136-80. After 1 g of phosphogypsum sand waste was dissolved in 30 ml of acid solution (conc. HCl:conc. HNO_3 = 1:1), elemental composition of dissolved solution were analyzed for various elements by Inductively Coupled Plasma-Atomic Emission Spectroscopy (ICP). Non-dissolved material was filtered and weighed after drying at 110 °C. The concentration of phosphorus in dissolved solution was measured using a Hach kit (Hach Co.).

Table 1. Chemical Analysis of Kazakhstan Sulfur and
Sulfur Polymer

Element	Concentration in Elemental Sulfur, ppm	Concentration in Sulfur Polymer, ppm
Ag	<0.5	<0.5
Al_2O_3	300	300
As	<1	<1
Ba	7	6
Be	<2	<2
Bi	<5	<5
Br	<0.5	<0.5
CaO	<100	<100
Cd	<0.5	<0.5
Co	0.8	1.1
Cr	6.9	11.9
Cs	<0.2	<0.2
Cu	5	6
Fe_2O_3	300	200
Hf	<0.2	<0.2
Ir	<0.001	<0.001
K_2O	<100	<100
MgO	100	400
MnO	<100	<100
Mo	<2	<2
Na_2O	<100	<100
Ni	3	3
P_2O_5	<100	<100
Pb	<5	<5
Rb	<10	<10
Sb	<0.1	<0.1
Se	<0.5	<0.5
SiO_2	1,400	2,700
Sr	<1	<1
Ta	<0.3	<0.3
TiO_2	<100	<100
V	<2	<2
W	<1	<1
Y	<2	<2
Zn	5	7
Zr	3	3
Sc	<0.01	0.01
La	0.2	0.2
Ce	1	2
Nd	<1	4
Sm	0.03	0.05
Eu	<0.05	<0.05
Tb	<0.1	<0.1
Yb	<0.05	0.06
Lu	<0.01	<0.01
U	<0.1	<0.1
Th	<0.1	0.1
C	<3	50,020

Table 2. Properties of Phosphogypsum Sand Waste

Property	Value
Density, g/cm^{3a}	2.57
Moisture Content, wt%	12.3
Acidity in Water, pH	5.1
Non-Dissolved Materials in Acid Solution, wt%[b]	7.4

[a]Density measured by Multi-Pycnometer after drying at 105 °C.
[b]Acid solution (conc. HCl:conc. HNO$_3$ = 1:1).

The properties of phosphogypsum sand waste are summarized in Table 2. The elemental composition of phospogypsum sand waste is presented in Table 3. Particle size distribution of waste is shown in Fig. 2.

The moisture content of the "as-received" phosphogypsum sand waste was 12.3 wt%. The waste was then dried by heating overnight in a convection oven at 110 °C, prior to encapsulation in SPC. Dried waste was encapsulated in SPC directly without any size screening or size reduction because the content of relatively large particles (>2 mm) was less than 4 wt%. The major elements of phosphogypsum sand waste were calcium, phosphorus, iron, and aluminum. Non-dissolved materials in acid solution may be silicon compounds. Hazardous (toxic) elements are not present in the waste in significant quantities. In addition to analysis for non-radioactive elements, radionuclides analysis was conducted qualitatively by counting 100 g of sample for 24 hours on an intrinsic Ge gamma-spectrometer (with a Canberra computer system). However, no evidence was found for the existence of any radionuclides other than background levels of natural uranium. Although analyses on elements and radionuclides revealed no toxic and

Table 3. Elemental Composition of Phosphogypsum Sand Waste

Element	Concentration, ppm
Al	4,754
Ba	210.4
Ni	8.9
Fe	5,390
Cr	5.1
Zn	27.3
Cu	32.7
Cd	—
Ca	45,770
Sr	2,515
Mg	377.6
Na	1,703
K	2,512
Rb	598.1
As	103.6
Se	9.7
Ti	52.4
Mn	24.5
Sb	68.8
Mg	421.3
Mo	7.2
Pb	1.0
V	1.2
P	9,350

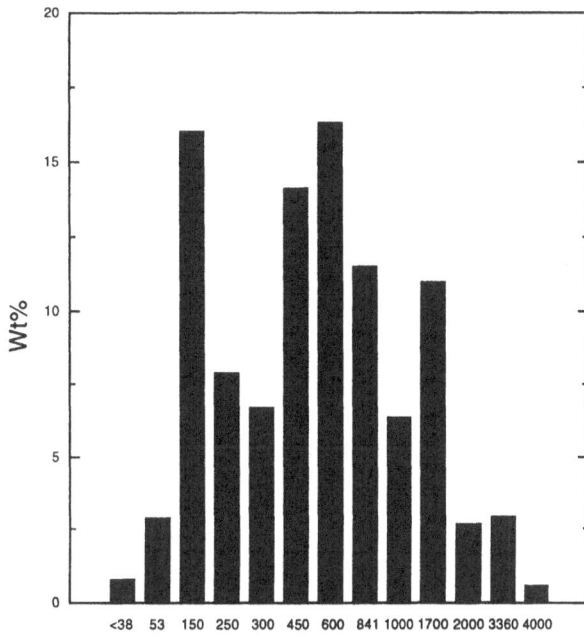

Figure 2. Particle Size Distribution of Phosphogypsum Sand Waste.

radioactive constituents, handling and disposal of this material may provide environmental concerns due to dispersibility.

6. DETERMINATION OF OPTIMUM LOADING WITH SURROGATE

Only a limited supply of the Kazakhstan phosphogypsum sand waste was available for this study. Therefore, waste form loading optimization was performed using calcium sulfate reagent as a surrogate for Kazakhstan phosphogypsum sand waste. Surrogate calcium sulfate was a fine powder of $CaSO_4.1/2H_2O$ (calcium hemihydrate). Its density and moisture content were $2.61\,g/cm^3$ and $5.6\,wt\%$, respectively. It was dried at $110\,°C$ prior to encapsulation. The density of the dry powder was $2.613\,g/cm^3$.

SPC waste forms were formulated using an electrically heated, stirred mixer by: (1) heating of the SPC at $130\,°C$ until the SPC has completely melted, (2) adding dry surrogate to molten SPC and mixing the constituents into a homogeneous slurry by mixer, and (3) pouring into a suitable mold and cooling to a monolithic solid. Formulations were prepared with increasing quantities of waste until the limits of processibility were reached.

SPC waste forms containing 10, 15, 20, 30, 40, 50, and 56 wt% surrogate calcium sulfate were prepared according to above mentioned steps. Above 50 wt% surrogate, the molten SPC and surrogate mixture was so viscous that it hardly flowed to the mold. Maximum loading of surrogate determined by the limit of processibility in this experiment was 56 wt%. The molten mixtures containing 30 and 40 wt% surrogate, respectively, had a good fluidity and homogeneity at $130\,°C$. The waste forms were also homogeneous after cooling. In the mixtures containing 10, 15 and 20 wt% surrogate, some of the particles of surrogate seemed agglomerate in the melted sulfur.

All waste forms prepared with the surrogate waste were immersed in water and were observed for changes in their physical appearance over a period of 90 days. SPC waste forms containing 50 wt% and 56 wt% surrogate cracked within 2 weeks and 1 week, respectively. The other waste forms maintained their original forms after 90 days.

As a result of processibilty and immersion tests, optimum loading of surrogate calcium sulfate was determined at the range of 30–40 wt% in the SPC waste form.

7. WASTE FORM PERFORMANCE

7.1. Sample Preparation

Based on determination of optimum loading with surrogate, SPC waste forms containing 40 wt% Kazakhstan phosphogypsum sand waste were prepared by the same steps and conditions described above. The molten mixture of SPC and waste was poured into a PVC tube measuring 1.6″ in diameter by 3′ in height. The same molten mixture was also poured into another PVC tube measuring 1″ in diameter by 1′ in height. After cooling to ambient temperature, they were cut into specimens measuring 1.6″ in diameter by 3.2″ in height and 1″ in diameter by 1″ in height, respectively, within 7 days. Eight specimens measuring 1.6″ in diameter by 3.2″ in height and five specimens measuring 1″ in diameter by 1″ in height were obtained. Of the eight specimens measuring 1.6″ in diameter by 3.2″ in height, four specimens were used in the compressive strength testing and another three specimens were used in water immersion testing. The last specimen was cross-sectioned for observation. Three of five samples measuring 1″ in diameter by 1″ in height were used in an accelerated leach test (ALT) and the other two samples were archived. SPC waste forms containing 40 wt% surrogate calcium sulfate were prepared in the same fashion. Compressive strength testing, water immersion testing, and ALT were performed on SPC waste form specimens containing surrogate as well as on SPC waste form specimens containing actual phosphogypsum sand waste.

7.2. Homogeneity

A thorough homogenization of waste within the solidified waste form is desirable so that the waste form will exhibit uniform performance characteristics. As a measure of homogeneity, the apparent density of waste form was obtained for all specimens by dividing the weight of the waste form by the volume calculated from the form dimensions. Apparent densities for the specimens measuring 1.6″ in diameter by 3.2″ in height are presented in Table 4.

SPC waste form specimens containing phosphogypsum sand waste have similar apparent densities of approximately $1.89 \, g/cm^3$, regardless of their level in the mold. Waste form specimens containing surrogate have almost the same apparent density of $2.06 \, g/cm^3$ at any level within two molds. This is indicative of good homogeneity within SPC waste form specimens. Particles of phosphogypsum sand and surrogate calcium sulfate were observed to be homogeneously dispersed within the SPEC. No aggregation and settling were observed in the samples.

7.3. Compressive Strength

Compressive strength testing was performed in accordance with the standard method ASTM C-39 (Compressive Strength of Cylindrical Concrete Specimens).[11] Results are summarized in Table 5.

Table 4. Apparent Densities of SPC Waste Forms Containing 40 wt% Waste

Sample Number	Apparent Density of SPC Waste Form, g/cm^3	
	Phosphogypsum Sand Waste	Surrogate Calcium Sulfate
#1	1.908	2.068
#2	1.893	2.052
#3	1.897	2.070
#4	1.925	2.061
#5	1.901	2.056
#6	1.898	—
#7	1.855	—
#8	1.890	—
#1[a]	—	2.051
#2[a]	—	2.062
#3[a]	—	2.060
#4[a]	—	2.066
Mean[b]	1.895 ± 0.017	2.060 ± 0.005

*The larger sample number it has, the closer it was to the bottom of the mold.
**Same sample number for each waste does not mean that the samples were taken from same level of the mold, but refers to the relative level in the mold for each waste.
[a]Replicate specimens obtained using another PVC mold with the mixture in the same batch.
[b]Error expressed at the 95% confidence limit.

The neat SPC forms containing no waste tested for this study had a mean compressive strength of 3,616 psi. This compressive strength is higher than those of neat SPC forms (approximately 1,800–2,600 psi) previously reported.[4,9] While the compressive strength of the waste form containing 40 wt% surrogate increased to 5,134 psi by the addition of surrogate waste to the SPC, the compressive strength of the SPC waste form containing phosphogypsum sand waste was 3,470 psi, similar to that of neat SPC forms. The SPC waste form containing phosphogypsum sand waste had more voids than that of surrogate. Voids in the SPC waste form may be decreased by careful sample preparation. If the voids decrease, the compressive strength of the waste form will increase. Although the compressive strength of the SPC waste form containing phosphogypsum sand waste

Table 5. Compressive Strengths of SPC Waste Forms

Waste Type and Loading	Compressive Strength, psi	Compressive Strength, Mpa
Neat[a,b]	3,616	24.9
Phosphogypsum Sand Waste,[c] 40 wt%	3,470 ± 287	23.9 ± 2.0
Surrogate calcium sulfate,[d] 40 wt%	5,135 ± 632	35.4 ± 4.4

[a]Result reflects mean value for two specimens.
[b]Standard deviation for two specimens: ±286 psi (±2.0 Mpa). Error expressed at 95% confidence limit for two replicate specimens: ±2,571 psi (±17.7 Mpa).
[c]Mean value and error expressed at 95% confidence limit for four replicate specimens.
[d]Mean value and error expressed at 95% confidence limit for five replicate specimens.

was not higher than those of the neat SPC form and the SPC waste form containing surrogate, the SPC waste form containing phosphogypsum sand waste possessed excellent mechanical integrity.

7.4. Water Immersion Testing

Water immersion testing was performed on three replicate specimens for SPC waste forms containing 40 wt% phosphogypsum sand waste and 40 wt% surrogate calcium sulfate, respectively. Each specimen was immersed completely in demineralized water at ambient temperature (25 °C) and was examined periodically. Upon completion of 90 days, they were removed from the water and checked for variation in weight and dimension. Then, compressive strength testing was conducted according to ASTM C-39. All specimens had negligible changes in weight and dimensions after 90 day water immersion. Their compressive strength was also retained after water immersion. Results are summarized in Table 6. Improvements of compressive strength may be attributed to experimental scatter associated with the limited sample population.

7.5. Accelerated Leach Test

Leach testing for the SPC waste form containing phosphogypsum sand waste was conducted in accordance with Accelerated Leaching Test (ALT), developed at BNL and adopted by ASTM as Standard Test Method C-1308-95.[12,13] Three replicate specimens were leach tested at 25 °C. Each specimen was leached in a leachant volume that was 100 times the geometric surface area of the specimen. In this experiment, 2,900 ml of distilled water was used in leachant for each specimen because each specimen had a surface area of 29 cm². It was changed twice on the first day, and then daily for 11 days. Leachate were analyzed for major elements of phosphogypsum sand waste, e.g., Ca, Al, Fe, Na by ICP. Phosphorous in the leachate was checked by Hach kit.

Concentrations of Al, Fe, Na, and phosphorous in the leachate were under, or near, the detection limit for all leachates so they were omitted from leach rate calculations. Calcium leach rates were calculated using a computer program that was developed to accompany the Accelerated Leach Test (the ALT program). Figure 3 shows leaching data and modeling curves for three replicate SPC waste forms containing 40 wt% phosphogypsum sand.

The release of calcium from SPC waste forms was very slow. Release data were in close agreement with those predicted by the diffusion model, indicating diffusion as the dominant leaching mechanism. The goodness-of-fit value of the diffusion model to the experimental data was less than 1% for all specimens. The average diffusion coefficient of calcium was 2.02×10^{-10} cm²/sec. It is interesting to note that release of calcium from the SPC waste form was controlled by diffusion while previous studies indicate the release of calcium from hydraulic cement waste form is probably controlled by solubility.[13] Table 6 shows the projected cumulative fraction release of calcium from the SPC waste form containing 40 wt% calcium as a function of waste form size after 10 years, 50 years, 100 years, and 300 years, using the average diffusion coefficient of 2.02×10^{-10} cm²/sec at 25 °C obtained from this experiment. This indicates clearly that the SPC waste form is stable and releases calcium very slowly. Releases of radioactive and hazardous contaminants are therefore also expected to be low.

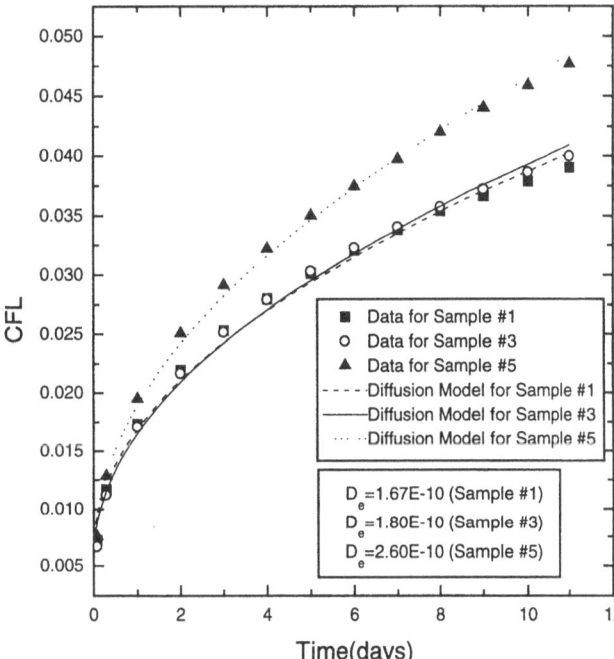

Figure 3. Leaching Data and Diffusion Model Curves for Three Replicate SPC Waste Forms Containing 40 wt% Phosphogypsum Sand Waste at 25°c.

8. SUMMARY AND CONCLUSIONS

Large quantities of by-product sulfur with little commercial or social benefit are generated in Kazakstan and are currently being treated as waste. Conversion of this material to sulfur polymer will facilitate beneficial use of the material for treatment of hazardous, radioactive and mixed wastes and for various construction applications. By-product sulfur produced at Tengizchevroil oil production facilities in the Republic of Kazakstan was shipped to the U.S. and was successfully converted to SPC as evidenced by the presence of 5 wt% carbon detected in the SPC product. Small quantities of chromium were also detected in the Kazakhstan sulfur and SPC product (approximately 6 ppm and 12 ppm, respectively), but neither material failed TCLP testing for hazardous waste.

The Kazakstan SPC was then used to conduct a treatability study for the encapsulation of phosphogypsum waste generated in Kazakstan as a result of fertilizer production. Process development activities resulted in maximum waste loadings of 56 wt%,

Table 6. Projected Cumulative Fractional Release of Calcium from the SPC Waste Form Containing 40 wt% Phosphogypsum Sand Waste

	Projected Cumulative Fractional Release, %[a]			
Dimension of Waste Form	10 years	50 years	100 years	300 years
57 cm in I. D. × 70 cm in height	2.8	6.2	8.7	14.6
1 m in I. D. × 1 m in height	1.7	3.8	5.3	9.2
2 m in I. D. × 2 m in height	0.86	1.9	2.7	4.6

[a]Projection was made by the diffusion model using a diffusion coefficient of 2.02×10^{-10} cm^2/sec measured at 25°C.

but optimal mixing and pouring was limited to mixtures containing 40 wt% waste. At 40 wt% waste loading, homogeneity was demonstrated by the close agreement of apparent density measurements throughout the waste forms. Performance testing of SPC waste forms included compression, water immersion, and accelerated leaching. Compressive strength for neat Kazakhstan SPC waste forms was about 1.5 times greater than previous data for domestic SPC. The compressive strength of SPC waste forms containing phosphogypsum sand waste was 3,470 psi, similar to that of neat SPC forms. Although the compressive strength of the SPC waste forms containing phosphogypsum sand waste was not higher than those of the neat SPC forms and the SPC waste forms containing surrogate, the SPC waste forms containing phosphogypsum sand waste possessed excellent mechanical integrity. Waste forms containing 40 wt% waste did not show any deterioration under saturated conditions in water immersion. Their compressive strength was also retained after water immersion. Accelerated Leach Testing revealed diffusion is the dominant leaching mechanism and projected leaching for a full size water form (2 m in diameter × 2 m in height) is <5% after 300 years.

REFERENCES

1. Sullivan T.A. and W.C. Mcbee, "Development and Testing of Superior Sulfur Concretes," RI-8160, Bureau of Mines, U.S. Dept. of the Interior, Washington, DC, 1976.
2. Mcbee, W.C., T.A. Sullivan, and B.W. Jong, "Modified Sulfur Cements for Use in Concretes, Flexible Pavings, Coatings and Grouts," RI-8545, Bureau of Mines, U.S. Dept. of the Interior, Washington, DC, 1981.
3. Colombo, P., P.D. Kalb, and J.H. Heiser, "Process for the Encapsulation and Stabilization of Radioactive Hazardous and Mixed Wastes," U.S. Patent No. 5,678,234, Oct. 14, 1997.
4. Kalb P.D. and P. Colombo, "Modified Sulfur Cement Solidification of Low-Level Waste, Topical Report," BNL-51923, Brookhaven National Laboratory, NY, October 1985.
5. Kalb P.D., J.H. Heiser III, and P. Colombo, "Modified Sulfur Cement Encapsulation of Mixed Waste Contaminated Incineration Fly Ash," Waste Management Vol. 11, Number 3, pp. 147–153, Pergammon Press, New York, 1991.
6. Adams J.W. and P.D. Kalb, "Thermoplastic Stabilization of a Chloride, Sulfate, and Nitrate Salts Mixed Waste Surrogate," Proceedings of the I&EC Special Symposium, Emerging Technologies for Hazardous Wastes, American Chemical Society, Atlanta, GA, September 19–20, 1995.
7. Kalb P.D., J.W. Adams, M.L. Meyer, and H.H. Burns, "Thermoplastic Encapsulation Treatability Study for a Mixed Waste Incinerator Off-Gas Scrubbing Solution," Stabilization and Solidification of Hazardous, Radioactive, and Mixed Wastes, ASTM STP1240, T.M. Gilliam, and C.C. Wiles, eds., American Society for Testing and Materials, Philadelphia, PA, 1995.
8. Kalb P.D., L.W. Milian, A.J. Grebenkov, and S.P. Rutenkroger, "Thermoplastic Process Treatability for Contaminated Hearth Ash from the Republic of Belarus," Proceedings of the 1996 International Conference on Incineration and Thermal Treatment Technologies, Savannah, GA, May 6–10, 1996.
9. Kalb P.D., J.H. Heiser III, R. Pietrzak, and P. Colombo, "Durability of Incineration Ash Waste Encapsulated in Modified Sulfur Cement," Proceedings of the 1991 Incineration Conference, Knoxville, TN, May 13–17, 1991.
10. U.S. Environmental Protection Agency, "Toxicity Characteristic Leaching Procedure," 40 CFR 261, Fed. Reg. 55, 11863, March 29, 1990.
11. ASTM, "Standard Method of Test for Compressive Strength of Cylindrical Concrete Specimens," C39-72, American Society for Testing and Materials, Philadelphia, PA, 1975.
12. Fuhrmann M., R.F. Pietrzak, J. Heiser III, E.M. Franz, and P. Colombo, "User's Guide for the Accelerated Leach Test Computer Program," BNL-52267, Brookhaven National Laboratory, Upton, NY, 11973, 1990.
13. ASTM, "Standard Method of Test for Accelerated Leach Test for Diffusive Releases from Solidified Waste and a Computer Program to Model Diffusive, Fractional Leaching from Cylindrical Waste Forms," ASTM C1308-95, American Society for Testing and Materials, West Conshohocken, PA, 1996.

17

ENCAPSULATION OF NITRATE SALTS USING VINYLMETHYLPOLYSILOXANE

Stephen Duirk and Christopher M. Miller

Department of Civil Engineering
210 Auburn Science and Engineering Center
University of Akron
Akron, Ohio 44325-3905

ABSTRACT

The Department of Energy has an immediate need for multiple technologies to transport and dispose of low level mixed waste. Solidification/stabilization (S/S) processes offer one possible solution that encapsulates sludge and dry hazardous wastes in a low permeability solid form. The advantage of vinylmethylpolysiloxane over other organic solidification processes is its ability to cure at ambient temperature. This paper examined compressive strength, metal leaching, and void area measurements of vinylmethylpolysiloxane-encapsulated waste as a function of waste loading (28–48 wt%) for three surrogate wastes having compositions similar to those found at national laboratories. Compressive strength was greater than 4,390 kPa for all but one sample (48 wt% waste load-High Chloride waste). Image analysis shows void area increases linearly with increasing waste load, attributed to an increasing total sample volume physically occupied by the waste. Leaching test results indicate metal specific behavior and waste composition effects on Toxicity Characteristic Leaching Procedure (TCLP) concentrations. Cadmium diffusion from the encapsulated waste is slow (leach index >10 at all waste loads). Cost comparison of vinylmethylpolysiloxane to cement shows that it could be an economical waste disposal alternative.

1. INTRODUCTION

Low level mixed waste (LLMW) resulting from nuclear weapons research and dismantling activities is present in large quantities at several Department of Energy (DOE) national labs.[1] It is estimated that approximately 82,000 m^3 of LLMW is currently stored, and an additional 137,000 m^3 is expected to be generated over the next 20 years. The largest storage facilities are Hanford and Idaho National Engineering Laboratory (INEL)

Emerging Technologies in Hazardous Waste Management 8, edited by Tedder and Pohland
Kluwer Academic/Plenum Publishers, New York, 2000.

which contain 36,000 and 35,000 m^3 of LLMW, respectively.[2] This waste, which contains both hazardous and low-level radioactive materials, can be extremely toxic to humans and animals.

It is estimated that the federal government will spend between \$234 and \$289 billion on environmental remediation over the next 75 years at sites owned by the Departments of Defense, Energy, Interior, and Agriculture.[3] Encapsulation is one type of solidification/stabilization (S/S) technique that has been shown to effectively treat many difficult wastes in both liquid and solid form for disposal.[4,5] Encapsulation is a nondestructive process that involves mixing waste with reagents to create physical and/or chemical reactions.[6] S/S processes are designed to reduce leaching and mobility of contaminants, improve handling, and decrease the surface area over which loss of contaminants can occur.[7]

S/S processes can be grouped into inorganic processes (cement and pozzolanic binders) and organic processes (thermoplastic and thermosetting polymer binders). One organic process that has not been evaluated is a vinylmethylpolysiloxane (VMPS) based system, typically referred to as silicone or polysiloxane. Silicone polymers were first developed in 1950.[8] The silicon backbone and room temperature cure capability of silicones may offer potential advantages for binding hazardous waste because of favorable thermal, chemical, and mechanical properties. Furthermore, it is believed that waste loads greater than traditional solidification technologies could also be achieved.[9] The purpose of this paper is to present compressive strength, metal leaching, and void area measurements as a function of waste loading for three surrogate wastes having compositions similar to those found at national laboratories. One commercial vinylmethylpolysiloxane formulation (GE RTV-664a,b) was examined at variable waste loads (28–48 wt%). Compressive strength and heavy metal leaching results should provide insight into the potential and use of polysiloxane as a viable DOE waste management option.

2. EXPERIMENTAL

2.1. Materials and Chemicals

The water used in all experiments was de-ionized with a Barnstead NANOpure water system. All chemicals used in waste surrogates and reagent solutions were purchased through Fisher Scientific (Pittsburgh, PA), GFS Chemicals (Powell, OH), or Aldrich Chemical Co. (Milwaukee, WI). General Electric silicone compound and platinum catalyst (GE RTV-664a,b) were obtained from General Electric Silicones (Waterford, NY).

2.2. Surrogate Waste Compositions

Three surrogate wastes were studied, hereafter referred to as Nitrate Salt (NS), High Nitrate (HN), and High Chloride (HC). Their compositions are shown in Table 1. These wastes were modeled after salt materials found at the Idaho National Engineering Laboratory.[10] Surrogates were prepared in one kg batches by crushing and sieving the material so that it passed a 0.6 mm sieve, but was retained on a 0.3 mm sieve.

2.3. Encapsulation Procedure

Depending on the selected waste load, various masses of base material VMPS, catalyst (manufacture recommends 10 wt% catalyst to base material), and waste were

Table 1. Surrogate Waste Compositions

Constituent	Nitrate Salt (wt%)	High Nitrate (wt%)	High Chloride (wt%)
$NaNO_3$	52.9	—	—
KNO_3	34.7	—	—
Al_2O_3	0.2	5.8	12.1
NaF	1.4	—	—
Na_2SO_4	5.4	—	—
NaCl	3.1	—	9.5
$NaH_2PO_4 \cdot H_2O$	1.9	2.0	4.1
$Fe(OH)_3$	—	5.9	12.2
$Mg(OH)_2$	—	3.9	8.1
Portland Cement	—	2.0	4.1
MicroCel E	—	7.8	16.2
H_2O	—	13.7	28.4
$CaSO_4$	—	—	4.8
Metal Oxides	(mg metal/kg)	(mg metal/kg)	(mg metal/kg)
CrO_3	1,000	1,000	1,000
CdO	1,000	1,000	1,000
HgO	—	1,000	1,000
PbO	—	1,000	1,000

measured. Mixing took place in one gallon plastic buckets using a hand-held drill equipped with a stainless steel paddle. Vinylmethylpolysiloxane, referred to as polysiloxane, base material and waste were mixed together first for a minimum of five minutes, and then the catalyst was added. The combination was mixed for an additional five minutes and then transferred to plastic cylinder molds. Samples were placed under a hood and allowed to cure for a minimum of 24 hours before testing.

2.4. Toxicity Characteristic Leaching Procedure (TCLP)

A simulated Toxicity Characteristic Leaching Procedure Test was performed after USEPA SW-846 to examine metal leaching from encapsulated samples. The samples were cut into pieces that passed a 9.42 mm sieve and retained on a 4.75 mm sieve. Ten grams of cut sample was then placed in a 250 ml high density polyethylene bottle and 200 ml of pH 2.88 acetic acid extract fluid was added to each bottle. The extraction vessel was shaken for 18 hours on a Burrell wrist action shaker set at one (scale of 0 to 10). Solution suspensions were then filtered using 0.7 m Whatman borosilicate glass filters, acidified, and stored in high density polyethylene vials until analysis.

2.5. Accelerated Leach Test (ASTM C-1308)

In accordance with ASTM C-1308, "Accelerated Leach Test for Diffusive Releases from Solidified Waste", three cylindrical replicate samples of each waste load were prepared (approximate size: diameter-30 cm and height-40 cm). The cylindrical samples were suspended from the bottle cap of a 500 ml high density polyethylene container with Teflon netting. Leachate volume was specified and set to be ten times the surface area of the specimen (±2%). This method is a short-term test with frequent leachant changes to facilitate the accelerated release of soluble compounds. Each interval leachant was acidified and stored for future analysis.

2.6. Analytical Methods

Compressive strength was conducted according to ASTM D-695. Cylindrical samples with a 1:2 (diameter:height) aspect ratio were smoothed if necessary with a scroll saw and then placed in a Geotest instrument with a 906 kg rated loadcell. The load was applied until the sample failed or the maximum load was achieved. TCLP and Accelerated Leach Test metal samples were analyzed using a Perkin Elmer Plasma 400 inductively coupled plasma atomic emission spectrometer (ICP-AES) with manufacturer recommended wavelengths for each metal. Image analysis was used to determine encapsulated sample void area (due to both gas generation during curing and waste occupation). Three samples from a cylindrical specimen (top, middle, and bottom third of cylinder) were cut, briefly exposed to water to dissolve waste particles, and then allowed to dry prior to analysis. Sample images were captured using an Olympus SZH10 stereo microscope with a magnification of 8.75 and analyzed using Optimas version 6.1 image analysis software. All pH measurements were done with an Accument Basic pH meter after appropriate calibration.

3. RESULTS AND DISCUSSION

3.1. Compressive Strength

Compressive strength was measured to determine the mechanical integrity of the solidified waste form. Table 2 shows compressive strength results for each waste and the corresponding waste loads. The NS waste was tested at 28, 38, and 48 wt% waste load. NS waste encapsulated samples exceeded the maximum load of the testing apparatus, corresponding to a compressive strength greater than 4,390 kPa for all waste loads. HN and HC wastes were tested at waste loading of 28 and 48 wt%. HN encapsulated waste behaved similar to the NS and exceeded the maximum load of the testing apparatus. HC waste also surpassed the capacity of the testing apparatus at 28 wt% waste load, however, at 48 wt%, the specimen failed at 2,830 kPa. Therefore, HC encapsulated waste exhibited a decrease in strength with increasing waste load. This behavior is not completely unexpected with increasing waste load because the polymer system becomes more waste-like (i.e. brittle) resulting in reduced compressive load bearing capacity.

3.2. Image Analysis

Image analysis was performed on all three wastes to determine void area as a function of waste load. The NS waste was examined at 28, 38, and 48 wt% waste load while

Table 2. Compressive Strength as a Function of Waste Load

Waste	Waste Load (wt%)	Compressive Strength (kPa)
Nitrate Salt	28	>4,390
	38	>4,390
	48	>4,390
High Nitrate	28	>4,390
	48	>4,390
High Chloride	28	>4,390
	48	2,830

Table 3. Void Area Analysis as a Function of Waste Load

Waste Type	Waste Load (wt%)	Sample 1	Sample 2	Sample 3	Avg. ± STD Dev.
Nitrate Salt	28	14.2%	14.9%	13.9%	14.3 ± 0.51%
	38	17.1%	17.5%	17.7%	17.4 ± 0.31%
	48	22.3%	22.4%	21.3%	22.0 ± 0.61%
High Nitrate	28	18.6%	16.2%	16.4%	17.1 ± 1.3%
	48	22.1%	22.7%	20.2%	21.7 ± 1.3%
High Chloride	28	16.7%	21.9%	24.3%	21.0 ± 3.9%
	48	18.8%	21.0%	12.6%	17.5 ± 4.4%

the HN and HC wastes were examined at 28 and 48% waste load. The NS and HN wastes exhibited increasing void area with increasing waste load (Table 3). This trend is expected because increasing waste load results in a higher percentage of volume physically occupied by the waste. The HC waste, however, did not show any statistically significant difference for the two waste loads. HC waste was difficult to mix resulting in a non-homogeneous sample and highly variable measurements (Table 3). Image analysis of all samples also indicated that the waste is physically captured and not chemically bound by the polysiloxane.[10]

3.3. Heavy Metal Leaching

TCLP for NS, HN, and HC wastes was performed to evaluate heavy metal leaching from the encapsulated waste form. The NS waste was examined for leaching of cadmium and chromium at 28, 38, and 48 wt% waste load. Increasing waste load from 28 to 48 wt% increased cadmium leaching from 1.97 mg/L to 3.27 mg/L and chromium from 1.87 mg/L to 4.14 mg/L (Table 4). The increase in metal concentration for both cadmium and chromium is approximately proportional to waste load.

TCLP tests were also used to determine the observed diffusion behavior of cadmium from NS encapsulated waste as a function of waste load. Samples were taken at 0.08 (5 minutes), 1, 4, 8, and 18 hours. The five minute sample was taken to measure instantaneous dissolution of the waste exposed by the cut surface. Soluble cadmium concentrations were converted to cumulative fraction leached (CFL) for each time interval and plotted versus the square root of time (Fig. 1). The linearity of Fig. 1 indicates transport controlled dissolution of cadmium.[11] Depending on waste load, approximately 5–7% (i.e. CFL) of the total 13–15% leached in the first five minutes. Therefore, instantaneous

Table 4. TCLP Results as a Function of Waste Load

Waste	Waste Load (wt%)	Cd	Cr	Hg	Pb
Nitrate Salt	28	1.97	1.87	—	—
	38	2.45	3.01	—	—
	48	3.27	4.14	—	—
High Nitrate	28	0.30	0.34	ND*	ND*
	48	0.05	1.26	0.05	ND*
High Chloride	28	0.69	0.47	0.01	ND*
	48	0.18	0.64	0.01	ND*

ND*: Below detection limits (<0.01 mg/L).

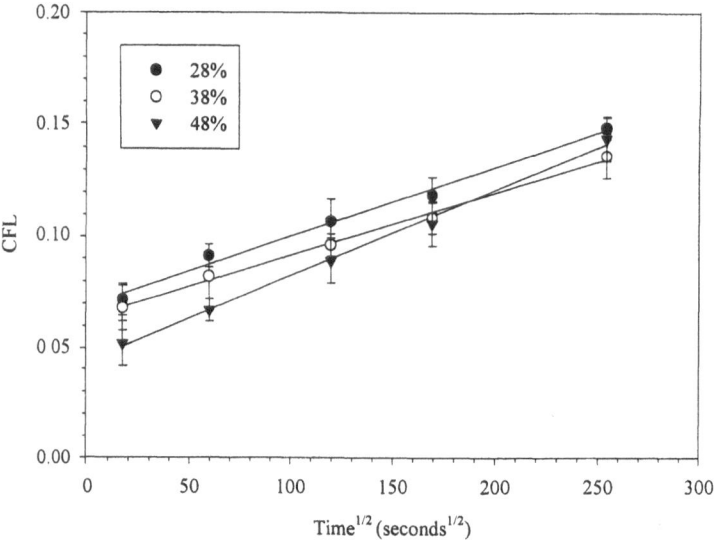

Figure 1. Cadmium leaching under TCLP conditions (pH = 2.88, 10 g sample, 200 ml water) as a function of waste load.

dissolution of salt waste exposed by the cut surface of the TCLP sample accounts for 38–47% of the total cadmium concentration measured during the TCLP test. Consequently, instantaneous dissolution of the salt waste at the cut surface is a significant fraction of the leachate concentration.

HN and HC wastes were TCLP tested at 28 and 48 wt% waste load. TCLP results are summarized in Table 4. At 28 wt% waste load, HN waste TCLP leachate showed cadmium and chromium concentrations of less than 0.4 mg/l and less than detectable mercury and lead concentrations (<0.01 mg/l). At 48 wt%, cadmium, mercury, and lead concentrations were all less than 0.1 mg/l, but chromium was significantly greater at 1.26 mg/l. Interestingly, the cationic metals (i.e. Cd and Hg) decreased with increasing HN waste load and the anionic metal (i.e. Cr) increased.

HC 28 wt% waste load TCLP leachate showed cadmium and chromium concentrations of less than 0.7 mg/l, 0.01 mg/l mercury, and a less than detectable lead concentration (<0.01 mg/l). At 48 wt%, cadmium decreased to 0.18 mg/L, mercury and lead were approximately the same level as 28 wt%, and chromium increased to 0.64 mg/L. Increasing chromium concentration and decreasing cadmium concentration with increasing HC waste load is the same trend observed with HN waste. This behavior is likely attributed to HC and HN waste properties. Both wastes contain Portland cement, synthetic calcium silicate (MicroCel E), and iron oxide. Portland cement and the synthetic calcium silicate appeared to lower cadmium leachate concentration by buffering the TCLP extract solution and chemical reaction during the extraction procedure. In fact, the final pH of HN and HC 28 wt% TCLP solutions was approximately 5 (and even greater at 48 wt%) compared to NS with a final pH = 4. As for chromium results, iron oxide is known to have a large sorption capacity for anions at pH < 7.8.[12] Therefore, HC waste with 12.2 wt% iron oxide would be expected to have a lower final chromium concentration compared to HN waste with 5.9 wt% iron oxide, and indeed this is what was observed (Table 4). NS waste (no iron oxide content) TCLP chromium concentrations were greater than both HN and HC waste, also consistent with iron oxide sorption of chromium. Finally,

Table 5. Cadmium Accelerated Leach Test Results as a
Function of Nitrate Salt Waste Load

Waste Load (wt%)	D_e (cm^2/s)	Leach Index $-\log D_e$
28	1.32×10^{-11}	10.88
38	3.09×10^{-11}	10.51
48	3.45×10^{-11}	10.42

because chromium solubility is less pH sensitive than cadmium in pH range 4–5, increasing chromium concentration with increasing waste load is consistent with NS waste results showing chromium TCLP concentration increases approximately proportional to waste load.

Accelerated leach tests were performed at 28, 38, and 48 wt% NS waste load to measure the effective diffusion coefficient (D_e) of cadmium at room temperature (~20 °C). D_e was determined by a computer program provided with ASTM C-1308 which minimizes the sum squared error of CFL measurements.[13] Table 5 summarizes the effective diffusion coefficient as a function of waste load. D_e increased slightly with increasing waste load from 1.32×10^{-11} cm^2/s at 28 wt% waste load to 3.45×10^{-11} cm^2/s at 48 wt% waste load. The effective diffusion coefficient is often converted into a leach index (LI) by equation 1.

$$LI = -Log(D_e).\qquad(1)$$

A LI greater than seven is often required for stabilized waste land disposal.[4] The LI for NS encapsulated waste was greater than 10 for all waste loads (Table 5), indicating slow diffusion and a waste form suitable for land disposal.

3.4. Cost Comparison and Considerations

Table 6 presents a cost comparison between waste encapsulation using GE RTV-664 and cement. Polysiloxane encapsulation is more cost effective at 38 and 48 wt% waste load than cement at 15 wt% (note: cement has difficulty curing at waste loads greater than 15 wt%). There is a savings of $12.58/kg waste at 48 wt%, which corresponds to a savings of greater than $20,000,000 at current waste inventory levels (note: the waste disposal

Table 6. Cost Comparison of GE RTV-664 and Cement Encapsulation of Salt Waste

	GE RTV-664			Cement
Waste Load	28%	38%	48%	15%
Salt Waste (kg)	1.0	1.0	1.0	1.0
Polysiloxane Base or Cement (kg)	2.25	1.41	1.02	5.70
Catalyst (kg)	0.22	0.14	0.10	—
Total (kg)	3.47	2.68	2.12	6.70
Bulk Density (kg/ft^3)	44.73	40.38	37.08	42.67
Volume (ft^3)	0.077	0.066	0.057	0.157
Disposal Cost ($300/ft^3)	$23.10	$19.80	$17.10	$47.10
Material Cost:	$43.73	$27.28	$19.71	$2.29
Base and Catalyst ($17.60/kg)				
Cement ($0.40/kg)				
Total Cost $/kg salt waste	$66.57	$47.08	$36.81	$49.39

cost, estimated at $300/ft^3, is based on mixed waste disposal at EnviroCare in Utah and does not include transportation costs). In addition, the polysiloxane encapsulated waste volume at 48 wt% is 64% less than cement at 15 wt% (0.057 vs. 0.157 ft^3) which would translate into lower transportation costs. Therefore, initial cost estimates show that polysiloxane could be an economical alternative to cement. Additives to the polysiloxane waste form that minimize heavy metal leaching such as cross-linking agents, however, need to be investigated. Cross-linking agents could coat the waste and chemically bind it into the polysiloxane system. Process and scaling issues to encapsulate large quantities of waste also need to be addressed.

ACKNOWLEDGMENTS

This research was supported by Orbit Technologies, Inc. and The Department of Energy (Contract #DE-AC07-94ID13223). No endorsement by granting agencies should be inferred. Tom Quick, Annabelle Foos, Wayne Butala, and Eric Wagner provided analytical assistance and General Electric donated the RTV-664 material.

REFERENCES

1. Kalb, P.D., Heiser, J.H., and Colombo, P. Polyethylene Encapsulation of Nitrate Salt Wastes: Waste Form Stability, Process Scale-Up, and Economics-Technology Status Topical Report, Department of Nuclear Energy. Brookhaven National Laboratory. 1991.
2. U.S. Department of Energy. Final Waste Management Programmatic Environmental Impact Statement For Managing Treatment, Storage, and Disposal of Radioactive and Hazardous Waste. Office of Environmental Management. 1997.
3. MacDonald, J.A. and Rao, P.S. Shift Needed to Improve Market for Innovative Technologies in Soil and Groundwater Cleanup, August/September, 1997.
4. LaGrega, M.D., Buckingham, P.L., and Evans, J.C. Hazardous Waste Management. McGraw-Hill, New York, 1994.
5. O'Brien & Gere Engineers, Inc. Innovative Engineering Technologies for Hazardous Waste Remediation. Van Nostrand Reinhold, New York, 1995.
6. Connor, J.R. Chemical Fixation and Solidification of Hazardous Wastes. Van Nostrand Reinhold, New York, 1990.
7. Freeman, H.M. Standard Handbook of Hazardous Waste Treatment and Disposal. McGraw-Hill, New York, 1989.
8. Klempner, D. and Frisch, K.C. Handbook of Polymeric Foams and Foam Technology. Hanser Publishers, Munich, 1991.
9. Franz, E.M., Heiser, J.M., and Colombo, P. Immobilization of Sodium Nitrate Waste with Polymers. Brookhaven National Laboratory. 1987.
10. Loomis, G., Miller, C.M., and Prewett, S. Mixed Waste Salt Encapsulation Using Polysiloxane—Final Report. Idaho National Engineering and Environmental Laboratory. 1997.
11. Batchelor, B. Leach Models for Contaminants Immobilized by pH-Dependent Mechanisms, Environmental Science and Technology, 32(11):1721–1728, 1998.
12. Stumm, W. and Morgan, J.J. Aquatic Chemistry. John Wiley & Sons, New York, 1996.
13. Fuhrmann, M., Heiser, J.H., Pietrzak, R., Franz, E., and Colombo, P. User's Guide for the Accelerated Leach Test Computer Program. Department of Nuclear Energy. 1990.

18

DEVELOPMENT OF AN ON-LINE ANALYZER FOR VANADOUS ION IN LOMI DECONTAMINATION

Sherman Kottle[1], Robert A. Stowe[2], and John V. Bishop[2]

[1]Lake Service Company
Lake Jackson, Texas
[2]VanTek Corporation
Midland, Michigan

ABSTRACT

An automatic, repetitive titrimeter was developed for the determination of vanadous ion concentration during chemical decontamination of nuclear power plants by means of the LOMI method. The LOMI (an acronym for low oxidation state, metal ion) treatment is used during refueling shut-downs to reduce the exposure of plant personnel to radiation by removing radioactive deposits which have accumulated in pumps and piping. A dilute aqueous mixture of vanadous formate and picolinic acid (2-pyridine carboxylic acid) reduces, complexes, and dissolves oxide deposits which contain radioisotopes which were produced by neutron activation of the products of corrosion and wear. The titrimeter makes it possible to follow depletion of the V(II) concentration as the process proceeds. The instrument automatically withdraws samples of the circulating chemicals, performs a titration with Fe(III), and determines the end point by means of oxidation-reduction electrodes. Means were found to secure rapid and meaningful electrode response. That made it possible to design and assemble an instrument which was successful in laboratory simulations and field use.

INTRODUCTION

In nuclear power plants, mechanical wear and corrosion liberate small quantities of metals into the heat transfer system which become radioactive as the result of neutron absorption. The radioisotopes so formed become part of a refractory layer of metal oxides on the inner surfaces of the stainless steel pipes and pumps. They produce a radiation field external to the equipment which must be minimized in order to minimize radiation exposure to plant workers. The radiation field is a major cost factor as well as

Emerging Technologies in Hazardous Waste Management 8, edited by Tedder and Pohland
Kluwer Academic/Plenum Publishers, New York, 2000.

a safety factor because it limits the working time permitted to operating personnel. It has become widely accepted to remove the radioactivity by chemical means during scheduled refueling shut-downs. The removed material is then concentrated by means of ion exchange to minimize the volume of radioactive waste for disposal. The importance of radiation field control and decontamination techniques has made it the subject of a series of conferences which are sponsored by EPRI (Electric Power Research Institute) and which are reported upon in a series of monographs available from them.[1]

One of the decontamination procedures which is widely used is the LOMI (acronym for low oxidation state, metal ion) process, in which a dilute aqueous mixture of vanadous formate and picolinic acid (2-pyridinecarboxylic acid) is circulated at elevated temperature through the system which is to be cleaned.[2] The coating which must be dissolved is composed primarily of iron, chromium, and nickel oxides, and their mixed oxides. It is desired to use chemicals for cleaning which are not aggressive to the materials of construction, yet will dissolve this very insoluble coating and the accompanying radioactive species. This is accomplished by a dilute mixture of vanadous formate and picolinic acid through a process of reduction and chelation. The dissolved metals are subsequently removed from the circulating solution for disposal by means of cation and anion resin beds.

The operators follow the course of the process by laboratory analysis of samples withdrawn from the circulating chemicals. The radioactivity and metals are assayed by scintillation spectroscopy and ICP spectrometry, respectively. There has been no rapid, convenient way for the operating personnel to monitor the vanadous ion concentration to insure that cessation of corrosion product and radioactivity removal is the result of successful completion of the cleaning process and is not the result of depletion of the reducing agent.

The laboratory method employed commercially for the analysis of vanadous formate solution relies on redox titration with visual end point detection.[3] Analysis for V(II) in decontamination is made difficult by the low concentration, in the range 0–500 ppm, and its reactivity with atmospheric oxygen, which makes securing samples and transporting them to the laboratory difficult.

The vanadous picolinate complex exhibits a molar absorbance of $3,800 \, M^{-1} cm^{-1}$ at 660 nm, which is more than adequate for colorimetric analysis in the LOMI range of 0–0.010 M.[4] We studied direct photometric measurement of the complex and found excellent analytical sensitivity. A prototype on-line colorimeter was evaluated at a nuclear plant. It was found not to be effective because of excessive interference by light-absorbing material of extremely small particle size which was present in the system.[5]

In the present work, we elected to take advantage of the strong reducing action of the V(II). We considered both direct measurement of the redox potential and the use of a redox titration. Noble metal oxidation-reduction electrodes have been shown to be responsive to the vanadous picolinate complex, but electrode equilibration in LOMI solutions was found to require up to 40 minutes in the absence of iron and 5–10 minutes when iron was present.[6] Our preliminary tests showed that equilibration was slow and the potentials were not entirely stable. It was necessary to overcome those limitations to make an online instrument successful.

EXPERIMENTAL PROCEDURES

The laboratory apparatus which was employed is illustrated in Fig. 1. Experiments were carried out in a closed, cylindrical vessel of 100 ml volume which was fabricated

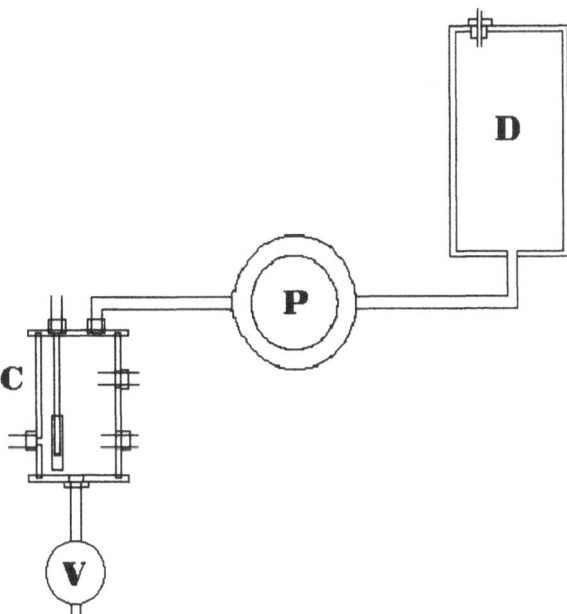

Figure 1. Laboratory ORP titration apparatus. C-acrylic vessel, D-inert gas-purged reagent tank, P-piston pump, V-solenoid drain valve.

from transparent acrylic plastic. Small polypropylene tubing fittings were used to introduce a platinum oxidation-reduction potential (ORP) electrode, a silver-silver chloride reference electrode, a pH probe, and to provide openings for introduction of argon as an inert gas purge, and for entry and removal of materials. Picolinic acid solution and the other reagents required were purged with argon before they were added to the titration vessel. The vanadous formate used was a commercial product approximately 0.25 M in V(II).[3] All other reagents were of standard analytical reagent grade.

Both the ORP electrode measurements and redox titrations were performed in this cell. To titrate, standardized ferric ammonium sulfate solution of the desired concentration was supplied from a storage bottle by means of a calibrated positive displacement pump (Model RHO, Fluid Metering, Inc., Oyster Bay, NY 11771). Simulated LOMI solutions were prepared in an inert gas-purged glove bag and were weighed in from a pre-purged hypodermic syringe. An electrically-operated valve was used to drain the titration vessel. After it was discovered that stirring afforded by bubbling argon through its contents was not adequate, magnetic stirring was added. The drain connection was relocated so that a PTFE-coated, X-shaped stirring bar could be rotated by a permanent magnet mounted on a 270 rpm motor located below the cell.

The potential difference between the platinum and reference electrodes was measured by means of an high input impedance digital voltmeter designed for pH and ORP control (Model MV7615, Phoenix Electrode Company, Houston, TX), which switched off the titrant pump at the desired ORP. The quantity of titrant introduced was calculated from the running time of the calibrated titrant pump and the ferric reagent concentration. The elapsed time and the ORP were recorded using a personal computer equipped with an analog to digital adapter. During the titration the titrant pump running time and ORP data were collected, while the progress of the titration was displayed on the computer monitor. Subsequently, the data were converted into both tabular and graphic form by importing them into a spread sheet file.

RESULTS

Preliminary measurements on simulated LOMI samples showed that directly reading the ORP was not adequate for the analysis because electrode equilibration was slow and could be erratic. Manual titrations performed by adding small aliquots of ferric ammonium sulfate reagent over a period of 5–10 minutes yielded ORP curves which exhibited the two end points expected for a 2-step oxidation, from V(II) to V(III), and then from V(III) to V(IV). However, continuous addition of reagent by means of a pump at rates in the range of 0.5–2 ml/min met with poor success (Fig. 2). The titration curves were distorted in shape, and the apparent end points were subject to shifts which depended on the rate of titrant addition. Since the titration vessel was transparent, it was possible to observe the end point visually as a change from the intense blue-black color of the vanadous complex to the light green of the vanadic complex. That end point, noted by the arrow in Fig. 2, was observed to occur before the rapid increase in ORP which accompanied the oxidation from V(II) to V(III). A second increase in ORP occurred, as expected for a second oxidation step, from V(III) to V(IV).[7] The second color transition,

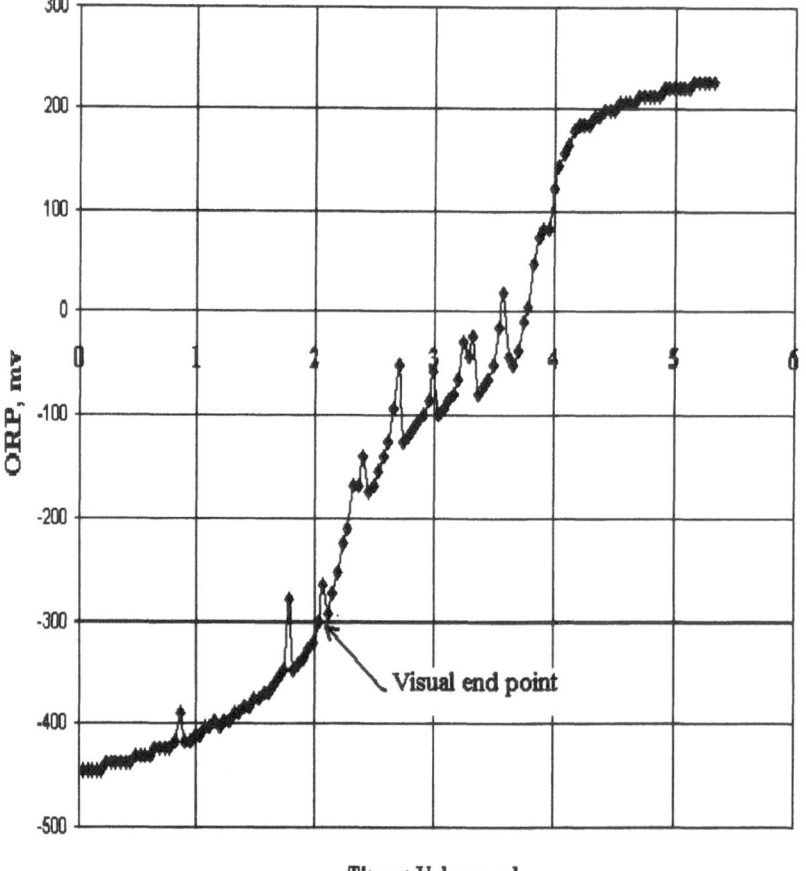

Figure 2. Rapid titration of 0.21 mmole vanadous picolinate complex with Fe(III) showing visual end point at 0.21 mmole and apparent ORP end points at 0.22 and 0.38 mmoles.

from light green to colorless, also preceded the apparent potentiometric end point. The lag of the ORP was a consequence of slow electrode response.

Up to this point, the experiments were performed at 4.5 pH with a picolinic acid/vanadium ([Pic]/[V]) molar ratio of 4.5/1, as is usually practiced in the LOMI process. In the course of investigating the experimental variables, it was observed that reducing the pH greatly speeded up the electrode response. Figure 3 shows the titration curves at 0.7 pH of samples which contained 0.003, 0.006 and 0.009 M vanadous formate. These titration curves exhibit the desired S-shape, and their inflection points occur at the correct values. Only a single potential increase is present. The vanadic to vanadyl oxidation is no longer observed. In addition, it was found that the visual end point, as evidenced by a change in color from blue to green, coincided with the ORP end point.

The rapidity of electrode response at low pH is confirmed by the increase in the noise present in the ORP data shown in Fig. 3. Since the design plans for the titrator called for detecting the point at which a selected ORP value was reached, it became essential to reduce the noise which would otherwise produce false triggering and poor reproducibility. Examination of numerous titration curves revealed that 2 sources of ORP fluctuations were present. Fast noise was found to be related to agitation of the cell contents by the purge gas, and slow noise was found to be related to strokes of the reagent pump. When the purge gas inlet was raised to a position above the liquid level in the titration cell and a magnetic stirrer was added, a major portion of the fast noise was eliminated. The slower noise component associated with pump strokes was then almost entirely eliminated by changing from a single-piston pump to a peristaltic pump (Model 7014 Masterflex Type LS, Cole-Parmer Instrument Company, Niles, IL 60714), in which three roller bearings operating on elastomer tubing created a more uniform flow of titrant.

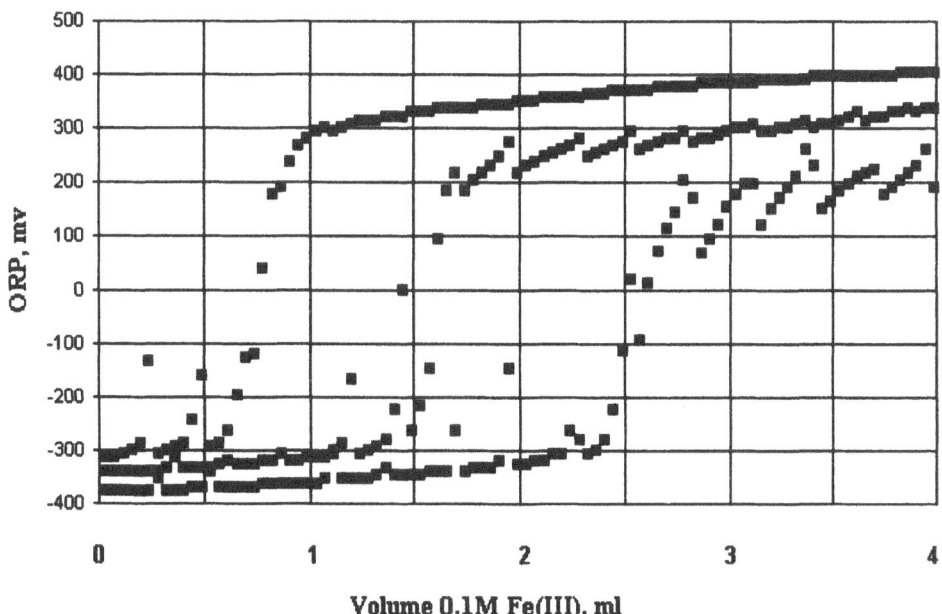

Figure 3. Titration of 3, 6, and 9 mM vanadous picolinate samples showing noise priduce by using inert gas purge for mixing and by pump strokes. Sample size 25 ml, 0.7 pH, [Pic]/[V] = 4.5, piston pump delivering 0.5 ml/min.

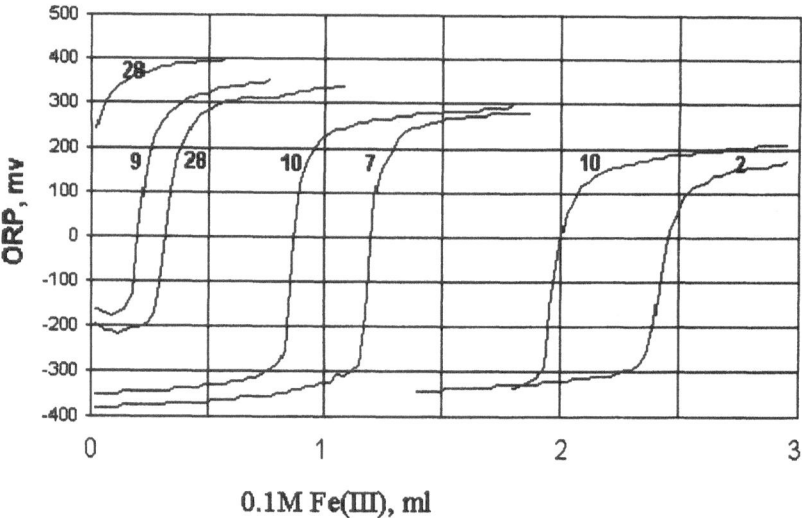

Figure 4. Automatic titration curves using peristaltic pump. Present: 0.009, 0.028, 0.036, 0.098, 0.133, 0.209, and 0.254 mmole V(II). [Pic]/[V] ratios shown adjacent to curves, sample volume 30 ml, pH 0.7, pumping rate 0.625 ml/min.

The effects achieved by employing a low pH and more uniform reagent flow are evident in the data of Fig. 4, in which are superimposed titration curves for 7 vanadous picolinate samples. The 30 ml samples were acidified with 8% sulfuric acid to approximately 0.7 pH, and were titrated automatically with 0.1 M ferric solution added at the rate of 0.625 ml/min. (The 3 ml mark on the abscissa represents 4.8 minutes reagent pumping time.) The ORP curves were symmetric and exhibited well-defined inflection points. For the samples shown in Fig. 4, the V(II) concentrations in millimoles/ liter were:

Given: 0.3   0.8   1.1   3.0   4.1   6.2   7.8
Found: 0   0.7   1.1   2.8   4.0   6.3   8.0

The data in Fig. 4 also show that the analytical results are unaffected by the relative picolinic acid and vanadium concentrations, the molar ratios of which are shown adjacent to each titration curve.

The variation of ORP with vanadous ion concentration is illustrated in Fig. 5. The ORP values obtained were not completely reproducible from day to day. That result, which was attributed to changes of the Pt electrode, was not further investigated, since it was expected to be of little effect when ORP changes were measured in the course of performing titrations. Shifts in ORP values at the equivalence points which are evident in Fig. 4 were more important, since they suggest that selection of a suitable end point ORP might not be feasible. The choice of a value of −100 mv for the end point produces an error of approximately −15% at a V(II) concentration of 1 mmole/liter. A correction can be applied, if it is desired to do so. For the leftmost curve in Fig. 4, the end point was reached within a few seconds of the onset of titrant flow, leading to an estimate of a lower limit of detection of approximately 0.1 mM V(II) for a 30 ml sample.

An automatic titrimeter was assembled using the components and arrangement shown in Fig. 6, where the solid lines indicate fluid paths and the dashed lines denote electrical connections. The 3 peristaltic pumps supply 25 ml of sample, 0.1 M ferric

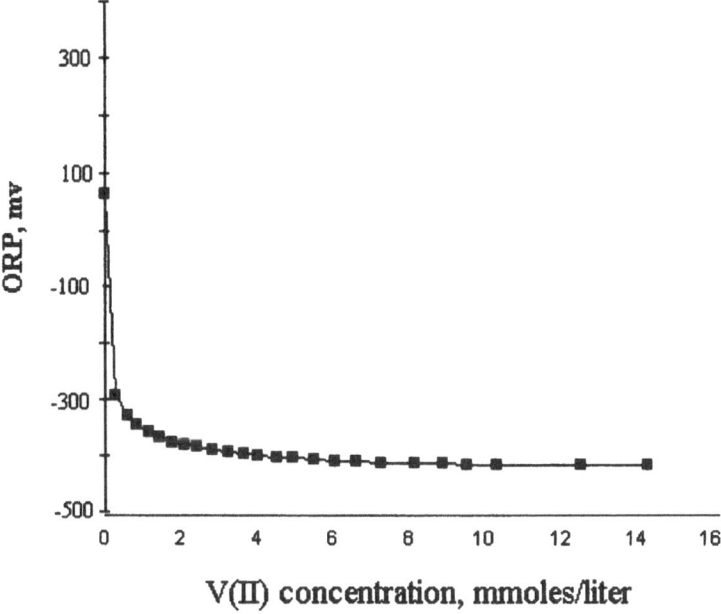

Figure 5. Dependence of oxidation-reduction potential (ORP) on vanadous picolinate concentration. pH 4.5, [Pic]/[V] molar ratio 4.5.

reagent, and 1 ml of sulfuric acid solution to the titration cell. The titration cell is essentially the same as the cell shown in Fig. 1, except for relocation of the drain and close coupling of a magnetic stirrer at the lower end. All of the sample stream components except the ORP electrodes and the drain valve are comprised of plastic or elastomer materials. The pumps and solenoid-operated drain valve are controlled by a Micrologic 1,000 programmable logic controller (PLC) (Allen-Bradley, Milwaukee, WI 53204). The PLC is programmed to provide timed outputs which govern the quantities of sample and of acid to be added to the titration cell. It also starts the ferric reagent pump and turns it off when the selected end point ORP has been reached. A Model DC1100 digital recorder

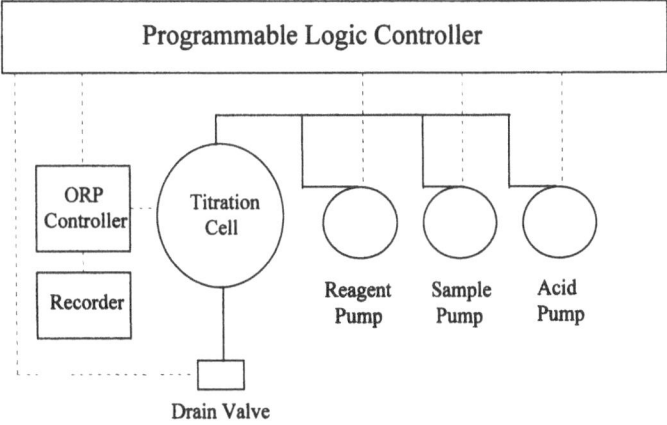

Figure 6. Titrimeter block diagram. Fluid paths solid lines, electrical paths dotted lines.

(Monarch Instrument Co., Amherst, NH 03031) is used for data acquisition. Following completion of a series of analyses, the recorded data may be downloaded to a computer from the recorder by means of a removable memory card. They may then be tabulated and graphed in the same manner as described above for the laboratory tests. A mode selector switch is incorporated in the instrument to permit recording either complete titration curves or acquiring only the reagent pump running times.

Under the direction of the PLC, the sample pump withdraws a 25 ml sample from plastic tubing which connects to the reactor piping and introduces it into the titration vessel. The acid pump operates at the same time to introduce 1 ml of 8% (v/v) aqueous sulfuric acid for pH adjustment. When sample and acid introduction into the titration cell are complete, the reagent pump is energized to add 0.1 M ferric ammonium sulfate reagent at the rate of 0.625 ml/min. The potential difference of the electrodes is sensed by the ORP controller. When the platinum electrode reaches a potential of -100 mv relative to the reference electrode, a relay contact in the controller stops the reagent pump. The running time of the reagent pump establishes the quantity of reagent consumed. The instrument cycle time is set to 6 minutes, which provides ample time to drain the vessel, to introduce the next sample, and to titrate. The cell contents are drained into a plastic bottle. In a nuclear plant LOMI run the drained material is on the order of 1 mc/ml in radioactivity. It is returned to the radioactive waste. Figure 7 shows the results of titrations of 20 known solutions during a series of trials under simulated LOMI conditions. The instrument response was linear, with a concentration precision of approximately 0.1 mM. When the titrimeter is used in LOMI decontamination, the recorder trace serves to provide an immediate and on-going indication of the vanadous ion concentration.

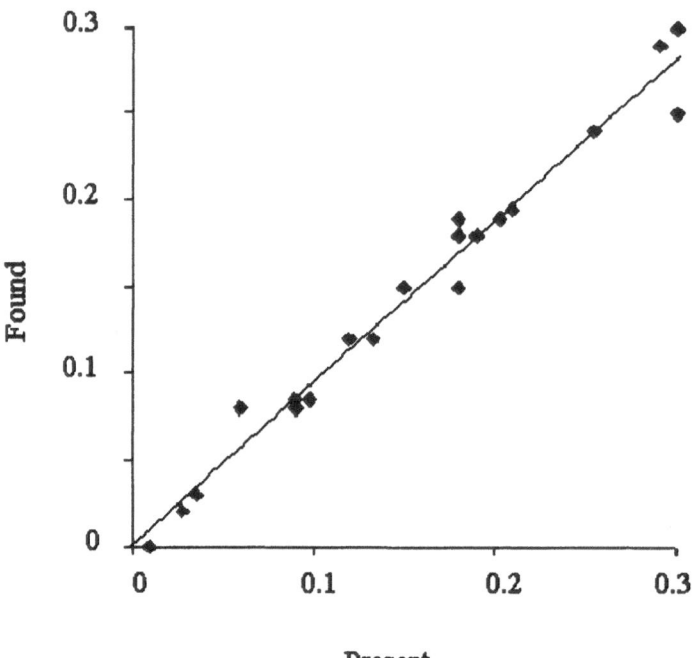

Figure 7. Results of decontammation simulations. Millimoles V(II) added and found.

DISCUSSION AND CONCLUSIONS

The slow electrode response effects which were anticipated were evident in the disparity between the visual and potentiometric end points and between rapid, automatic addition and slow, manual addition of titrant. Adjustment of the pH to a value <1 has two benefits. It reduces the ORP electrode equilibration time, making it practical to perform rapid on-line titrations. It also suppresses the oxidation of V(III) to V(IV), thereby providing an unambiguous result which permits more latitude in the instrument design.

The titrimeter provides adequate sensitivity, linearity, and precision for its intended application. When it is used in a LOMI decontamination, the recorder trace provides the vanadous ion concentration data at the work site. Its use can potentially reduce the need for additional LOMI treatments which are sometime used to insure that maximum radioactivity removal has been attained, thereby minimizing the volume of radioactive waste. Since LOMI treatments require times of the order of 2–8 hours, the 6-minute analysis cycle is sufficiently rapid. The sample system components are made of plastic and elastomer materials, making it possible to replace them at low cost in the case of excessive radioactive contamination. A completed titrimeter underwent a successful trial during a LOMI decontamination at the Brunswick-2 plant of the Carolina Power and Light Company in February, 1996, where its performance met the design criteria. Details of the trial have been reported.[8]

ACKNOWLEDGMENTS

The authors wish to thank the personnel of the Carolina Power & Light Company Brunswick nuclear power plant, and of PN Services of Richland, Washington, the decontamination contractor, for making it possible to conduct field evaluations. They also gratefully acknowledge support by Omni Tech International, Ltd., VanTek Corporation, and the Research Division of the U.S. Nuclear Regulatory Commission.

REFERENCES

1. Electric Power Research Institute, Palo Alto, California 94304.
2. D. Bradbury, et al., "Experience of Plant Decontamination with LOMI Reagents", in Water Chemistry of Nuclear Reactor Systems, British Nuclear Energy Society, London, 1980, 3, 203.
3. J.V. Bishop, S. Kottle, and R.A. Stowe, Analytical Methods for Vanadous Formate, Fifth EPRI Workshop on Chemical Decontamination, Charlotte, NC, June, 1993.
4. A.M. Lannon, et al., Inorg. Chem., 1984, 23, 4167.
5. J.V. Bishop, R.A. Dutcher, M.S. Fisher, S. Kottle, and R.A. Stowe, Continuous Spectrographic Analysis of Vanadous and Vanadic Ions, Omni Tech International, Ltd., U.S. Nuclear Regulatory Commission Report NUREG/CR-6047, October, 1993.
6. J.L. Smee and D. Bradbury, Low Picolinate LOMI—Update, Fifth EPRI Workshop on Chemical Decontamination, Charlotte, NC, June, 1993.
7. K.L. Chawla and J.P. Tandon, Tantala, 1965, 12, 665.
8. S. Kottle, R.A. Stowe, and J.V. Bishop, Continuous Analysis for Vanadium in LOMI Deontamination, Omni Tech International, Ltd., U.S. Nuclear Regulatory Commission Report NUREG/CR-6494, December, 1996.

19

MAGNETIC SEPARATION FOR NUCLEAR MATERIAL SURVEILLANCE

Laura A. Worl[1], David Devlin, Dallas Hill, Dennis Padilla, and F. Coyne Prenger

[1]NMT-11 MS-E505
Los Alamos National Laboratory
Los Alamos, NM 87545

ABSTRACT

A high performance *superconducting magnet* is being developed for particle retrieval from field collected samples. Results show that the ratio of matrix fiber diameter to the diameter of the captured particles is an important parameter. The development of new matrix materials is being pursued through the controlled corrosion of stainless steel wool, or the deposition of nickel dendrites on the existing stainless steel matrix material. We have also derived a model from a continuity equation that uses empirically determined capture cross section values. This enables the prediction of *high gradient magnetic separator* performance for a variety of materials and applications. The model can be used to optimize the capture cross section and thus increase the capture efficiency.

1. INTRODUCTION

Magnetic separation has received much attention recently due to industrial,[1-3] medical[4,5] and governmental[6-8] interest in selective separation technologies. Currently, the Department of Energy (DOE) is developing technologies for single particle capture of submicron *actinide* particles from forensic samples of interest.

High gradient magnetic separation (HGMS) utilizes large magnetic field gradients to effectively separate micron sized *paramagnetic particles*. Because actinide compounds are paramagnetic, HGMS can extract and concentrate micron sized actinides and actinide containing particulates with no sample destruction. Most host materials, such as air, water and organic matter are diamagnetic, which will allow for a physical separation of the paramagnetic actinides from these materials. Typically, HGMS separators consist of a high field solenoid magnet, the bore of which contains a fine-structured ferromagnetic matrix material. The matrix fibers locally distort the magnetic field

Emerging Technologies in Hazardous Waste Management 8, edited by Tedder and Pohland
Kluwer Academic/Plenum Publishers, New York, 2000.

generating high field gradients at the surface of the filaments. These areas then become trapping sites for paramagnetic particles and are the basis for the magnetic separation. Because of the commercial availability of 45-micron diameter stainless steel wool, HGMS of micron sized materials has been demonstrated in the kaolin clay industry,[9] for soil remediation,[10,11] and for water treatment.[12–14] The magnetic capture of smaller particles (submicron) can be obtained by increasing the magnetic force between the fibers and the paramagnetic particles. One approach towards increasing the magnetic force is to reduce the size of the matrix filament, which is believed to be the most effective method of improving separator efficiency. Ritter[15] and others[16] have examined submicron sized spheres for the collection of nanometer sized particles. In addition, fractionation methods for HGMS of nanometer particles is underdevelopment.[17,18] We are in investigating dendritic growth on surfaces, and controlled corrosion of stainless steel wool filaments to augment matrix surface topography in an effort to reduce the effective matrix capture diameter.

Much work[19–25] has been done to model the behavior of the paramagnetic particles as they interact with the magnetized matrix. Dynamic effects have been investigated by analyzing particle trajectories in homogeneous materials, and the influence of particle buildup on the matrix elements and their effect on the flow field have been studied. These studies, though requiring simplifying assumptions, provide useful insights into important aspects of this complex problem. In particular, they identify appropriate independent variables that form the basis for this investigation. The matrix materials currently used for HGMS are inhomogeneous and have a complex cross section. In addition, the paramagnetic particles are nonspherical and include a range of particle sizes. All of these factors discourage precise analytical treatment. Therefore, our approach is semi-empirical combining a mechanistic analytical model with empirical input from controlled experiments.

We have described previously, a *rate model*[26] that is derived from a force balance on individual paramagnetic particles in the immediate vicinity of a matrix fiber. The model assumes that if the magnetic capture forces are greater than the competing viscous drag and gravity forces, the particle is captured and removed from the flow stream. The rate model depends on the magnitude of the capture cross section (separation coefficient) as determined by the force balance on the particle. In this work we have expanded the model to include the loading behavior of particles on the matrix fibers using the continuity equation[19] and we utilize capture rate data from the experimental results. The resulting analytical model predicts the critical separation parameters for submicron single particle collection presented in this work. It thus provides a predictive tool for HGMS tests conducted here, and can be used for designing both prototype and full-scale units for specific applications.

2. EXPERIMENTAL

The magnet system used for the studies described here was a six inch warm-bore superconducting high gradient magnetic separator (manufactured by Cyromagnetics, Inc.) with an 8 tesla maximum field strength. The matrix material of extruded stainless steel wool was obtained from MEMTEC, Inc. The matrix body canister was designed and machined in our laboratories. The matrix body was packed to achieve a 94% void fraction. The experiments described here were conducted with CuO, Pr_2O_3 or PuO_2 and were used to define the significant separation and magnet design parameters. CuO and Pr_2O_3 were obtained from CERAC; PuO_2 was obtained from ongoing work in our laboratories.

Specific ranges of micron and submicron particles of Pr_2O_3 and PuO_2 were obtained by sedimentation and decantation procedures, relying on the relationship that exists between settling velocity and particle diameter in Stokes Law.[27,28] Particle sizes were verified by dynamic light scattering for particle size analysis.

For the HGMS experiments, a constant volume of solution was pumped through the magnetic matrix using a peristaltic pump at constant flowrate. Experimental variables that were systematically examined included applied magnetic field strength, flow velocity, carrier fluid and paramagnetic particle size. Typically, one to two passes through the matrix were performed using 100–200 ml of feed solution. Back-flushing of the magnetic matrix was done at zero magnetic field, with the flow direction reversed.

Feed and effluent samples were analyzed for contaminant concentration. Samples containing praseodymium were analyzed by ICP-MS where the detection limits are typically less than 1 parts per billion (ppb). In cases where a pre-concentration step is incorporated in the sample preparation, the detection limits can be more sensitive. Samples were analyzed for copper using a spectrophotometric method[29] where the detection limits are typically 2–5 parts per million (ppm). Plutonium samples were analyzed by liquid scintillation with a Packard Instrument for liquid scintillation analysis. For Pu-239 analysis, a detection limit of 6 parts per trillion (ppt) was obtained for samples with no pre-concentration steps. If lower detection limits are desired, a pre-concentration step in the sample preparation is required. In most cases, duplicate samples were analyzed to check consistency. Mass balances for the paramagnetic materials were determined for each HGMS test. Separation efficiencies were determined using the paramagnetic mass values determined by three different methods: (1) mass values from backwash and feed, (2) mass values from feed and effluent, and (3) mass values from backwash and effluent. The three methods would all give the same result for separation efficiency if the mass balance error was zero. The data is summarized below with the separation values calculated from the effluent and feed values, method (2).

Controlled corrosion studies were conducted on MEMTEC 430 series stainless steel wool. The material was cleaned by an ethanol bath prior to etching. The material was etched using HCl acid solution to obtain a general corrosion pattern. The material was exposed to a 2.9 M HCl solution for a brief time period, rinsed with de-ionized water and dried. The material was weighed and on average a forty to fifty percent weight loss was obtained. The material was characterized after corrosion by SEM.

Also, in a parallel approach, nickel dendrites were deposited on the extruded steel wool matrix by chemical vapor deposition of $Ni(CO)_4$ described previously.[30] The reactor is a hot-wall reactor as opposed to the previous cold-wall system. This type of system is easier to scale for deposition on large mats. The extreme toxicity of nickel carbonyl required special precautions in the vapor deposition system. All the valves were pneumatically controlled so the reactor could be activated or shut down remotely. Vacuum lines and gas flush systems were also remotely controlled. The system was operated in a hood within a room, which was also operated on its own exhaust system at a negative pressure with respect to the building. Mass flow controllers regulated argon carrier gases. The nickel carbonyl flow was regulated using a calibrated rotometer. The exhaust gases were passed through a hot scrubber at 500 °C to consume unused nickel carbonyl leaving the reaction zone. During operation the lines were heated to 40 °C to avoid condensation of carbonyl. The sample was placed in the middle of the reaction chamber. Processing parameters were adjusted to optimize the nickel dendrite distribution. The material was characterized by SEM.

3. RESULTS/DISCUSSION

A series of HGMS experiments were performed using particulate oxides of copper, praseodymium or plutonium suspended in a liquid. The experimental data were correlated with the analytical model. The experimental results and the correlated model results are discussed below. The model validation has been demonstrated previously with a systematic series of soil remediation tests.[26]

3.1. Model Formulation

The mathematical model is based on a geometry comprised of a highly porous bed of ferromagnetic parallel cylinders describing the matrix elements. These elements are arranged in a regular staggered manner and are orientated at right angles to both the flow and magnetic field directions. The application of HGMS involves passing a slurry of the contaminated sample through a magnetized volume. In the separation, there are several key independent experimental variables affecting the performance of the magnetic separation system. Variable ranges that we have studied are listed in Table 1. We have defined the variables as either material characteristics or operating parameters. The significance of these parameters has been discussed elsewhere.[26]

The model is formulated from the continuity equation and an expression for the capture rate of the paramagnetic particles. We start with the continuity equation[19] for the matrix given by

$$\frac{\partial c}{\partial t} + \frac{(1-\varepsilon_o)}{\varepsilon_o}\rho m\frac{\partial n}{\partial t} + \frac{U_o}{\varepsilon_o}\frac{\partial c}{\partial z} = 0 \tag{1}$$

Where, c is the contaminant concentration, ε_o is the matrix void fraction, ρ_m is the matrix packing density, n is the mass of particles captured per unit mass of packing, U_o is the superficial velocity, t is time and z is axial distance through the matrix. This formulation describes the distribution of particles in a magnetic filter and can be used to generate breakthrough curves for the filter.

Table 1. Independent Variables for Magnetic Separator

Material Characteristics	Separator Parameters
Particle Size: 0.2–5.0 μm	Matrix Element Size: 25 μm
Paramagnetic Particle Concentration: 50–0.3 ppm	Matrix Element Spacing: 80 μm
Solids Concentration: <1 wt %	Magnetic Field Strength: 2.0–6.5 T
Magnetic Susceptibility* (×10⁶): 200–2,400	Matrix Material: 430 Stainless Steel
Carrier Fluid: water or dodecane	Residence Time: 2–30 s
Slurry pH: 7 (water)	Superficial Velocity: −0.5–2.0 cm/s

*SI units.

The rate of particle capture given by the partial derivative of n with respect to t is related to the capture cross section, λ of the magnetic separator. This relationship is given by eq 2, where D_m is the diameter of the matrix wire.

$$\frac{\partial n}{\partial t} = \frac{4U_o \lambda c}{\pi D_m \rho_m \varepsilon_o} \left[1 - \frac{n}{n_t} \right] \tag{2}$$

We determine the capture cross section empirically from experiments and have correlated the results.[26] In summary from the previous work, we found that the capture cross section is significantly less that one, which implies that the particles can be swept around the matrix elements by the fluid flow even if the particles occupy an upstream capture zone defined by the projected area of the matrix element. The capture cross section was found to be proportional to the magnetic force and the particle diameter and inversely proportional to the matrix element diameter and the matrix element spacing. This indicates that the improved capture of submicron particles can be achieved by decreasing the matrix element diameter and the spacing between the matrix elements.

3.2. Single Particle Collection

Assuming a spherical particle with a known density, the concentration of a specific diameter particle can be calculated. The single particle concentrations for plutonium oxide at various particle diameters are shown in Table 2. For a concentration of 4 parts per billion (ppb) of a 0.8 micron particle, there would be approximately 1300 *plutonium oxide* particles in 1 ml of solution. In the initial feed concentrations for our experiments, we typically use starting plutonium concentrations in the range of 2–15 ppb and an initial volume of 100 ml. This would translate to approximately 6,500 particles with a 0.8 micron diameter in the feed solution. In these applications, we attempt to extract all the 6,500 particles to get below the single particle threshold.

3.2.1. Pr_2O_3. HGMS experiments on praseodymium oxide in water were conducted in order to optimize the system parameters for single particle collection. These experiments were performed with small particulate Pr_2O_3, which serves as a surrogate for plutonium and can be detected at less than 1 ppb levels by ICP-MS. In our experiments, the three particle size ranges that were studied include 0.2–0.8, 0.8–2 and 2–5 μm sized particles. For the single particle threshold for Pr_2O_3 of these particle size ranges we must be below the 1 ppb level for 7 μm sized particles and below the 1 parts per trillion (ppt) level for 0.7 μm sized particles. With Pr_2O_3 particles, we were able to approach single particle capture because ppb effluent concentrations were achieved.

The results for the *magnetic separation* of Pr_2O_3 particles in water are shown in Table 3. The variables that were examined include particle size, applied magnetic field

Table 2. The calculated number of PuO_2 particles in concentrations of 4 parts per billion to 3 parts per trillion

[PuO_2] (ppb)	5 μm Particle Count	0.8 μm Particle Count	0.2 μm Particle Count
4	5	1,300	83,330
0.006	0.008	2	125
0.003	0.004	1	62

Table 3. The separation of submicron Pr_2O_3 particles in water with HGMS

Run#	Particle Size (μm)	Applied Magnetic Field (T)	Superficial Velocity (cm/s)	Matrix Length (cm)	Res. Time (s)	[M]i[1] (ppm)	[M]f[2] (ppm)	Sep. Eff. Pass One[3] (%)
1207	0.2–0.8	6.5	1	15	15	0.437	0.001	99.77
1207A	0.2–0.8	6.5	0.5	15	30	0.846	0.0003	99.97
1208	0.8–0.2	6.5	1	15	15	0.832	0.002	99.76
1209	2.0–5.0	6.5	1	15	15	2.372	0.0047	99.80
1214A	0.2–0.8	2	1	15	15	1.89	0.005	99.74
1215	2.0–5.0	2	1	15	15	5.932	0.002[4]	99.97
1216	0.2–0.8	6.5	1	15	15	2.92	0.01	99.66

1) [M]i is the concentration of the paramagnetic material in the initial feed stream. 2) [M]f is the concentration of the paramagnetic material in the final effluent stream. 3) Separation effectiveness after one pass through the separator, determined from specific mass (corrected for sample volume) of praseodymium in the influent and effluent streams. 4) Data point is below the detection limits for the series of experiments from 1214–16.

strength and matrix residence time. Exceptionally high recovery was observed in all tests. The single particle concentration was approached in the two tests with 2–5 μm sized particles (test #1209 and #1215), where concentration of Pr_2O_3 in the effluent was 4.7 and 2 ppb, respectively. It is also expected that the magnetic separation of larger particles (greater than submicron) is more efficient due to the increased magnetic forces. Tests with submicron material were an order of magnitude above the single particle threshold limit. From these experiments, the most significant parameter appears to be the matrix residence time. Run 1207A showed an exceptionally high separation with the 0.2 to 0.8 μm material with a 30 second residence time. This indicates that for the best separation, a long matrix or low superficial velocity may be required to assure submicron particle capture.

The data are correlated with the model in Fig. 1. The open circle curve is the predicted effectiveness versus particle diameter; the closed circles are the measured data points. From our previous work, we have demonstrated that the model correlates with

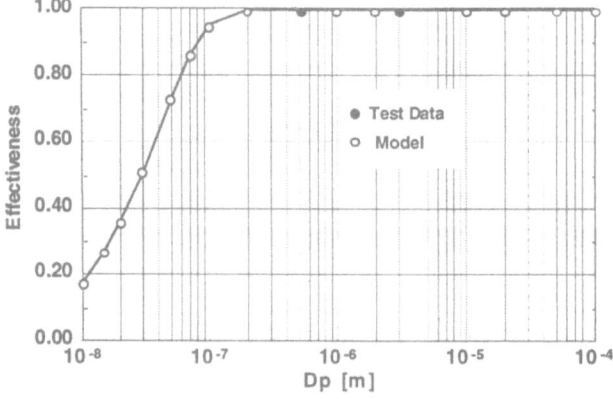

Figure 1. The effect of particle size (Dp) on the HGMS performance for Pr_2O_3 is plotted. The applied magnetic field strength is 6.5 T, and the superficial velocity is 1 cm/s. The open circles are the predicted effectiveness; the closed circles are the experimental data points. For the experimental results, the mean particle size is determined from particle size distribution data.

experimental results.[26] From the current data we are able to confirm that the model correlates with the separation efficiencies obtained with rare earth materials.

3.2.2. PuO₃. The single particle capture limit of plutonium oxide varies slightly from praseodymium due to the density differences of the compounds. The concentration of one $0.8\,\mu m$ particle of plutonium oxide is 3 ppt (for Pu-239 this translates as 0.2 picoCurie/mL). We must be below this limit for submicron single particle collection.

To obtain a concentration of less than $3 \times 10^{-11}\,M$ (~3 ppt) of plutonium oxide, the small but finite solubility of PuO_2 in water demanded that we examine alternate carrier fluids. Though the PuO_2 solubility constant is very low, in slightly basic conditions, plutonium (IV) in water solutions at pH 8 can be found at concentrations up to $10^{-7}\,M$.[31,32] Preliminary HGMS tests were conducted in pH 8 water but the PuO_2 solubility prevented complete plutonium removal from the feed samples. For the remaining HGMS tests, we have examined dodecane as a carrier fluid because of the plutonium oxide insolubility in long chain hydrocarbons. The results are illustrated in Fig. 2. Plutonium oxide particles sized from 0.2–0.8 microns were removed below 6 ppt. This value is the detection limit of the current analytical method (liquid scintillation counting) without a pre-concentration step in the sample preparation. The separation efficiencies for these tests were over 99%. For PuO_2 there are predicted to be 2 particles in one ml of solution at the 6 ppt

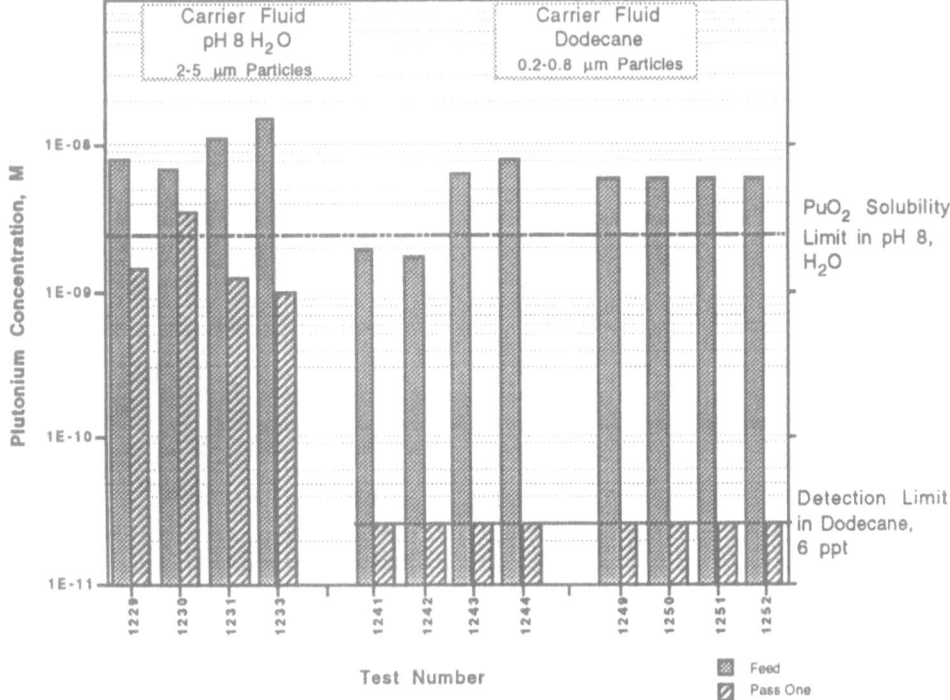

Figure 2. The bar graph depicts the plutonium concentration in the HGMS feed and effluent from pass one for 12 HGMS experiments. Experiment numbers 1229–1232 were conducted in pH 8 water with a 2–5 μm PuO_2 particle size range. Experiments 1241–1252 were conducted in dodecane as a carrier fluid with a 0.2 to 0.8 μm PuO_2 particle size range.

concentration (see Table 2). Yet under these conditions for the smaller 0.2 μm particles, over 125 particles still remain in one ml the effluent liquid.

In early experiments, HGMS data collected in water were corrected for plutonium solubility through a series of standard filtration steps. These aqueous based data are correlated with the model in Fig. 3. The open point curve is the predicted effectiveness versus particle diameter; the closed points are the experimental data. The data are in good agreement with the model as we have demonstrated previously in extensive testing on soils.[26]

3.3. Matrix Material

The experiments discussed here used a matrix material based on extruded stainless steel fibers. Figure 4 shows a scanning electron micrograph of the extruded material. The multiple ridges in the material form a continuum of sharp axial edges that generate large magnetic field gradients and serve as trapping sites for the submicron particles. The extruded sample developed for this program contains a much higher density of edge locations compared with traditional HGMS shaved stainless steel wool materials. The extrusions produce small matrix fiber diameters and high matrix loading capacities.

In previous tests the matrix capacities for both the shaved stainless steel wool and expanded metal was determined to be 0.05 of the calculated surface area of the fibers. This fraction we refer to as the load factor. If the load factor were 1, then all of the calculated surface area in the matrix would be available for particle capture. Breakthrough occurs in these materials when 5% of the total estimated surface area of the matrix fibers is covered with paramagnetic particles. Significantly, with the extruded stainless steel wool no breakthrough was observed in several of the tests. In fact, 25 liters of solution containing 60 grams of 5 μm CuO was passed through the matrix material without reaching saturation. From this information a lower limit of the matrix capacity can be determined for the extruded stainless steel material. Figure 5 shows model calculations of the matrix

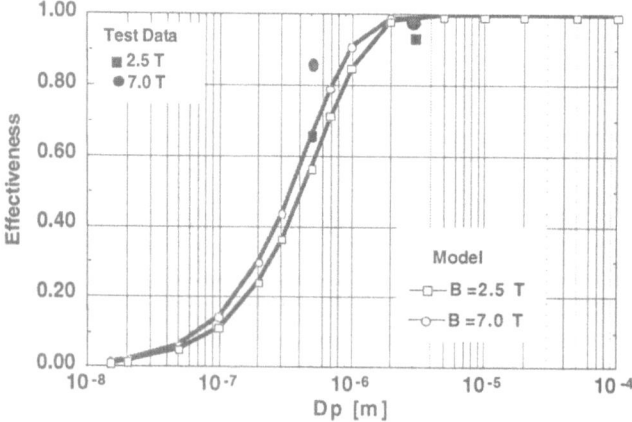

Figure 3. The calculated effect of particle size (Dp) on the HGMS performance for PuO_2 in pH = 8 water is plotted. The data was corrected for plutonium solubility. The experimental data points were determined in HGMS runs 1226, 1227, 1230 and 1231. The applied magnetic field strength is 7.0 T (circles), or 2.5 T (squares); the superficial velocity is 1 cm/s. The open points are the predicted effectiveness; the closed points are the experimental data points.

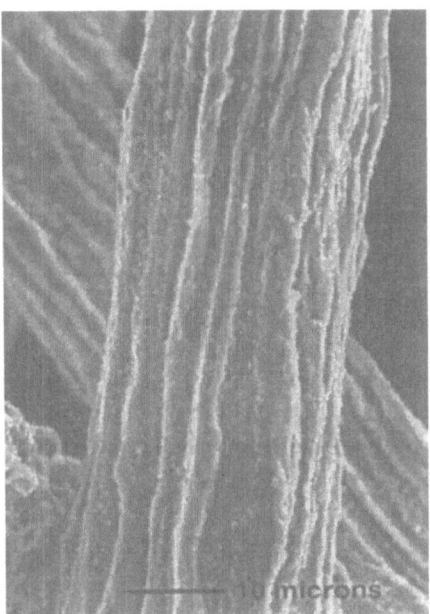

Figure 4. Scanning electron micrograph of extruded stainless steel matrix material.

capacity for 25 µm extruded stainless steel matrix material as a function of slurry volume required for breakthru. The curves show the effect of capturing two different sized particles. For 5 µm CuO, the lower limit for the matrix capacity is 0.60 of calculated surface area for the extruded material (corresponding to the 25 L of slurry volume). This shows that the capacity of the extruded matrix is an order of magnitude higher than the conventional matrix. The significant increase in the matrix capacity for the extruded wool material can be attributed to the deeply grooved surface of the fibers (see Fig. 4) creating large surface areas and numerous edge capture sites. The high edge density contributes

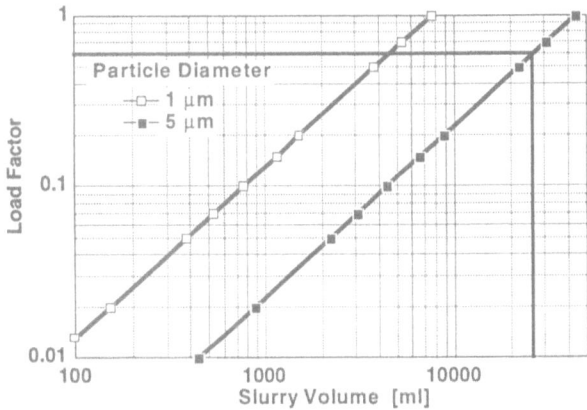

Figure 5. Model calculations of the matrix capacity for 25 µm extruded stainless steel matrix material. The curves show the effect of capturing two paramagnetic particle sizes. From the experimental data, 25 liters of 5 µm CuO was passed through the matrix with no observed saturation. The lower limit of the matrix capacity is expected to be 0.6 from the plot.

to higher local field gradients and increased surface area, which may cause the captured particles to adhere in multiple layers.

The effect of *matrix* diameter was examined experimentally by determining HGMS effectiveness for four different matrix materials with different cross sectional areas. The correlation of the matrix cross sectional areas and the HGMS results for submicron particles is shown in Fig. 6. The qualitative exponential relationship indicates that for selective small particle capture, very fine matrix fibers are required. Previous work has shown that the high gradient magnetic force reaches a maximum for a ratio of fiber diameter radius to particle radius (**p/f**) equal to approximately three.[33] It has also been experimentally confirmed that in the range of **p/f** from 1 to 10, the change in the separation efficiency of capture is insignificant.[34] We have experimentally demonstrated this effect, that as the matrix fiber diameter approached the particle diameter then the separation efficiency was enhanced. The analytical model also predicts this trend. Figure 7 shows the predicted HGMS effectiveness for matrix wire diameters of 5, 15 and 25 µm. For particles below 0.2 µm, HGMS separation is much more effective with smaller wire diameters. For the same matrix size the effect of changing the applied magnetic field from 7 T to 2.5 T is also shown in the figure. This change is not nearly as significant as the effect of reducing the matrix size. At 2.5 T the matrix material is nearly saturated magnetically, and any further increase in the applied field has a small effect on the separation. Increasing the matrix residence time will also increase the separation effectiveness but at extremely low superficial velocities, the resulting particle settling can be a significant problem.

To greatly enhance performance and to achieve a single particle collection threshold, it has become apparent through model predictions and experimental data that the matrix fiber diameter is a crucial separation variable. Tests have been initiated to develop new matrices with submicron sized fiber diameters. Two different approaches are being studied to generate these materials. One approach is controlled corrosion studies on

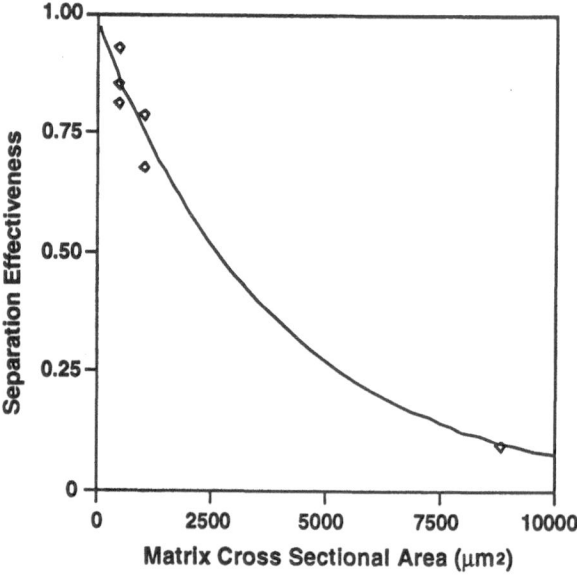

Figure 6. Experimental results showing the effect of the matrix cross sectional area for submicron particle capture on the separation effectiveness.

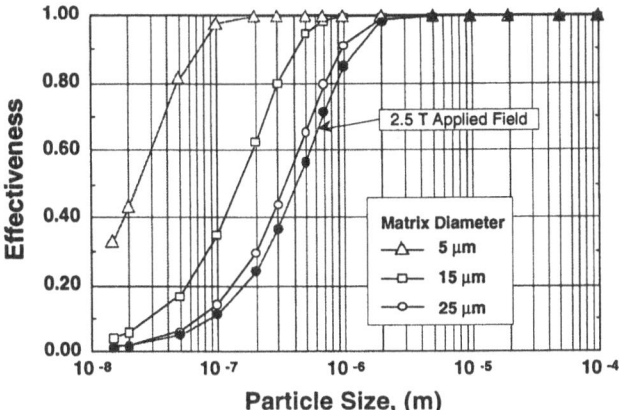

Figure 7. Model calculations for HGMS separation effectiveness for three matrix fiber diameters. The prediction is based on an applied field of 7.0 T (open points) or 2.5 T (closed points); the superficial velocity is 1 cm/s; the carrier fluid is dodecane.

25 μm sized drawn stainless steel wool material, and the second approach is the growth of micron sized nickel dendrites on stainless steel wool fibers by chemical vapor deposition. Controlled corrosion studies on stainless steel wool fibers have been conducted under mildly corrosive conditions so that a general corrosion pattern is obtained. Matrix material has been prepared that has a 20–80% weight loss depending on the acid contact time. The material that has an 80% weight loss has some fibers approaching 5 μm in diameter, with many fibers close to 20 μm. Enough material will be prepared for performance assessment with HGMS experiments.

Nickel dendrites were deposited on the stainless steel matrix material using methods described previously, but with the distinction of using a hot wall reactor.[30] Nickel dendrite structures were uniformly deposited on the surface throughout the wool structure. The length of the dendrites is 1–2 orders of magnitude smaller than the diameter of the wool fiber substrate. The resulting matrix material has a very high surface area containing fine points that may greatly enhance the high field gradients. In addition to magnetic characterization of the material, HGMS tests on these materials are planned. This material has the potential to advance HGMS towards sub-nanometer particle collection. We are currently evaluating the HGMS performance of the new materials.

4. SUMMARY

We have derived a model based on the continuity equation that uses empirically determined capture cross section values. This enables the prediction of high gradient magnetic separator performance for a variety of materials and applications. The model can be used to optimize the capture cross section and thus increase the capture efficiency. Results show that maximum separation effectiveness is obtained when the matrix fiber diameter approaches the diameter of the particles to be captured.

Experimentally, we obtained a single particle capture limit with 0.8 μm PuO$_2$ particles with dodecane as a carrier fluid. The finite solubility of plutonium in water prevented the complete removal of the contaminants when using water as the carrier fluid.

The development of new *matrix materials* is being pursued through the controlled corrosion of *stainless steel wool*, or the deposition of *nickel dendrites* on the existing stainless steel matrix material. The new materials are promising for the submicron collection of *paramagnetic particles*. HGMS experiments on the new materials are planned.

REFERENCES

1. Svoboda, J. *Magnetic Methods for the Treatment of Minerals*; Elsevier: New York, 1987.
2. J. Lyman in *High-Gradient Magnetic Separation, Unit Operations for Treatment of Hazardous Industrial Wastes*, D.J. De Renzo, ed. Noyes Data Corp., Park Ridge, New Jersey, 1978, pp. 590–609.
3. Arvidson, B.R. and Henderson, D. *Minerals Engineering*, **1997**, *10*, 127.
4. Richards, A.J., Roath, O.S., Smith, R.J.S., and Watson, J.H.P. *IEEE Trans. On Magnetics*, **1996**, *32*(2), 459.
5. Pourfarzaneh, M., Snavely, K., and Lawlor, J. *Genetic Engineering News*, **1995**, *15*, (13).
6. Avens, L.R., Hill, D.D., Prenger, F.C., Stewart, W.F., Tolt, T.L., and Worl, L.A. "Process to Remove Actinides from Soil Using Magnetic Separation," The Regents of the University of California, Office of Technology Transfer, Alameda, CA, United States (U.S. Corporation), US Patent 5538701 960723, 1996.
7. Nuñez, L., Buchholz, B.A., and Vandegrift, G.F. *Sep. Sci. & Tech.*, **1995**, *30*, 1455–1472.
8. Ritter, J.A., Zamecnik, J.R., Hutson, N.D., Smith, M.E., and Carter, J.T. *Wat. Sci. Tech.* **1992**, *25*, 269.
9. Iannicelli, J. *Clays and Clay Minerals*, **1976**, *24*, 64–68.
10. Schake, A.R., Avens, L.R., Hill, D.D., Padilla, D., Prenger, F.C., and Worl, L.A. Conference Proceedings, Conference: 207. Spring National Meeting of the American Chemical Society (ACS), San Diego, CA (United States), 13–18 Mar 1994.
11. Coe, B.T., Gerber, R., and Witts, D. *IEEE Trans. On Magnetics*, **1998**, *34*(4) 2126–2128.
12. Yiacoumi, S., Rountree, D.A., and Tsouris, C. *J. of Colloid and Interface Science*, **1996**, *184*, 477–488.
13. Emory, B.B. *IEEE Trans. On Magnetics*, **1981**, *Mag-17*, 3296.
14. Kolm, H., Kelland, J., and Kelland, D. *Sci. Am.*, **1975**, *46*.
15. Ebner, A.D., Ritter, J.A., and Ploehn, H.J. *Separation and Purification Technology*, **1997**, *11*, 199.
16. Haque, M.F., Aidun, R., Moyer, C., and Arajs, S. *J. Appl. Phys.* **1988**, *63*, 3239.
17. Tsukamoto, O. and Ohizumi, T. *IEEE Transactions on Appl. Superconductivity*, **1995**, *5*(2), 311–313.
18. Kelland, D.R. *IEEE Trans. On Magnetics*, **1998**, *34*(4), 2123–2125.
19. Akoto, I.Y. *IEEE Transactions on Magnetics*, **1977**, *Mag-13*(5), 1486.
20. Friedlander, F.J., Takayasu, M., Rettig, J.B., and Kentzer, C.P. *IEEE Transactions on Magnetics*, **1978**, *Mag-14*(6), 1158–1164.
21. Lawson, W.F., Simons, W.H., and Treat, R.P. *Journal of Applied Physics*, **1977**, *48*(8), 3213–3224.
22. Luborsky, F.E. and Drummond, B.J. *IEEE Transactions on Magnetics*, **1976**, *Mag-12*(5), 463–465.
23. Nesset, J.E. and Finch, J.A. Industrial Applications of Magnetic Separation, Conference Proceedings, pp. 188–196, July 1978.
24. Stekly, Z.J.J. and Minervini, J.V. *IEEE Transactions on Magnetics*, **1976**, *Mag-12*(5) pp. 474–479.
25. Watson, J.H.P. *J. Applied Physics*, **1973**, *44*(9) 4209–4213.
26. Prenger, F.C., Stewart, W.F., Hill, D.D., Avens, L.R., Worl, L.A., Schake, A., deAguero, K.J., Padilla, D.D., and Tolt, T.L. *Adv. in Cryogenic Engineering*, **1994**, *34*, 485.
27. Tanner and Jackson, *Soil Science Society Amer. Proc.* **1947**, *12*.
28. *Methods of Soil Analysis*, *Part 1*, Klute, A. ed., American Society of Agronomy, Inc. Soil Science Society of America, Inc. Madison, Wisconsin, 1986, Chapter 15.
29. Stephens, B.G., Felkel, H.L., and Spinelli, W.M. *Anal. Chem.* **1974**, *46*, 692.
30. Grimmer, D.P., Herr, K.C., and McCreary, W.J. *J. Vac. Sci. Technol.*, **1978**, *15*, 59.
31. Allard, B. in *Actinides in Perspective*, Edelstein, N.M. Ed. Proceedings of the Actinides—1981 Conference, Pergamon Press, New York, 1982, 553–580.
32. Delegard, C.H. *Radiochim Acta*, **1987**, *41*, 11–21.
33. Oberteuffer, J.A. *IEEE Trans. Mag.*, MAG-10 (1974), 223.
34. Collan, H.K., Jantunen, J., Kokkala, M., and Ritvos, A. *IEEE Trans. Mag.*, **MAG-14**, (1978), 398.

20

TRANSITION METAL CATALYSTS FOR THE AMBIENT TEMPERATURE DESTRUCTION OF ORGANIC WASTES USING PEROXYDISULFATE

G. Bryan Balazs, John F. Cooper, Patricia R. Lewis, and Martyn G. Adamson

Lawrence Livermore National Laboratory
Livermore, California 94550

1. ABSTRACT

Destruction of the organic components of hazardous or mixed waste has been demonstrated in a process known as Direct Chemical Oxidation (DCO). This technology, developed over the last six years at Lawrence Livermore National Laboratory (LLNL), is an aqueous-based process which mineralizes almost all organics through oxidative destruction by peroxydisulfate. The process typically operates at 80–100 °C but there are obvious advantages to lowering this operating temperature to near ambient for the treatment of volatile materials, or for situations where heating the waste is impractical (e.g., large area decontamination of soil). For the purposes of quantifying transition metal catalytic activity in a prototype DCO system, the Destruction and Removal Efficiencies (DRE's) for the oxidation of several model organic substrates in acid peroxydisulfate solutions have been measured. Results are presented for the enhancement of destruction rates of ethylene glycol, 1,3-dichloro-2-propanol, tributyl phosphate, and a "real world" waste surrogate using catalysts such as ionic Ag, Cu, Co, and Fe.

2. INTRODUCTION

Direct Chemical Oxidation (DCO) is a non-thermal, ambient pressure, aqueous-based technology for the oxidative destruction of the organic components of hazardous or mixed waste streams. The process has been developed for applications in waste treatment and chemical demilitarization and decontamination at LLNL since 1992, and is applicable to the destruction of virtually all solid or liquid organics, including: chloro-solvents, oils and greases, detergents, organic-contaminated soils or sludges, explosives, chemical and biological warfare agents, and PCB's.[1-15]

Emerging Technologies in Hazardous Waste Management 8, edited by Tedder and Pohland
Kluwer Academic/Plenum Publishers, New York, 2000.

DCO uses solutions of the peroxydisulfate ion (typically sodium or ammonium salts) to completely mineralize the organics to carbon dioxide and water. The net waste treatment reaction is (Eq. 1):

$$S_2O_8^{2-} + \{\text{organics}\} \Rightarrow 2HSO_4^- + \{CO_2, \text{ inorganic residues}\} \tag{1}$$

Peroxydisulfate is one of the strongest chemical oxidants known (oxidation potential is +2.05V), and is exceeded in oxidative power only by fluorine, ozone, and oxyfluorides. The oxidation potential of peroxydisulfate is high enough to oxidize nearly any organic substance.[16,17]

While many oxidants exhibit a redox potential capable of broad-spectrum organic oxidation, peroxydisulfate uniquely combines a high oxidation potential with a rapid, nucleophilic charge-transfer capability. Oxidation occurs principally through the formation of the sulfate radical anion $SO_4^{\cdot-}$, following mild thermal (70–100 °C) or UV activation of peroxydisulfate solutions(Eq. 2):[18-25]

$$S_2O_8^{2-} \rightleftharpoons 2SO_4^{\cdot-} \tag{2}$$

The subsequent reaction of the sulfate free radial with the organic and with water results in a cascade of active oxidants including organic free radical fragments and hydroxyl free radicals. The decomposition of peroxydisulfate produces a number of intermediate oxidizers including peroxymonosulfate (a strong industrial bleach), hydrogen peroxide, and nascent oxygen bubbles.

The oxidant ammonium or sodium peroxydisulfate is sufficiently stable at or slightly below room temperature to be stored almost indefinitely as a solid or a wet slurry for months. This being the case, the process of waste destruction can be decoupled in time and place from the generation of the peroxydisulfate oxidant. The oxidant becomes reactive only at elevated temperatures or through contact with transition metal catalysts. This allows the oxidant to be slowly produced and stockpiled for use in intermittent waste treatment campaigns of short duration.

The expended oxidant may be electrolytically regenerated to minimize secondary waste. The ammonium (or sodium) hydrogen sulfate produced as a byproduct of the organic waste oxidation process is relatively non-hazardous, and may be incorporated in a subsequent inorganic treatment step, or recycled in a flowing electrolyte cell. If recycled, an industrial cell may be used which employs a platinum or glassy carbon anode, an inert graphite cathode, and a porous ceramic separator to prevent cathodic reduction of the product. The anodic reaction is (Eq. 3):

$$2NH_4HSO_4 \rightleftharpoons (NH_4)_2S_2O_8 + 2H^+ + 2e^- \quad \text{(anode)} \tag{3}$$

while the cathodic half reaction is the reduction of water to form hydrogen gas.[26-28] For such cells, this gas is best immediately oxidized to water (in a catalyzed bed), and the water internally recycled. Commercial catalysts are available and used for this purpose. In specialized applications where the production of hydrogen gas is not desirable (such as in a confined space), the cathodic reaction can be replaced with oxygen reduction using a porous gas diffusion electrode. This modification should reduce the cell voltage by 1 volt (about 20%).

The DCO process normally operates at 80–100 °C, a heating requirement which increases the difficulty of surface decontamination of large objects or, for example, treatment of a wide area contaminated soil site. Alternatively, the radical-generation process shown in Eq. 2 may be accelerated at near ambient temperatures (20–50 °C) through the use of a catalyst such as metallic platinum, or with dissolved silver, iron, or copper ion catalysts.

Available literature data includes the catalysis of peroxydisulfate oxidation of organics by a number of transition metal catalysts.[18–20,29–34] These catalysts include the ions of Cu,[29–31,34] Ag[18–20,32–34] and Fe.[34] In addition, a number of organic substrates were studied, including alcohols, aldehydes, and ethers. In general, the catalysts act to accelerate the decomposition of peroxydisulfate into the sulfate free radical anion (Eq. 2 above) which leads to a higher rate of organic oxidation by this aggressive oxidant.

A generalized rate equation applicable to all combinations of catalyst, organic substrate, and oxidant concentrations is not easily formulated. In addition, the actual mechanism is often dependent on the particular catalyst, making a universal statement about transition metal catalysis impossible. For example, Fe(III) appears to accelerate some reactions while inhibiting others. However, the general rate equation for the reduction of peroxydisulfate (Eq. 1 above) follows the form (Eq. 4):

$$d\left[S_2O_8{}^{2-}\right]\Big/dt = -k\left[S_2O_8{}^{2-}\right]^x[\text{catalyst}]^y \qquad (4)$$

where x is some value between 1/2 and 3/2 and y is some value between 0 and 3/2, with negative values possible in some cases. Note that this rate equation is independent of the organic concentration, i.e., the rate of organic oxidation is limited only by the rate of the formation of the sulfate free radical anion (assuming no mass transport limitations).

3. EXPERIMENTAL

Experiments were conducted at a laboratory scale (1.0 liter reactor), using acidified (0.1 M H_2SO_4) solutions of 1.0 M $(NH_4)_2S_2O_8$. Catalyst ions were added in the form of the sulfate salts of the appropriate transition metal, except in the case of Ag which was by addition of the nitrate salt. Solutions were put into lightly stoppered 250 mL glass flasks immersed in a circulating water bath at the appropriate temperature. Four organic substrates were tested: ethylene glycol, 1,3-dichloro-2-propanol, tributyl phosphate, and the hydrolysis product of 1,1,1-trichloroethane.* Initial organic loading was between 500 and 6,000 ppm total carbon, and the catalyst concentrations were varied between 1 and 1,000 ppm (with the exception of additional Fe tests at 5,900 ppm). Destruction and Removal Efficiencies (DRE's) were determined by Total Organic Carbon (TOC) analysis of the resultant solutions at the end of the experiment. Determination of peroxydisulfate concentrations was done by quenching the reaction in an ice bath, followed immediately by an iodometric titration (peroxydisulfate reacts with iodide to form iodine).

4. RESULTS

In order to assess the effect of several transition metal ions on the DCO system at normal operating temperatures, the reduction of peroxydisulfate and the mineralization of the organic were studied at 80 °C. Measurements were made of both the instantaneous

---

*1,1,1-trichloroethane was hydrolyzed in a prior experiment in 1.0 M NaOH at 100°C to form a mixture of water soluble, less volatile organic compounds.

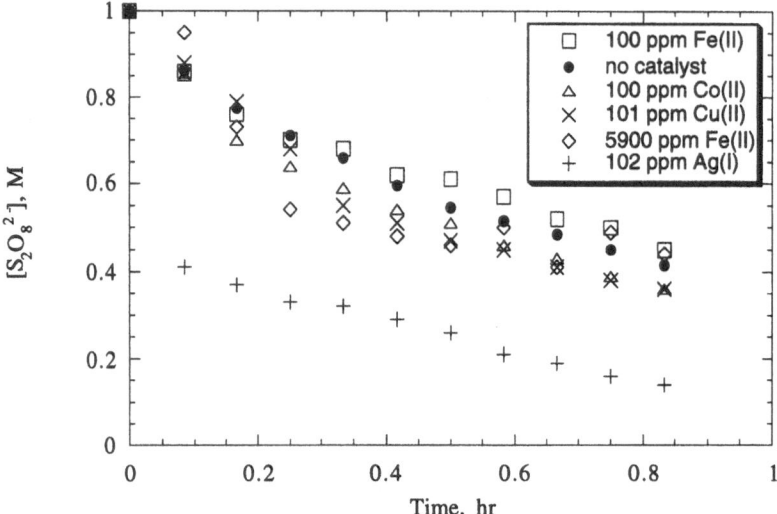

Figure 1. Peroxydisulfate concentration as a function of time with various catalysts and an initial carbon loading of 1,400 ppm, at 80 °C. Initial $[S_2O_8{}^{2-}]$ is 1.0 M in 0.1 M H_2SO_4.

concentration of peroxydisulfate (by titration) and the conversion of organic to carbon dioxide (by on-line IR gas analysis). Ethylene glycol and the hydrolysis output of chloro-solvent hydrolysis were used as substrates; several typical transition metals were tested with each.

Figure 1 shows the decrease in peroxydisulfate concentration as a function of time during the oxidation of ethylene glycol. Each of the catalysts systems, except for 100 ppm Fe(II), showed some catalytic activity for the decomposition of peroxydisulfate, with the silver system being the most pronounced.

Figure 2 shows the output of carbon dioxide obtained during each of seven runs, with and without the catalysts as noted, during the oxidation of ethylene glycol at 80 °C.

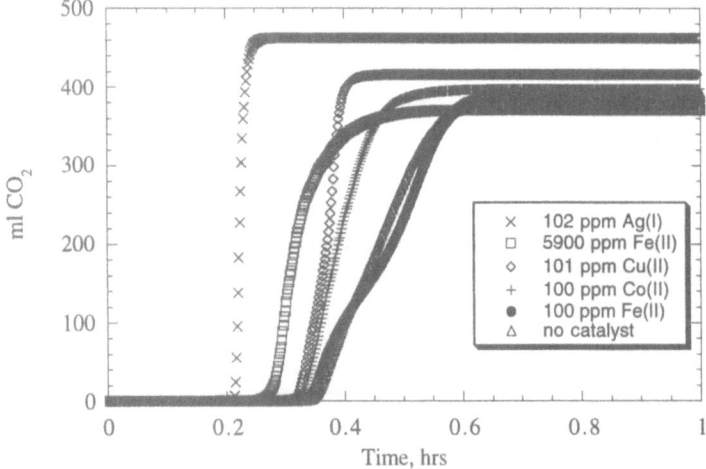

Figure 2. Evolution of carbon dioxide as a function of time with various catalysts and an initial carbon loading of 1,400 ppm, at 80 °C. Initial $[S_2O_8{}^{2-}]$ is 1.0 M in 0.1 M H_2SO_4.

The expected output based on the amount of ethylene glycol initially added was 420 ml CO_2 with a system measurement precision of ±10%. All of the results fall within this range, but the silver system is noticeably different, both in terms of a higher CO_2 output as well as a much steeper rise. Although the latter result is expected due to the catalytic activity, a reason for the former observation is not known.

Previous work on the effect of transition metal catalysts on the rate of oxidation of organic substrates by peroxydisulfate[18–20,29–34] focused primarily on the decomposition of the peroxydisulfate, and not on the mineralization of the organic. For the current work, it was felt that a more realistic approach with regard to waste treatment by DCO was to measure the overall Destruction and Removal Efficiencies (DRE's) on a matrix of catalysts, temperatures, and organic substrates. In addition, most of this current work is focused on 20–50 °C as this regime is closer to temperatures likely to be encountered when using DCO to treat waste under close to ambient conditions. No additional oxidant was added in most cases, except as noted, in order to more closely simulate a "one-pass" treatment approach.

Results for ethylene glycol are shown in Figs. 3–5, while those for 1,3-dichloro-2-propanol are showed in Figs. 6 and 7. Tributyl phosphate is only partially soluble in water, and results with this organic substrate gave inconsistent results. The initial conditions of the tributyl phosphate experiments were such that the added level of organic was 1,700 ppm as carbon, but TOC results on this initial solution gave results of 200–300 ppm as carbon. Indeed, a second phase was observed on the walls of the reaction flasks and thus the data are judged to be invalid.

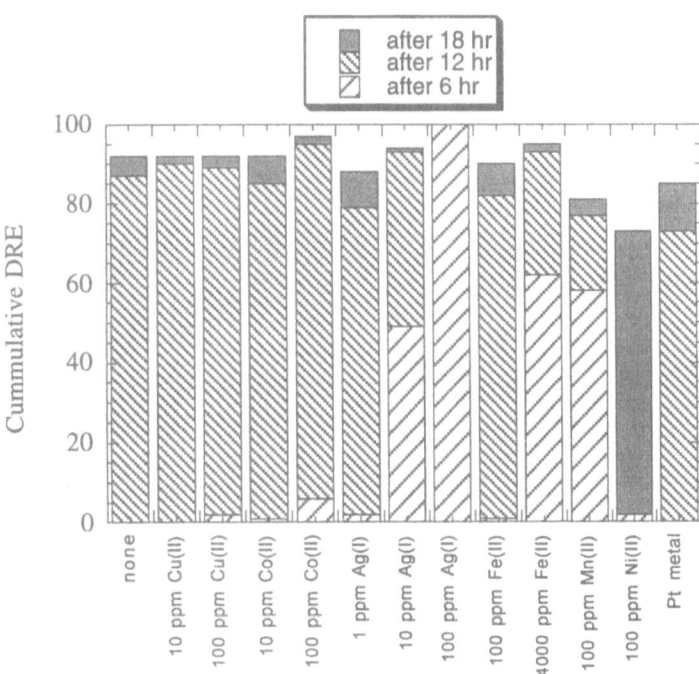

Figure 3. Cumulative Destruction and Removal Efficiency (DRE) for ethylene glycol (1,800 ppm as carbon) at 50 °C with various catalysts.

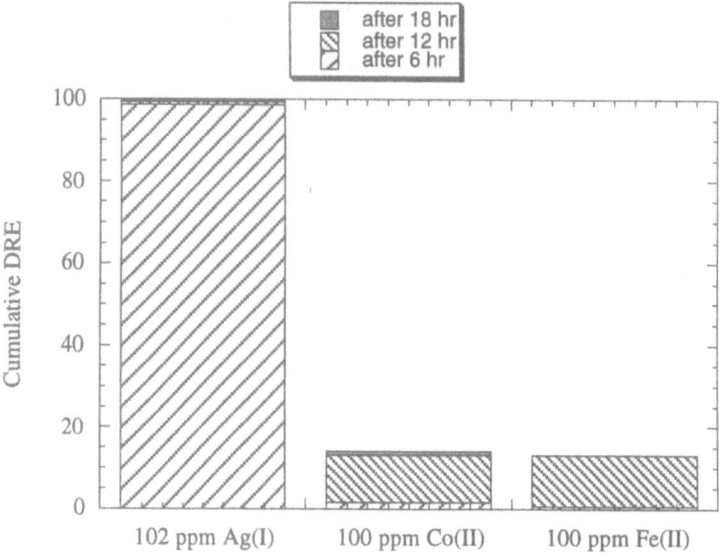

Figure 4. Cumulative Destruction and Removal Efficiency for ethylene glycol (1,800 ppm as carbon) at 35 °C with various catalysts.

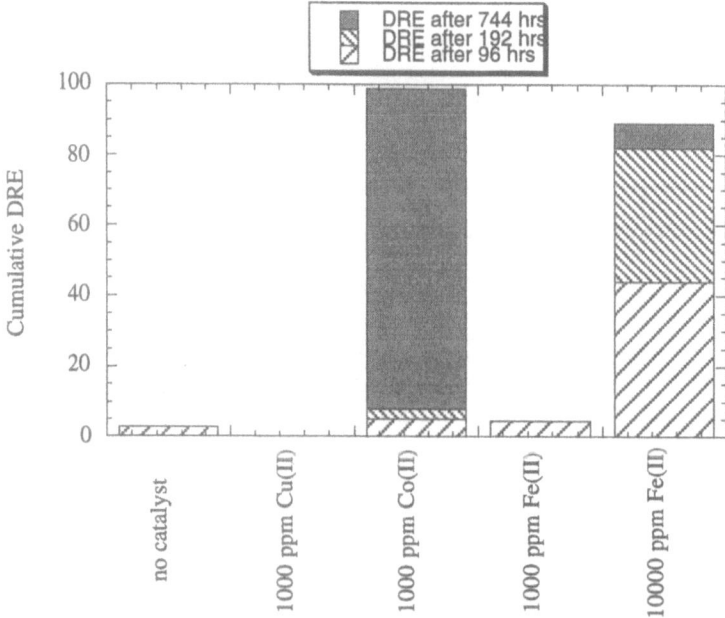

Figure 5. Cumulative Destruction and Removal Efficiency for ethylene glycol (1,800 ppm as carbon) at 20 °C with various catalysts.

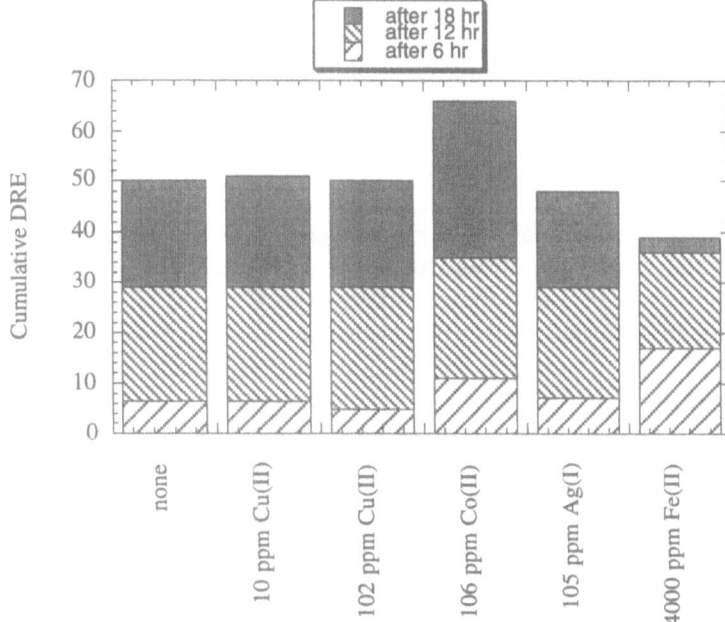

Figure 6. Cumulative Destruction and Removal Efficiency for 1,3-dichoro-2-propanol (1,700 ppm as carbon) at 50 °C with various catalysts.

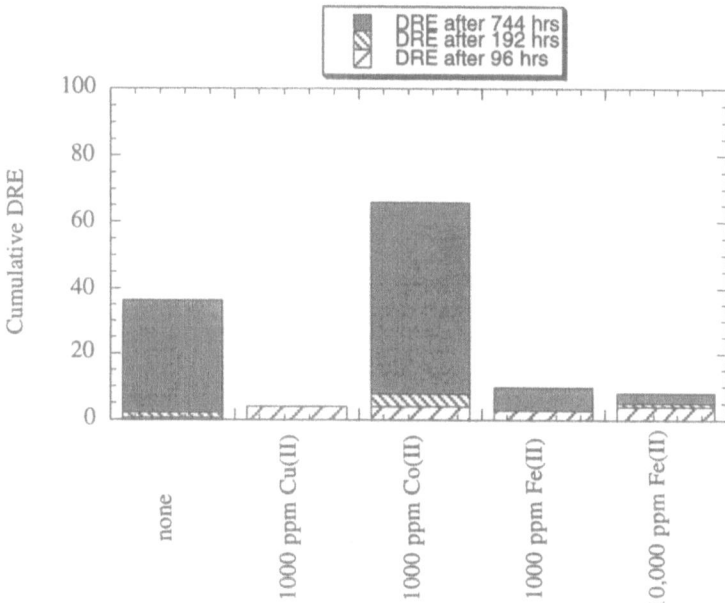

Figure 7. Cumulative Destruction and Removal Efficiency for ethylene glycol (1,700 ppm as carbon) at 20 °C with various catalysts.

5. CONCLUSIONS

As shown in Figs. 1 and 2, the concentration of peroxydisulfate decreases rapidly at 80 °C due to its reduction by the organic and, to a lesser extent, water. All of the added transition metals tested had some effect (with the exception of low concentrations of iron), and this effect was towards an acceleration of the peroxydisulfate decomposition rate. In the case of ethylene glycol as the added organic substrate (Fig. 1), silver ion showed the highest perturbation, which is expected based on the fact that silver showed the highest catalytic effect in the destruction tests. The effect of silver was not nearly as pronounced in tests where chloride ion was present. However, based on results obtained with and without catalysts at 80 °C, peroxydisulfate does not appear to be a very aggressive oxidant at concentrations of less than about 0.3 M. Catalysts did not change this cutoff value appreciably. Thus, any DCO waste treatment system would have to keep the oxidant concentration greater than 0.5 M for optimal efficiency.

During the experiments, it was noted that additional heat outputs were obtained at 80 °C with the catalysts, and the amount of extra heat correlated well with the acceleration in peroxydisulfate decomposition/organic destruction. As noted in the organic destruction tests detailed in the following section, silver exhibited the most pronounced effect. As the oxidative destruction of organics is exothermic, any organic waste treatment system using peroxydisulfate *must* take into account the extra heat output encountered when transition metal ions (especially silver) are present. As an additional safety note, the combination of an aggressive oxidant, the possible presence of chlorine species when treating chloro-organics, an exothermic reaction, and the possibility of reaction catalysis depending on the species present in the waste stream mandate that reactor material, sizing, throughputs, and heat generation be taken into consideration. In addition, because of the release of carbon dioxide gas, the system design and materials must incorporate means to prevent pressure buildup.

It was noted that, when treating chlorinated organics, catalysts also affected the production of chlorine gas. However, this effect followed no logical pattern; in some cases the chlorine production (observed by watching the reactor offgas trap color) was immediate and quantitative and in other cases it was minimal.

For the organic destruction tests on surrogate waste materials shown in Figs. 3–7, all of the transition metals studied exhibited some effect on the rate of oxidation of the organic substrates by peroxydisulfate, although the magnitudes varied greatly. This is not surprising, as peroxydisulfate has a sufficient oxidation potential to oxidize all of the transition metals to their highest valent state, and the participation of these higher valence states in organic oxidations is well-known.[35-38] It was generally the case that the transition metals which exhibited the most facile organic oxidation capability (Ag(II), Co(III), Fe(III)) also exhibited the most pronounced catalytic effect in these studies. Note that this order of catalytic effectiveness is surprisingly similar to the results obtained on Mediated Electrochemical Systems as described in Refs 35–38. Although an all-encompassing maxim for the rate enhancements obtained with a specific catalyst is not possible, a general rule of thumb is that the effectiveness of the catalysts followed the general order Ag >> Co > Cu, Fe > Pt. Individual results varied depending on the temperature regime, the organic substrate, and the matrix. This has been noted before in the literature.[18-20,29-34] The above results lead to the conclusion that catalysis of organic destruction by peroxydisulfate is merely due to the formation of another, favorably solvated transition metal cation with a high oxidation potential.

Complete destruction of the organics tested was achievable with silver as a catalyst, even at 20 °C in a relatively short time, and cobalt, iron or copper showed some enhancement although a much longer time was required. This result is backed by previous literature studies as discussed above. However, it was also noted in this current study that different organic substrates exhibited different behavior with respect to catalysis.

In all the cases studied, the enhancement of the organic destruction rate was a function of the catalyst concentration, and this is reflected in the literature as well (Eq. 4). Thus, higher rates of organic destruction can be achieved with higher catalyst concentrations. Due to the copious data available in the literature, a systematic reinvestigation of the order of the catalyst concentration on the rate equation was not attempted.

Although rate equations such as that shown in Eq. 4 would predict that destruction of organics might be faster with higher oxidant concentrations, the present work was not sufficiently detailed to either bolster or refute this statement. Additional oxidant made only a slight difference in DRE's, at least at close to ambient conditions. However, with higher organic loadings it may be necessary to periodically replenish the oxidant in waste treatment systems. See also the following section regarding oxidant concentration.

During catalyzed oxidations, there is an incubation period during which the carbon content of the solutions is unchanged from the initial value. This is presumably reflective of the fact that oxidation is occurring, but has not proceeded to the stage of evolving carbon dioxide from the solution and thus the carbon content of the solution remains unchanged from the original value. This incubation period was not noticed with silver, as oxidations with this catalyst proceeded much faster than the sampling timescales at the temperatures tested.

In waste systems containing chloride ion, or containing organic chlorine, silver ion is ineffective as a catalyst. AgCl, being highly insoluble, immediately precipitates thus removing the active catalyst. Along these same lines, silver will be ineffective as a catalyst in systems where the pH is sufficiently high to result in the precipitation of silver hydroxide. In fact, this latter characteristic would presumably be extendable to all potential transition metal catalysts in systems at high pH.

This work was performed under the auspices of the U.S. Department of Energy by the Lawrence Livermore National Laboratory under contract number W-7405-ENG-48.

REFERENCES

1. Bryan Balazs, John Cooper, Peter Hsu, Pat Lewis, and Martyn Adamson, "Direct Chemical Oxidation: Destruction of chlorinated solvents by aqueous peroxydisulfate," poster presentation, SERDP Symposium, Washington, D.C., December 3–5, 1997.

2. Peter Hsu, John Cooper, Bryan Balazs, Pat Lewis, and Martyn Adamson, "Process Engineering of Direct Chemical Oxidation for the Treatment of Insoluble Chlorinated Solvents in Waste Streams," UCRL-JC-128142abs., presented at AIChE Spring National Meeting, New Orleans, LA, March 8–12, 1998.

3. Bryan Balazs, John Cooper, Peter Hsu, Pat Lewis, Joe Penland, Larry Finnie, and Martyn Adamson, "Integrated Direct Chemical Oxidation System for the Non-Thermal Destruction of Chlorosolvents," WM '98 Proceedings, Tucson, AZ, March 1–5, 1998, UCRL-JC-128131.

4. John F. Cooper, Francis Wang, Roger Krueger, and Ken King, "Destruction of organic wastes with electrolytic regeneration of the oxidant," Proc. Sixth International Conference on Radioactive Waste Management and Environmental Remediation, Singapore, October 12–16 1997. UCRL-JC-121979 rev 2 July 1997.

5. John F. Cooper, G. and Bryan Balazs, "Demonstration of Omnivorous Non-thermal Mixed Waste Treatment: Direct Chemical Oxidation of Organic Solids and Liquids using Peroxydisulfate," Final Report to Mixed Waste Focus Area, Oct. 1997.

6. J. Cooper, F. Wang, B. Balazs, R. Krueger, P. Lewis, T. Shell, J. Farmer, and M. Adamson, "Destruction of organic wastes by ammonium peroxydisulfate," presented at AIChE Spring National Meeting (March 9–13, 1997), Houston, TX, UCRL-MI-126794.

7. Cooper, J.F., Wang, F., Balazs, G.B., Shell, T., Krueger, R., Lewis, P., and Adamson, M., "Omnivorous Waste Destruction Process Using Uncatalyzed Peroxydisulfate Solutions," poster presentation at WM '97 conference, Tucson, Arizona, March 2–6, 1997, UCRL-MI-126764.

8. John F. Cooper, F. Wang, J. Farmer, R. Foreman, T. Shell, and K. King, "Direct Chemical Oxidation of Hazardous and Mixed Wastes" (*Proc. of the Third Biennial Mixed Waste Symposium*, American Society of Mechanical Engineers; Aug. 1995), Lawrence Livermore National Laboratory Report UCRL-JC-120141 March 28, 1995.

9. John F. Cooper, F. Wang, J. Farmer, M. Adamson, K. King, and R. Krueger, "Direct Chemical Oxidation: Peroxydisulfate destruction of organic wastes," *Proc. World Environmental Congress, International Conference and Trade Fair*, Page 219, London Ontario, Sept. 17–22, 1995.

10. F. Wang, J.F. Cooper, J. Farmer, M. Adamson, and T. Shell, "Destruction of ion exchange resins by wet oxidation and by direct chemical oxidation—a comparison study," *Proc. World Environmental Congress, International Conference and Trade Fair*, p. 206; London Ontario, Sept. 17–22, 1995.

11. John F. Cooper, Francis Wang, Roger Krueger, Ken King, Joseph C. Farmer, and Martyn Adamson, "Destruction of organic wastes by ammonium peroxydisulfate with electrolytic regeneration of the oxidant," LLNL Internal Report, September 1995. UCRL-121979 Rev 1, October 10, 1995.

12. John F. Cooper, Roger Krueger, and Joseph C. Farmer, "Destruction of VX by aqueous-phase oxidation using peroxydisulfate: Direct chemical oxidation," (*Proc. Workshop on Advances in Alternative Demilitarization Technologies*, pp. 429–442, Reston VA September 2–7 1995, published by SAIC Aberdeen, MD).

13. John F. Cooper, Francis Wang, Roger Krueger, Ken King, Thomas Shell, Joseph C. Farmer, and Martyn Adamson, "Demonstration of omnivorous non-thermal mixed waste treatment: Direct chemical oxidation using peroxydisulfate," First Quarterly Report to Mixed Waste Focus Group, SF2-3-MW-35 October–December 1995; UCRL-ID-123193, February 1996.

14. John F. Cooper (with Joseph C. Farmer, Francis T. Wang, and Martyn G. Adamson, "Alternatives to Incineration," (Proc. ASME/EPRI Radwaste Workshop, New Orleans, LA, July 24–26 1996.)

15. John F. Cooper, Francis Wang, Thomas Shell, and Ken King, "Destruction of 2,4,6-trinitrotoluene using ammonium peroxydisulfate," LLNL Report UCRL-ID-124585, July 1996.

16. W.M. Lattimer, "The oxidation states of the elements and their potentials in aqueous solutions," Prentice Hall, NY.

17. "Peroxides and Peroxy Compounds, Inorganic," Encyclopedia of Chemical Technology, V. 17, ed. Kirk Othmer.

18. D.A. House, "Kinetics and mechanism of oxidations by peroxydisulfate," *Chem. Rev.* 62, 185, (1961).

19. Francesco Minisci, Attillio Citterio, and Claudio Giordano, "Electron-transfer processes: peroxydisulfate, a useful and versatile reagent in organic chemistry," *Acc. Chem. Res.* v. 16 27 (1983).

20. Gary R. Peyton, "The free-radical chemistry of persulfate-based total organic carbon analyzers," *Marine Chemistry* 41, 91–103 (1993).

21. Robert E. Huie and Carol L. Clifton, "Temperature Dependence of the Rate Constants for Reactions of the Sulfate Radical, SO_4^-, with Anions," *J. Phys. Chem.*, 94 (1990) 8561.

22. Neta, P., Huie, R.E., and Ross, A.B., "Rate constants for reaction of inorganic radicals in aqueous solution," *J. Phys. Chem. Ref. Data*, 17:1027–1284, 1988.

23. I.M. Kolthoff and I.K. Miller, "The chemistry of persulfate. I. The kinetics and mechanism of the decomposition of the persulfate ion in aqueous medium," *J. Am. Chem. Soc.*, p. 3055 July (1951).

24. P.D. Goulden and D.H.J. Anthony, "Kinetics of uncatalyzed peroxydisulfate oxidation of organic material in fresh water," *Anal. Chem.* 50(7) 953 (1978).

25. Model 700 TOC Total Organic Carbon Users Manual, (O.I. Analytical, Inc.; Graham road at Wellborn Rd. PO Box 2980; College Station TX 77841-2980).

26. Walter C. Schumb, Charles N. Satterfield, and Ralph L. Wentworth, Hydrogen Peroxide, (Reinhold Publishing, Inc., New York, 1955).

27. Wolfgang Thiele and Hermann Matschiner, "Zur elektrosynthese von Wasserstoffperoxid und Peroxodisulfaten, Teil I. Wasserstoffperoxid und Peroxodischwefelsauer," *Chem. Techn.* v. 29(3) p. 148 (1977).

28. "Uses of Persulfate," FMC, Inc. Buffalo NY; 1951; 1960.

29. T.L. Allen, "The oxidation of Oxalate ion by Peroxydisulfate," *J. Am. Chem. Soc*, 73, 3589 (1951).

30. D.I. Ball, M.M. Crutchfield, and J.O. Edwards, "The Mechanism of the Oxidation of 2-Propanol by Peroxydisulfate Ion," *J. Org. Chem.*, 25, 1599 (1960).

31. A.R. Gallapo and J.O. Edwards, "Kinetics and Mechanisms of the Spontaneous and Metal-Modified Oxidations of Ethanol by Peroxydisulfate Ion," *J. Org. Chem.* 36, 4089 (1971).

32. C. Walling and D.M. Camaioni, "Role of Silver (II) in Silver-Catalyzed Oxidations by Peroxydisulfate," J. Org. Chem., 43, 3266 (1978).

33. M.G. Ram Reddy, B. Sethuram, and T.N. Rao, "Kinetics of Decomposition of Peroxydisulphate in the Presence of Mannitol Catalyzed by Ag^+," Z. Phys. Chem., 256, 875 (1975).

34. V.N. Kislenko, A.A. Berlin, and N.V. Litovchenko, "Kinetics of Oxidation of Organic Substances by Persulfate in the Presence of Variable Valence Metal Ions," Kinet. Catal. (Trans. of Kinet. Katal.), 37, 767 (1996).

35. Joseph Farmer, "Electrochemical Treatment of Mixed and Hazardous Wastes," (Chapter in "Environmental Oriented Electrochemistry," ed. Cesar A.C. Sequeira, Elsevier Science Publishers B.V., Amsterdam, The Netherlands; 1994).

36. Zoher Chiba, Bruce Schumacher, Patricia Lewis, and Laura Murguia, "Mediated electrochemical oxidation as an alternative to incineration for mixed wastes" (Proc. Waste Management 95 Symposium, Tucson, Arizona ,March 1, 1995).

37. Joseph C. Farmer, Francis T. Wang, Patricia R. Lewis, and Leslie Summers, "Destruction of chlorinated organic by cobalt(III)-mediated electrochemical oxidation," *J. Electrochem Soc.* v. 139(11) 3025 (1992).

38. "Destruction of Hazardous and Mixed Wastes using Mediated Electrochemical Oxidation in a $Ag(II)/HNO_3$ Bench Scale System," Balazs, B., Chiba, Z., Hsu, P., Lewis, P. Murguia, L., Adamson, Martyn, ICEM '97 Proceedings (Singapore, Oct. 12–16, 1997), UCRL-126754.

CONTRIBUTORS

Martyn G. Adamson
Lawrence Livermore National Laboratory
Mail Code L-092
Livermore, CA 94550
TN: 925-423-5403
Fax: 925-423-4897
e-mail: balazsl@llnl.gov

D. D. Adrian
Department of Civil & Environmental Engineering
Louisiana State University
Baton Rouge, LA 70803

Akram N. Alshawabkeh
Department of Civil & Environmental Engineering
Northeastern University
Boston, MA

*Thomas Andrews
In-Situ Oxidative Technologies, Inc.
Lawrenceville, New Jersey
TN: 609-275-8500 ext. 120
Fax: 609-275-9608

G. Bryan Balazs
Lawrence Livermore National Laboratory
Mail Code L-092
Livermore, CA 94550
TN: 925-423-5403
Fax: 925-423-4897
e-mail: balazsl@llnl.gov

J. K. Bewtra
Civil & Environmental Engineering
College of Engineering & Science
University of Windsor
Windsor, Ontario, Canada N9B 3P4

John V. Bishop
VanTek Corporation
Midland, Mi

N. Biswas
Civil & Environmental Engineering
College of Engineering & Science
University of Windsor
Windsor, Ontario, Canada N9B 3P4

Kevin C. Bower
Department of Civil Engineering
210 Auburn Science and Engineering Center
University of Akron
Akron, OH 44325-3905
TN: 330-972-5915
Fax: 330-972-6020
e-mail: cmmiller@uakron.edu

G. Breitenbeck
Department of Agronomy
Luoisiana State University
Baton Rouge, LA 70803

Ting-Chien Chen
Department of Environmental Engineering and Health
Tajen Junior College of Pharmacy
En-Pu Hsiang, Pingtung
Taiwan

John F. Cooper
Lawrence Livermore National Laboratory
Mail Code L-092
Livermore, CA 94550
TN: 925-423-5403
Fax: 925-423-4897
e-mail: balazsl@llnl.gov

David Devlin
Los Alamos National Laboratory
NMT-11 MS-E505
Los Alamos, NM 87545
TN: 505-665-7149
Fax: 505-6655-4459
e-mail: lworl@lanl.gov

Stephen Duirk
Department of Civil Engineering
210 Auburn Science and Engineering Center

University of Akron
Akron, OH 44325-3905
TN: 330-972-5915
Fax: 330-972-6020
e-mail: cmmiller@uakron.edu

R. S. Dyer
U.S. Environmental Protection Agency
Office of International Activities
Ronald Reagan Building
1300 Penn Avenue
Washington, DC 20004

Stefan Foy
In Situ Research Group
Départment de géologie et de génie géologique,
Université Laval, Sainte-Foy
Quebec, Canada G1K 7P4

*R. J. Gale
Department of Chemistry
Louisiana State University
Baton Rouge, LA 70803
TN: 504-388-3361
Fax: 504-388-3458

Pierre J. Gélinas
In Situ Research Group
Départment de géologie et de génie géologique,
Université Laval, Sainte-Foy
Quebec, Canada G1K 7P4

Richard S. Greenberg
In-Situ Oxidative Technologies, Inc.
Lawrenceville, New Jersey
TN: 609-275-8500 ext. 120
Fax: 609-275-9608

Alain Hébert
In Situ Research Group
Départment de géologie et de génie géologique,
Université Laval, Sainte-Foy
Quebec, Canada G1K 7P4

Dallas Hill
Los Alamos National Laboratory
NMT-11 MS-E505
Los Alamos, NM 87545
TN: 505-665-7149

Fax: 505-6655-4459
e-mail: lworl@lanl.gov

*P. K. Andrew Hong
Department of Civil & Environmental Engineering
University of Utah
Salt Lake City, UT 84112
TN: 801-581-7232
Fax: 801-585-5477
e-mail: hong@civil.utah.edu

*Maria Inman
Faraday Technology, Inc.
315 Huls Drive
Clayton, OH 45315
TN: 937-836-7749
Fax: 937-836-9498
e-mail: faratech@erinet.com

A. Jackson
Department of Civil & Environmental Engineering
Louisiana State University
Baton Rouge, LA 70803

*Chad T. Jafvert
Purdue University
School of Civil Engineering
1284 Civil Engineering Bldg.
West Lafayette, IN 47907
TN: 765-496-2196
Fax: 765-496-1107
e-mail: jafvert@ecn.purdue.edu

Weimin Jiang
Department of Civil & Environmental Engineering
University of Utah
Salt Lake City, UT 84112
TN: 801-581-7232
Fax: 801-585-5477
e-mail: hong@civil.utah.edu

Prasad K. C. Kakarla
In-Situ Oxidative Technologies, Inc.
Lawrenceville, New Jersey
TN: 609-275-8500 ext. 120
Fax: 609-275-9608

*P. D. Kalb
Department of Advanced Technology

Environmental & Waste Management Group
Building 830, P.O. Box 5000
Upton, NY 11973-5000
TN: 516-344-7644
Fax: 516-344-4486
e-mail: kalb@bnl.gov

Sherman Kottle
Lake Service Company
Lake Jackson, TX
TN: 409-297-5256
e-mail: lakesvc@computron.net

René Lefebvre
In Situ Research Group
INRS-Géoressources
2535 boulevard Laurier
C.P. 7500
Sainte-Foy, Quebec, Canada
G1V 4C7
TN: 418-654-2604
Fax: 418-654-2615

Patricia R. Lewis
Lawrence Livermore National Laboratory
Mail Code L-092
Livermore, CA 94550
TN: 925-423-5403
Fax: 925-423-4897
e-mail: balazsl@llnl.gov

Chelsea Li
Department of Civil & Environmental Engineering
University of Utah
Salt Lake City, UT 84112
TN: 801-581-7232
Fax: 801-585-5477
e-mail: hong@civil.utah.edu

Michele E. Lindsey
Department of Chemistry
University of New Orleans
New Orleans, LA 70148
TN: 504-280-6323
Fax: 504-280-6860
e-mail: mtarr@uno.edu

*Krishnanand Maillacheruvu
Department of Civil Engineering & Construction

Bradley University
Peoria, IL
TN: 309-677-3764
Fax: 309-677-2867
e-mail: kmaillac@bradley.edu

Abdul Majid
Institute for Chemical Process & Environmental Technology
National Research Council of Canada
Montreal Road Campus
Ottawa, Ontario, K1A OR6, Canada
TN: 613-990-1769
Fax: 613-941-1571
e-mail: bryan.sparks@nrc.ca

*Richard Martel
In Situ Research Group
INRS-Géoressources
2535 boulevard Laurier
C.P. 7500
Sainte-Foy, Quebec, Canada
G1V 4C7
TN: 418-654-2604
Fax: 418-654-2615

W. R. Michaud
U.S. Environmental Protection Agency
Office of International Activities
Ronald Reagan Building
1300 Pennsylvania Avenue
Washington, DC 20004

D. A. Michels
Chemistry & Biochemistry
College of Engineering & Science
University of Windsor
Windsor, Ontario, Canada N9B 3P4
TN: 519-253-3000 ext. 3526
Fax: 519-973-7098
e-mail: taylor@uwindsor.ca

L. W. Milian
Department of Advanced Technology
Environmental & Waste Management Group
Building 830, P.O. Box 5000
Upton, NY 11973-5000
TN: 516-344-7644
Fax: 516-344-4486
e-mail: kalb@bnl.gov

*Christopher M. Miller
Department of Civil Engineering
210 Auburn Science and Engineering Center
University of Akron
Akron, OH 44325-3905
TN: 330-972-5915
Fax: 330-972-6020
e-mail: cmmiller@uakron.edu

H. R. Monsef
Chemistry & Biochemistry
College of Engineering & Science
University of Windsor
Windsor, Ontario, Canada N9B 3P4
TN: 519-253-3000 ext. 3526
Fax: 519-973-7098
e-mail: taylor@uwindsor.ca

Debbra L. Myers
Mitisubishi Silicon America
South Campus
3990 Fairview Industrial Drive. SE
Salem, OR 97302

E. Oszu-Acar
Electrokinetics Inc.
11552 Cedar Park Avenue
Baton Rouge, LA 70809

Dennis Padilla
Los Alamos National Laboratory
NMT-11 MS-E505
Los Alamos, NM 87545
TN: 505-665-7149
Fax: 505-6655-4459
e-mail: lworl@lanl.gov

J. H. Pardue
Department of Civil & Environmental Engineering
Louisiana State University
Baton Rouge, LA 70803

Robert W. Peters
Energy Systems Division
Argonne National Laboratory
Argonne, IL 60439-4815

*Frederick G. Pohland
Department of Civil & Environmental Engineering

University of Pittsburgh
Pittsburgh, PA 15261
TN: 412-624-1880
Fax: 412-624-0135
e-mail: pohland@engrng.pitt.edu

F. Coyne Prenger
Los Alamos National Laboratory
NMT-11 MS-E505
Los Alamos, NM 87545
TN: 505-665-7149
Fax: 505-6655-4459
e-mail: lworl@lanl.gov

M. F. Rabbi
Department of Civil & Environmental Engineering
Louisiana State University
Baton Rouge, LA 70803

Liz A. Ramer
Department of Civil & Environmental Engineering, 116 ERF
University of Iowa
Iowa City, IA 52242

Annie Roy
In Situ Research Group
INRS-Géoressources
2535 boulevard Laurier
C.P. 7500
Sainte-Foy, Quebec, Canada
G1V 4C7
TN: 418-654-2604
Fax: 418-654-2615

Nathalie Roy
In Situ Research Group
INRS-Géoressources
2535 boulevard Laurier
C.P. 7500
Sainte-Foy, Quebec, Canada
G1V 4C7
TN: 418-654-2604
Fax: 418-654-2615

Laurent Saumure
In Situ Research Group
Départment de géologie et de génie géologique,
Université Laval, Sainte-Foy
Quebec, Canada G1K 7P4

Jerald L. Schnoor
Department of Civil & Environmental Engineering, 116 ERF
University of Iowa
Iowa City, IA 52242

R. K. Seals
Department of Civil & Environmental Engineering
Louisiana State University
Baton Rouge, LA 70803

*Bryan D. Sparks
Institute for Chemical Process & Environmental Technology
National Research Council of Canada
Montreal Road Campus
Ottawa, Ontario, K1A OR6, Canada
TN: 613-990-1769
Fax: 613-941-1571
e-mail: bryan.sparks@nrc.ca

Robert A. Stowe
VanTek Corporation
Midland, MI

Timothy J. Strathmann
Purdue University
School of Civil Engineering
1284 Civil Engineering Bldg.
West Lafayette, IN 47907
TN: 765-496-2196
Fax: 765-496-1107
e-mail: jafvert@ecn.purdue.edu

*Matthew A. Tarr
Department of Chemistry
University of New Orleans
New Orleans, LA 70148
TN: 504-280-6323
Fax: 504-280-6860
e-mail: mtarr@uno.edu

E. Jennings Taylor
Mitisubishi Silicon America
South Campus
3990 Fairview Industrial Drive. SE
Salem, OR 97302

*K. E. Taylor
Chemistry & Biochemistry
College of Engineering & Science

University of Windsor
Windsor, Ontario, Canada N9B 3P4
TN: 519-253-3000 ext. 3526
Fax: 519-973-7098
e-mail: taylor@uwindsor.ca

*Phillip L. Thompson
Department of Civil & Environmental Engineering, Room 524 ENGR
Seattle University
Seattle, WA 98122
TN: 206-296-5521
Fax: 206-296-2173
e-mail: thompson@seattleu.edu

Radisav D. Vidic
Department of Civil & Environmental Engineering
University of Pittsburgh
Pittsburgh, PA 15261
TN: 412-624-1880
Fax: 412-624-0135
e-mail: pohland@engrng.pitt.edu

Richard J. Watts
Department of Civil & Environmental Engineering
Washington State University
Pullman, Washington

*Laura A. Wori
Los Alamos National Laboratory
NMT-11 MS-E505
Los Alamos, NM 87545
TN: 505-665-7149
Fax: 505-6655-4459
e-mail: lworl@lanl.gov

S. P. Yim
Korean Atomic Energy Research Institute
Nuclear Environment Management Center
150 Dukjin-dong
Yusong-gu, Taejon 305-600
Korea

Pu Yong
Department of Civil & Environmental Engineering, 116 ERF
University of Iowa
Iowa City, IA 52242

Chengdong Zhou
Mitsubishi Silicon America
South Campus
3990 Fairview Industrial Drive. SE
Salem, OR 97302

INDEX